£55.00

£35.00

W9-ADY-092

The Mouse

The Mouse

Its Reproduction and Development

ROBERTS RUGH

Formerly College of Physicians and Surgeons
Columbia University, New York

OXFORD NEW YORK TOKYO

OXFORD UNIVERSITY PRESS

1990

Oxford University Press, Walton Street, Oxford OX2 6DP

Oxford New York Toronto
Delhi Bombay Calcutta Madras Karachi
Petaling Jaya Singapore Hong Kong Tokyo
Nairobi Dar es Salaam Cape Town
Melbourne Auckland
and associated companies in
Berlin Ibadan

Oxford is a trade mark of Oxford University Press

Published in the United States
by Oxford University Press, New York

Copyright © by Burgess Publishing Co. 1968
First published by Burgess Publishing Co.

Published by Oxford University Press 1990

All rights reserved. No part of this publication may be reproduced,
stored in a retrieval system, or transmitted, in any form or by any means,
electronic, mechanical, photocopying, recording, or otherwise, without
the prior permission of Burgess Publishing Co.

This book is sold subject to the condition that it shall not, by way
of trade or otherwise, be lent, re-sold, hired out or otherwise circulated
without the publisher's prior consent in any form of binding or cover
other than that in which it is published and without a similar condition
including this condition being imposed on the subsequent purchaser.

British Library Cataloguing in Publication Data
Rugh, Roberts
The mouse
1. Mice & rats
I. Title
ISBN 0-19-854277-1

Library of Congress Cataloging in Publication Data
(Data available)

Printed in Great Britain by
Butler and Tanner Ltd, Frome, Somerset

Preface

This mouse embryology text was inevitable. The authorship was pure chance. During the last 19 years in Radiobiological Research for the Atomic Energy Commission, we have provided control material for hundreds of thousands of mouse embryos and fetuses subjected to X rays. The controls have gradually accumulated until we now have a complete and detailed series of normal embryos, whole and sectioned sagitally, frontally and transversely. Since the early stages of development were found to be particularly radiosensitive, some studies began with fertilization and terminated with the pre-parturition 18-day fetus. A variety of stains and techniques was devised to bring out certain morphological effects, among which the most graphic were probably those in which the skeleton was cleared.

After its initial cleavage the embryo seems to accelerate in divisions and after the onset of organogenesis and differentiation the mosaic of development changes so rapidly that it became imperative that we time the pregnancies very accurately. As it was known that ovulation generally occurs during the night, earlier studies dealt with pregnancies resulting from overnight exposure of females to sexually mature males. The criterion of successful mating was not the vaginal smear but the vaginal plug, a coagulum of semen which causes the sperm to be retained within the female genital tract. But, since a 5 P.M. to 9 A.M. exposure of females introduces a time variable of 16 hours in development, we were forced to shorten the exposure time of the females to 45 minutes, from 8:00 A.M. until 8:45 A.M., and to search for vaginal plugs immediately thereafter. It is now believed that, although ovulation likely occurs between midnight and 2 A.M., a 45 minute period of early morning mating produces embryos of more precisely determined age than those produced by any other technique. Thus the material here presented is based upon studies of the CFI-S mouse under the most rigidly controlled conditions.

The technical preparation of embryos and slides began with Miss Erica Grupp with whom I have published papers in teratology. It was continued energetically with Mrs. Ludmila Skaredoff, who has had years of experience in histopathological laboratories. Much of the data was collected, and certainly most of the actual handling and direct observations on embryos and fetuses was done by Miss Marlis Wohlfromm, and some skeletal data was collected by Miss Lyse Duhamel. Other members of the research staff made important contributions.

Photographs of whole embryos and fetuses, and even of sections, are valuable, but since they are largely two-dimensional, they do not always include certain details. These can be shown only in drawings. Mrs. Rhoda Van Dyke, who has an M.S. in biology, has taught embryology, and is also an accomplished artist, has made practically all of the new drawings for this book. Because she has a first-hand knowledge and understanding of the subject, her drawings are neither rigidly presented nor "touched up" for artistic purposes. They are based on projections of microscopic sections at various magnifications but with all proportions maintained. Since I believe so strongly in accurately illustrated teaching, I have been most fortunate to have Mrs. Van Dyke's assistance.

The question might arise as to why rat development was not included in this book, or a section on abnormalities of mouse development. There are basic differences between mouse and rat development, particularly with respect to time but also with respect to certain organogenies, so that coverage of both within a single volume would be impractical. There are, of course, also minor differences in development among the many species of mice, but the species chosen here is a very satisfactory representative of all. A section on anomalies of development might shift the emphasis from the normal to the abnormal. The vast majority of fertilized mouse eggs develop normally and for the research worker and student of embryology an accurate and detailed description of normal development is most desirable. After all, normality is in itself an exciting revelation. Since I have concerned myself during the last two decades with radiation effects on the mammalian embryo and fetus, a separate treatise on this subject will follow.

As both teaching aids and ready references, particularly for courses in embryology and the research laboratory, I have included a complete bibliography of titles on normal development of the mouse, and a glossary of embryological terms. In a few instances the titles deal with related rodents but are listed because there is no parallel reference on the mouse or because they describe special techniques that would be applicable to the mouse. The completeness of the bibliography is due to the help of Dr. Joan Staats of the Jackson Memorial Laboratory, Bar Harbor, Maine, and Suzanne Kriss of Medlars Service of the National Library of Medicine, Department of Health, Education, and Welfare, Bethesda, Maryland. The glossary is a composite from my other books, corrected and increased from the literature pertaining to the mouse.

As a teacher of embryology I have been dissatisfied with the use of the pig as a representative of the mammals, largely because it is available only after most of the organs have been formed (*i.e.*, 4 mm stage). I believe that the mouse will supplant the pig for teaching purposes within a few years. The mouse may be examined alive and studied from fertilization through cleavage, blastulation, implantation, germ layer development, and organogenesis with a minimum of expense and at any laboratory in the world. It is hoped that this book will be a beginning, a foundation upon which others can build even a more thorough description of the normal development of the mouse.

November 1967   R. R.

Table of Contents

Chapter

INTRODUCTION

The mouse is proving to be a boon to embryological research and hence to a better understanding of normal mammalian embryology. It is available the world around; it is easy to feed, raise, mate and handle; and it has a short gestation period (18 to 21 days) and a long period of reproductive activity (from 2 to about 14 months of age) during which it can produce over 10 litters and 100 offspring. Furthermore, strains or species may be crossed, either naturally or artificially, to produce hybrids that exhibit typical heterosis with longer natural and reproductive lives, healthier and larger litters, and better resistance to disease than most pure bred strains.

The subject chosen as representative of the species is the inbred CFI-S mouse strain. Its members have a life span of $2\frac{1}{2}$ to 3 years, and, if kept at uniform temperature (around 72° F.), with regulated humidity, and protection from drafts and from invasion by common mouse diseases, they will remain healthy in their caged environment for their lifetime. The food is generally a standardized synthetic diet, supplemented during pregnancy and lactation with rolled oats and whole wheat bread. Water must always be available, as it is as important as food for survival. When there is evidence of systemic infection, the drinking water may be treated with an antibiotic such as streptomycin, chloromycin, or tetracycline. Other deterrents to infectious conditions may be neomycin, sulfamerazine, terramycin, tylan (tylocentrartrate), polyotic, piperizine and PRL. Sterilization of cages and drinking bottles should be routine and regular.

Large mice appear to have a longer mean life span than small mice, even within a litter, and hybrids have the longest of all in life span and in survival. The CFI-S as an inbred strain is among the best available in this respect. Genetic constitution therefore plays a part in the life of the mouse.

The production of the CFI-S mouse* used in these embryological studies begins with the young which are delivered by hysterectomy-hysterotomy to become the F_1 axenic generation. After weaning and confirmation of their germfree status, the surgically-acquired young are transferred to a large breeding isolator. The F_1 mice are associated with a defined autochthonous flora, indigenous to the mouse, and are maintained as gnotobiotes. Here they are bred and their offspring, the F_2 gnotobiote generation, is derived. The F_2 mice are in turn transferred in disposable, germfree transporters into breeding rooms where they are maintained and bred under rigidly-controlled, micro-barrier conditions. The food, bedding, and water are sterilized, filter caps control airborne contaminants and technicians observe strict sanitary precautions.

---

* From Carworth Farms, New City, N. Y. This strain of mouse may be considered as quite representative of most mice, in its embryological development.

These F_2 mice become the breeders of the primary colony producing the F_3 mice. These F_3 primary-colony mice are moved on to become mass production breeding colonies, comprising the F_4 specific-pathogen-free (SPF) generation. Some of the F_3 generation of mice return to the cycle to become the parent generation from which another F_1 generation of axenic mice is derived. In this way colonies of mice with controlled and indigenous flora are provided, giving a uniformity which is so desirable for research. Such mice seem to be an improved product in terms of litter size and health.

Although we cannot simply extrapolate findings from mouse embryos to human embryos or vice versa, there is sufficient similarity in the development of these divergent mammals that findings are at least qualitatively suggestive. When comparable developmental stages of mouse and man are placed side by side, only the well informed can distinguish between them. Moreover, comparison of organ primordia and differentiation in the two forms yields a graph (Otis and Brent 1952) emphasizing the likeness. In summary, then, the accessible, abundant, and prolific mouse provides embryonic material for study of normal development and for basic experimentation leading to more critical and urgent studies on man, with some suggestion as to outcome.

A. GESTATION PERIOD: The gestation period is usually 19 days, although the range, in different strains is 18 to 21 days, with variations occurring even in mice simultaneously mated. The differences are sometimes related to litter size, not to the number of corpora lutea left in the ovary but to the total number (volume) of conceptuses. Apparently the total mass of fetal and placental tissue rather than the number of implants affects the length of the gestation period, so that heavier fetuses are associated with shorter periods. It seems, therefore, that a humoral factor, possibly from the placentae or the fetuses, may regulate the duration of pregnancy. There is some evidence that within a given litter the mean weight of the fetuses and their placentae decreases from the ovarian toward the cervical end of the uterus.

Intrauterine development of all vertebrates falls naturally into three consecutive phases: (1) Primary development from fertilization through germ layer derivation, (2) Basic organogenesis, and (3) Tissue differentiation with functional maturity and organismic integration (through nervous and circulatory systems, primarily). During the first, but more specially the second phase, the embryo is particularly labile so that congenital anomalies can result from extrinsic imposed hazards. The first period in the mouse lasts about 6 days, and the second about 6 more, although some organ systems are not fully differentiated until after birth (*e.g.*, cerebellum, gonads). After about day 13 gross congenital anomalies cannot be imposed because the major organ systems have been developed.

Parturition generally occurs between midnight and 4 A.M. in these nocturnal animals, but may occur at any time. Estrus begins again in 12 to 20 hours, depending upon the time of delivery with respect to the diurnal light cycle. Successful matings can then resume, proving that lactation is not a deterrent to ovulation in the mouse. When a lactating mouse becomes pregnant, however, its gestation period may be up to 2 weeks longer than usual. The extension of the gestation period may also be due to delays in implantation.

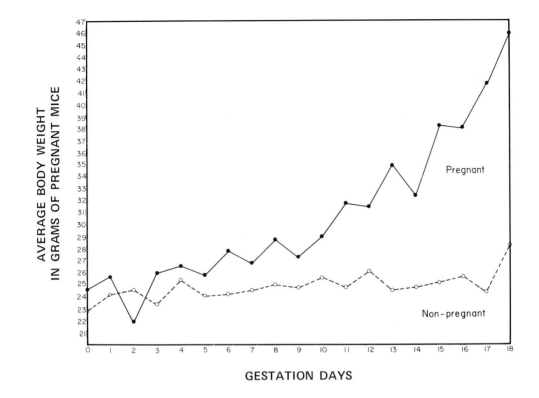

GESTATION DAYS

B. HEMATOLOGICAL AND WEIGHT CHANGES DURING NORMAL PREG-
NANCY IN THE MOUSE:
A detailed study of 15 pregnant mice tested daily for blood and weight changes revealed that there was little change in the white cell count, but the hemoglobin dropped appreciably at about the sixth day of pregnancy to a low of 11.3 (compared with 14.97 for the non-pregnant controls), with a corresponding drop in red cell count from 9,561,017 at conception to 7,593,333 at the time of delivery. During pregnancy the platelet count increased a bit at times, especially on day 16, but this may not be statistically significant. The total neutrophil count fluctuated, but did not deviate seriously from the controls. Thus, it appears that the hemoglobin and red cell counts were the most definitely affected. In addition the total body weight was increased from 24.56 grams to a maximum of 45.7 grams just before birth. Even after birth the weight did not return to the pre-pregnancy level, remaining around 6 grams in excess for a period of a month. This may be due largely to lactation. By the time of weaning the mother's hemoglobin and red cell counts, but not body weight, had returned to normal.

C. LITTER SIZE:
Factors in litter size include (1) the strain of the mouse, (2) whether it is the first or a later pregnancy, (3) the age, and (4) the health and vigor of the mother. When a female is pushed to maximum fetal production, it seems that her uteri become physiologically exhausted. In the CFI-S strain litters may range in size from 1 to 19 with the average between 10 and 11. The first litter is generally the smallest (average of 8.4) and the fourth or fifth the largest (average of 10.28), with a plateau maintained to 6 or 7 months of age thereafter and then showing a slow decline in

TABLE 1

HEMATOLOGICAL CHANGES DURING GESTATION

G.D.	#	AV. WT.	AV. HGB.	AV. WBC.	AV. RBC.	AV. PLAT.	TOT. NEUT.	STATS.	SEGS.	EO.	LYMPH.	MON.	D	T
CONTROL	21	24.92	14.97	8,980	8,923,273	1,299,127	26.6	0.19	20.5	2.9	65	4.7	.17	
0.0	59	24.56	15.6	7,671	9,561,017	1,293,051	33.1	0.30	32.1	4.3	60	2.2	.1	
1.0	15	25.56	14.57	8,373	9,120,000	1,367,333	22.1	0	22.1	3.5	69	5.1	.5	
2.0	14	22.0	14.8	9,571	9,270,000	1,282,143	16.9	0.14	16.9	4.5	72	6.0	.6	.07
3.0	15	25.93	15.1	9,800	9,372,000	1,372,000	20.7	0.13	20.6	2.6	71	4.8	.3	
4.0	15	26.57	15.5	8,546	9,492,000	1,490,000	22.0	0.26	21.8	2.4	71	3.8	.12	.6
5.0	15	25.80	15.1	8,840	9,405,333	1,270,000	22.0	0.2	21.8	3.8	69	4.6	.3	.06
6.0	15	27.76	14.3	8,466	8,866,666	1,392,000	14.5	0.26	24.3	2.8	67	5.2	.2	
7.0	15	26.71	14.1	7,866	8,694,666	1,352,666	27.0	0.46	26.5	4.1	64	4.8	.3	
8.0	15	28.7	13.6	8,120	8,064,000	1,145,000	40.4	0.46	39.9	3.4	51	5.1	0.06	.06
9.0	15	27.33	12.9	9,706	8,029,333	1,329,333	33.5	0.33	33.2	3.1	57	5.4	.6	.2
10.0	15	29.02	13.3	9,533	7,942,666	1,246,000	31.6	0.33	31.4	3.7	58	6.3	.3	.06
11.0	15	31.84	12.66	8,240	7,420,666	1,200,000	25.6	0.13	25.5	2.6	63	7.8	.4	
12.0	15	31.44	12.5	7,480	7,401,333	1,253,333	30.8	0.26	30.6	2.3	60	6.9		
13.0	15	34.76	12.76	7,120	7,646,666	1,371,333	26.6	0.20	26.4	2.4	65	.05	.13	.66
14.0	15	33.4	11.56	8,720	7,049,333	1,480,666	25.2	0.13	25.1	1.8	66	7.2	.13	
15.0	15	38.28	12.01	7,285	7,248,000	1,243,000	23.3	0.26	23.6	2.0	69	6.1	.06	
16.0	15	38.1	12.1	8,200	7,454,666	1,546,666	29.2	0.06	29.1	1.6	63	4.73		
17.0	15	41.68	10.83	7,293	6,717,333	1,440,666	32.8	0.26	32.5	1.7	62	3.4	.06	
18.0	15	46.1	11.07	7,028	7,196,000	1,446,000	34.8	0.2	34.6	1.5	59	5.1	.13	
19.0	3	45.7	11.3	7,833	7,593,333	1,326,666	22.6	0	22.6	3.0	68	6.3		

G.D. — gestation day
Av. Wt. — average weight
Av. HGB — average hemoglobin
Av. WBC — average number of white blood cells
Av. RBC — average number of red blood cells
Av. Plat. — average number of platelets
Tot. Neut. — total neutrophils

Stats. — stabophils
Segs. — segmentals
Eo. — eosinophils
Lymph. — lymphocytes
Mon. — monocytes
D — double nucleated cells
T — triple nucleated cells

(Data collected by Csilla Somogyi)

litter size. The mating and breeding period lasts from 2 to 14 months of age, and even longer in males. A count of the corpora lutea is a clue to the number of ova ovulated but not, of course, to the number of implantations or to the number of fetuses that reach term. These numbers are almost invariably less than the number of corpora lutea. Abnormal offspring rarely exceed 3% in this strain. Embryos implanted nearest the cervix appear to be resorbed more frequently than others and their mean weight is less, and those nearest the oviduct seem, at times, to be slightly more advanced in development. This may be related to the average larger placenta found toward the oviduccal end.

TABLE 2

REPRODUCTIVE PERFORMANCE OF VIRGINS AND MULTIPARA FEMALES

	VIRGINS (3—5 months)	VIRGINS (7—9 months)	MULTIPARAS (7—9 months)	EX-BREEDERS (10—12 months)
Number of litters	164	168	128	92
Average implantations	11.42	11.04	12.36	10.37
Average litter size	9.48	9.01	10.35	7.63
Normal offspring %	83.03	81.57	83.70	73.65
Stunted offspring %	1.97	3.17	2.52	6.36
Dead fetuses %	1,31	1.02	1.07	1.88
Resorbed in utero %	33.18	13.90	12.57	17.89
Anomalous %	0.48	0.32	0.12	0.19
Males %	53.80	52.60	50.90	51.70
Females %	46.20	47.40	49.10	48.30

Virgin mice, whether young or older, do not have as many implantation sites as do multipara mice of 7 to 9 months of age, and the ex-breeders have the smallest average number of implantation sites. This is probably because of physiological fatigue since these ex-breeders had been kept pregnant rather constantly for some 10 months. Likewise, the ex-breeders of 10—12 months of age had the highest percentage of stunted offspring.

D. SEX RATIO: In data from approximately 200,000 births of normal (control) mice the ratio of males to females was close to 52% to 48%. Theoretically, if all fertilized ova resulted in living offspring, the ratio would be 1:1 since the heterologous spermatozoa determine sex and spermatogenesis yields equal numbers of (dimorphic) male and of female producing spermatozoa. The 1:1 ratio represents an over-all mean probability and never applies to numbers less than 1,000, certainly not to single matings. The slight imbalance in favor of males may be related to the slightly smaller Y chromosome, as compared with the X chromosome, in the male-producing spermatozoon giving it an infinitesimally lighter chromosome burden to carry toward the matured ovum waiting to be fertilized in the ampulla of the oviduct.

TABLE 3

SIZE, NORMALITY, AND SEX RATIOS OF SUCCESSIVE LITTERS

	FIRST LITTER	SECOND LITTER	THIRD LITTER	FOURTH LITTER
Total litters	350	350	350	350
Total offspring	2940	3416	3560	3591
Average litter size	8.40	9.76	10.17	10.28
Total males	1601	1748	1878	1903
Total females	1339	1669	1682	1688
Ratio: males/females	54.5/45.5	51.2/48.8	52.8/47.2	53.0/47.0
Abnormals:				
Eaten by mother	0.47%	0.48%	0.63%	0.70%
Dead at birth	1.33	0.71	0.93	0.73
Persistent amnions	0.33	1.40	1.41	1.51
CNS anomalies	0.10	0.08	0.08	0.32
Other anomalies	0.06	0.0	0.05	0.0
Total not normal	2.29%	2.67%	3.10%	3.26%

The 350 females used in this study were the same for the four pregnancies. It will be noted that with each pregnancy the litter size was increased from an average of 8.40 to 10.28 for the fourth litter. The sex ratios varied slightly, but in all instances males were in excess of females. Total anomalies also increased with each successive litter, but never exceeded 3.26%. Among the 13,508 offspring examined 52.6% were males and 47.4% were females.

E. THE NEO-NATAL MOUSE: The mouse is born hairless, with its eyes and ears closed, and the vagina covered by a membrane. They may be handled gently from the beginning with the mother apparently unconcerned. After 2 or 3 days hair appears, and after 3 or 4 days the ears open. Sex can be determined at birth, or even earlier, by the spatial relations of the genital papilla or opening and anal opening and after 9 or 10 days the mammary nipples of the female can be identified. By 17 days the first maturation spindles are seen in the ovarian ova, the mice are very active, and begin the weaning process. Weaning may be forced at 3 to 4 weeks, but a young mouse is usually healthier if weaned naturally somewhat later. The eyes open at about 2 weeks, the ears grow rapidly, and the rather abundant hair undergoes its first moult. The vagina opens at about 35 days, and the first estrus follows. Such weanling females are generally reluctant to accept the male in mating, and even if mating does occur, it may not result in fertilization, implantation, or the production of viable fetuses. The first healthy mating takes place when a female is at least 7 to 8 weeks of age.

F. SEX DISTINCTION: The distance between the anus and the genital papilla or opening is greater in a male mouse than in a female. The anus and clitoris are roughly one-half to two-thirds as far apart as the anus and penis. The male penis may of course be withdrawn into the scrotal sac and into the body cavity. Adult mice exhibit other distinguishing secondary sex characters. Generally males are larger and heavier than females of similar age. At 2 months males average 27.8 grams and at 24 months 34.8 grams in total weight. Females at the same ages average 22.4 grams and 29.8 grams. A female is generally slenderer than a male, unless she has been multiparous, in which case she tends to be fatter. She also has more prominent mammary glands, whether or not she is lactating. Usually males are more aggressive, and females are docile.

REPRODUCTIVE SYSTEMS OF ADULT MICE

It is logical for us to begin our study with the matured reproductive organs of the adult mouse from which the germ cells are derived, and then to proceed to the normal development from fertilization to birth.

THE ADULT MALE

The male reproductive system consists of the testes and enclosing scrotal sac, epididymis and vas deferens, remnants of the embryonic excretory system that function in sperm transport, accessory glands, the urethra and the penis. Except for the urethra and penis, all these structures are paired.

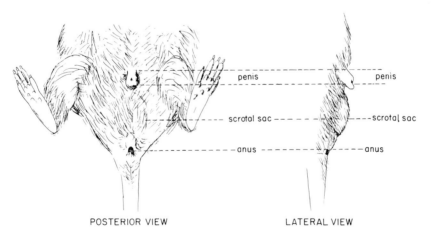

POSTERIOR VIEW   LATERAL VIEW

EXTERNAL GENITALIA OF ADULT MALE MOUSE

A. TESTES: Each testis is covered with a fibrous connective tissue, the tu-
nica albuginea, from which thin partitions, or septa, project into the organ to divide it into lobules, containing many convoluted tubules. These tubules are called seminiferous because within them are produced all the functional germ cells of the male. The region from which the tunica projects into the testis, and at which the testicular arteries enter, is known as the hilus. The arteries nourish every part of the testis, and then they connect with the testicular veins which leave it by the hilus.

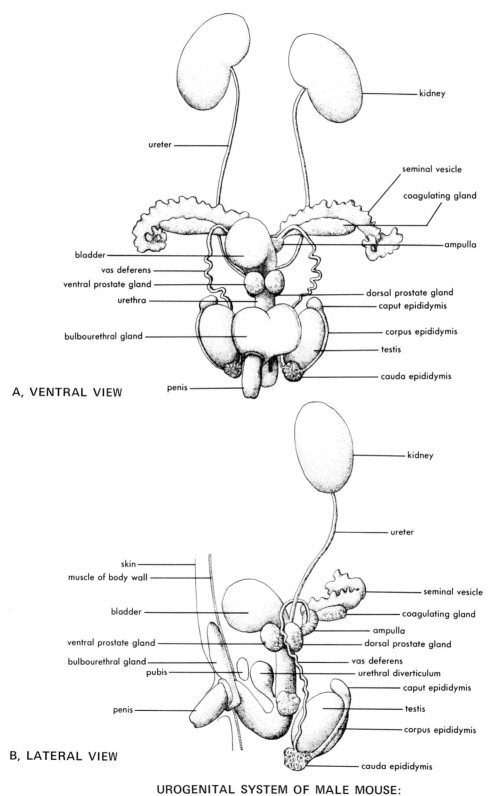

A, VENTRAL VIEW

B, LATERAL VIEW

UROGENITAL SYSTEM OF MALE MOUSE:

(From R. Rugh, "Vertebrate Embryology", Harcourt, Brace & World, Inc., New York, 1964.)

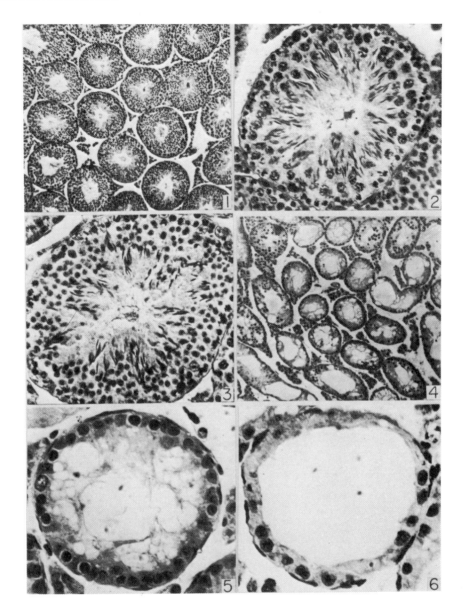

THE MOUSE TESTIS

Fig. 1 — Low power view of tubules within the testis as they are seen in transverse section. Note sparse interstitial tissue.

Fig. 2 — Enlarged view of single seminiferous tubule in section, one in which there are abundant large primary spermatocytes and some mature spermatozoa.

Fig. 3 — Tubule of sexually mature mouse showing all stages of spermatogenesis.

Fig. 4 — Low power view of section through testis of aged male, in which spermatogenesis is almost non-existent.

Fig. 5 — Seminiferous tubule of aged male, showing persistent Sertoli cells but no spermatogenesis. Coagulum within the tubule.

Fig. 6 — Tubule of aged male with most of the Sertoli cells also gone.

The seminiferous epithelium of the tubules lies against a basement membrane that is surrounded by thin fibrous connective tissue. Between the tubules is interstitial stroma, consisting of clumps of Leydig or interstitial cells and rich in blood and lymph. The interstitial cells of the testis have large round nuclei, each with one or more nucleoli containing coarse granules. Their cytoplasm is eosinophilic. It is believed that the interstitial tissue elaborates the male hormone testosterone. The seminiferous epithelium does not consist exclusively of spermatogenic cells, but also has sustaining nutritive or nurse (Sertoli) cells, found nowhere else in the body. Sertoli cells are attached by their bases to the basement membrane and project toward the lumen of a seminiferous tubule. They are elongate cells with large oval nuclei that may appear to be indented. Within the nucleus of a Sertoli cell is a compound or multiple nucleolus, one part composed of a central acidophilic body and the other of two or more peripheral basophilic bodies. The Sertoli cell may assume several forms, depending upon its activity. In the resting state it is closely associated with the basement membrane to which it is attached, and its oval nucleus is parallel to that membrane. As a supporting cell for the metamorphosis of a spermatid to a spermatozoon and the temporary retention of the mature spermatozoa, it is elongate, pyramidal, and its nucleus lies perpendicular to the basement membrane. The cytoplasm near the lumen generally contains the heads of many mature spermatozoa, the tails of which lie free within the lumen.

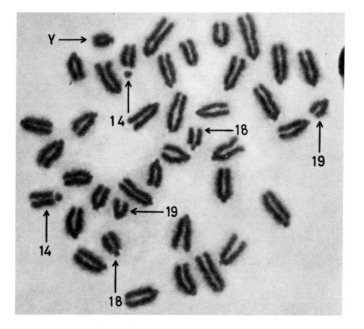

Chromosomes at mitotic metaphase in a cell from the spleen of a male CBA mouse. Arrows indicate the Y chromosome and the three pairs of autosomes that can be distinguished morphologically. The chromosomes of pair 19 are the shortest autosomes and commonly have a prominent proximal secondary constriction like those of pair 14. The X chromosome cannot be distinguished morphologically at this stage. Air-dried preparation. Stained in lactic-acetic orcein.

(Courtesy Drs. C. E. Ford and E. P. Evans)

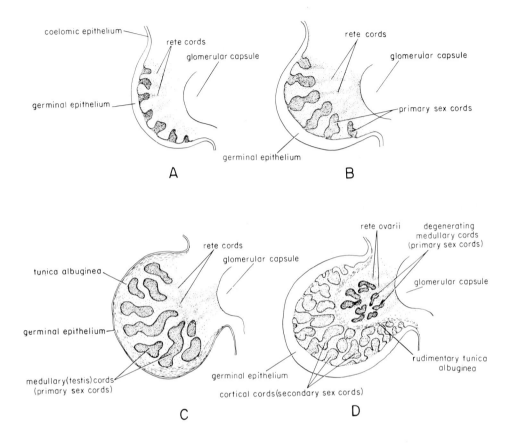

EMBRYOGENESIS OF THE MAMMALIAN GONAD

A — Origin of primary sex cords (medullary cords) from the germinal epithelium.

B — The gonad at an indifferent stage of sexual differentiation. There are well developed primary sex cords for the presumptive male or the medullary component, with the germinal epithelium representing the cortical component.

C — Differentiation of the testis, with further development of the primary sex cords and reduction of the germinal epithelium and development of the tunica albuginea.

D — Differentiation of an ovary which consists in the reduction of the primary sex cords to the medullary cords of the ovary, whereas the cortex is formed by continued development of the cortical cords from the germinal epithelium.

Redrawn from R. K. Burns in "Sex and Internal Secretions", 1961 (W. C. Young, Ed.)

B. SPERMATOGENESIS: The potential (primordial) germ cells of the male appear at about 8 days gestation, at which time there may be only about 100. They are the ancestors of all the millions of spermatozoa to be produced by the male (and there is no positive evidence of any other source for spermatozoa). They are first seen in the extra-gonadal endodermal yolk sac epithelium near the base of the allantois, far from their ultimate destination. Paired genital ridges arise independently at 9 days gestation, adherent to the paired mesonephroi, toward which the primordial germ cells migrate by ameboid movement. Since the germ cells are richly endowed with alkaline phosphatase which supplies the energy for their movement through the embryonic tissues, and which is not found in any other cells of the embryo, they are readily identifiable by appropriate staining techniques. Their route on days 9 and 10 is from yolk sac epithelium to hindgut endoderm to dorsal mesentery to root of the mesentery to coelomic angles during which passage many degenerate while others multiply and eventually migrate (at 11 or 12 days gestation) to the genital ridges. By that time the number of survivors has increased to probably 5,000 and identification of the testis can usually be made. Until that time gonad development in the two sexes is so similar that sex differentiation is not possible. In the testis subsequent proliferation and differentiation are medullary, while in the ovary they are cortical. Histological sex differentiation proceeds very rapidly. In cases of genetic sterility loss of germ cells seems to occur during this passage from the extra-gonadal region of origin toward the genital ridges.

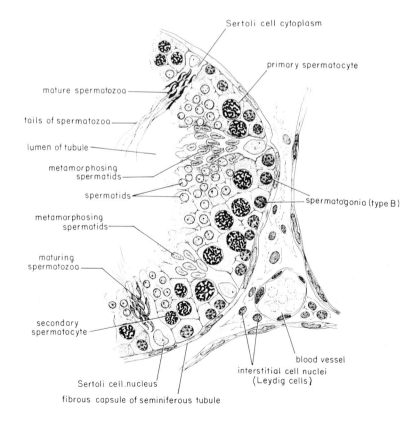

MATURATION OF SPERMATOZOA IN THE MOUSE

Toward the latter part of fetal life the mitotic activity of the primordial germ cells in the genital ridges declines, and some of the cells begin to degenerate by 19 days gestation. Shortly after birth larger cells, the spermatogonia are seen. Thereafter spermatogonia may be seen in the mouse testes throughout life. There are three types: type A, intermediate, and type B.

Type A is the ancestral stem cell: it is capable of mitosis or of giving rise to the other types and ultimately to spermatozoa. Type A spermatogonia are the largest and contain fine dust like particles of nuclear chromatin, and a single eccentrically placed chromatin nucleolus. Their metaphase chromosomes are long and slender. They may give rise, through intermediate spermatogonia to type B spermatogonia which are smaller, more numerous, and contain nuclear chromatin in coarse flakes or clumps on or near the inner surface of the nuclear membrane. There is a centrally placed plasmosome-like nucleolus. The metaphase chromosomes are usually short, rounded, and bean shaped. Type B spermatogonia may divide to give rise to more type B cells or change into primary spermatocytes, farther from the basement membrane. It has been estimated that the time lapse from spermatogonial metaphase to early meiotic prophase is 3 to 9 days, that diakinesis to second metaphase takes 4 days or less, and diakenesis to immature spermatozoa takes 7 or more days. Therefore, the time lapse from spermatogonial metaphase to immature spermatozoa is at least 10 days.

Type A cells first appear 3 days after birth. As they increase in number, the primordial germ cells, from which they are derived and which normally lie next to the basement membrane, decrease in number. Meiotic divisions in the testes begin about 8 days after birth.

The first indication that a type B spermatogonium will metamorphose into a primary spermatocyte is that it enlarges noticeably and moves away from the basement membrane. The primary spermatocyte divides into two smaller secondary spermatocytes, which in turn divide into four spermatids. They undergo a radical metamorphosis into an equal number of mature spermatozoa, losing most of their cytoplasm and changing form characteristically.

Between the primary and secondary spermatocyte stages, the chromatin material must divide. Premeiotic synthesis of DNA (deoxyribonucleotide) occurs in primary spermatocytes during the resting phase and is terminated just before the onset of meiotic prophase, with an average duration of 14 hours. No DNA synthesis occurs in the later stages of spermatogenesis. The first spermatocyte division reduces its volume, either without altering its quality, (equational division), or by actually separating members of allelomorphic pairs

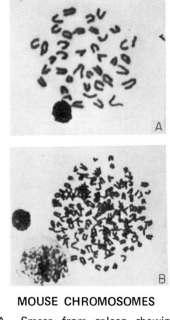

MOUSE CHROMOSOMES

A — Smear from spleen showing diploid chromosome number
B — Smear from bone marrow showing tetraploid chromosome number.

1.   Spermatogonial resting nucleus

2.   Early reproductive phase

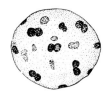

3.   Early prophase

4.   Late prophase

5.   Metaphase

6.   Early differentiation

7.   Early differentiation showing
      prominent reticulum

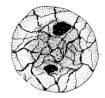

8.   Early differentiation
      showing orientation of reticulum

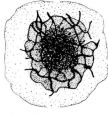

9.   Prophase-like stage

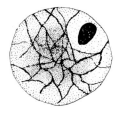

10.   Early spermatocyte

CYTOLOGIC CHANGES IN SPERMATOGONIAL NUCLEI IN MOUSE
(Redrawn from Fogg and Cowing, Exp. Cell Research, Vol. 4, 1953)

CYTOLOGIC CHANGES IN SPERMATOGONIAL NUCLEI IN MOUSE

Fig. 1 — Resting spermatogonium with finely granular chromatin mesh-work evenly distributed within the nucleus, a faintly basophilic nucleoplasm, and a well-defined nuclear membrane.

Fig. 2 — Spireme is lost, and the granular chromatin is assuming the form of compact bodies. These bodies tend to orient themselves peripherally but are not attached to the nuclear membrane. Nucleoli are still present.

Fig. 3 — The chromatin clumps have increased in number but not in size. This is a transitional phase, the orientation of the chromatin clumps tends to be peripheral and remain so until the beginning of nuclear breakdown in late prophase. The nuclear membrane in this figure seems to be fading out, the nucleus appears to be enlarging and the nucleoplasm is less definitely stainable.

Figs. 4 & 5 — These represent typical spermatogonial mitotic activity.

Fig. 6 — One of the first of the phases of differentiation, the nuclear reticulum is accentuated and clearly apparent, but the nucleolus, nuclear membrane, and staining capacity of the nucleoplasm are similar to the resting nucleus.

Fig. 7 — The next step in nuclear differentiation, with a more prominent linin network, slight increase in nuclear size, a reduction in the number of large-size chromatin staining particles, loss of morphological integrity of the nucleolus, and a decrease in the staining affinity of the nucleoplasm.

Fig. 8 — A further stage in nuclear differentiation showing an increase in the size of the reticulum and a gradual loss of the nucleolus. The nuclear membrane remains intact but the general picture of the nucleus shows wide variations at this time.

Fig. 9 — This is a key phase in the differentiation process, and may be confused with the late prophase. The reticulum is thick, more or less oriented in a mass, and there is no demonstrable nucleolus. Often the chromatin is massed to one side, leaving a non-stainable nuclear area and a faintly stainable nuclear membrane.

Fig. 10 — This is an early spermatocyte, with an enlarged reticulum which is breaking up into units, the nuceolus has reappeared and is surrounded by a clear zone, and the nuclear membrane is intact. The nucleus has increased in size and continues until the primary spermatocyte is formed.

(From L. C. Fogg and R. F. Cowing, 1953: Exp. Cell Research 4:107-115)

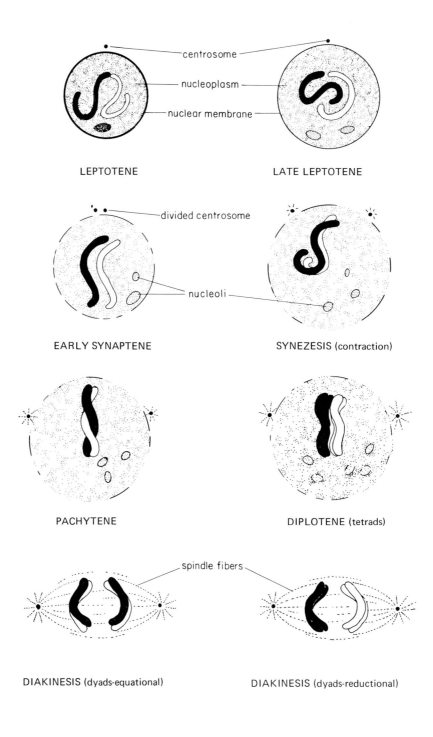

centrosome

nucleoplasm

nuclear membrane

LEPTOTENE

LATE LEPTOTENE

divided centrosome

nucleoli

EARLY SYNAPTENE

SYNEZESIS (contraction)

PACHYTENE

DIPLOTENE (tetrads)

spindle fibers

DIAKINESIS (dyads-equational)

DIAKINESIS (dyads-reductional)

MEIOSIS

(reductional division). The equational or quantitative division is essentially mitotic whereas the reductional or qualitative division, is characteristic only of meiosis, never seen in any other tissue or organ. Whichever type of division occurs, the other follows, making the secondary spermatocyte truly haploid (divided both qualitatively and quantitatively). There are 20 tetraploid chromosomes prior to the first spermatocyte division, and a spermatid has one-fourth of those, or 20 chromosomes, the haploid number. Should the reductional division not take place, there would be a doubling of the chromosome number with each generation.

Following is a schematic representation of meiosis (maturation)as illustrated by spermatogenesis. The following estimates have been made for the various stages of spermatogenesis (Oakberg 1957).

STAGE	DURATION: HOURS
Spermatogonia	
Type A	Always present
Intermediate	27.3
Type B	29.4
Primary spermatocytes	
Resting preleptotene	31.0
Leptotene	31.2
Zygotene	37.5
Pachytene	175.3
Diplotene	21.4
Diakenesis + metaphase	10.4
Secondary spermatocytes	10.4
Spermatids	229.2

Note that in diakenesis the chromosomes may be aligned in either of two ways. One of these results in their separation as pairs (equational), and the other in the separation of the members of allelomorphic pairs (reductional) of chromosomes.

The transformation of a spermatid into a spermatozoon involves no divisions. Most of the cytoplasm disappears carrying with it certain residual bodies. (See figure on page 18.)

The cytoplasm of the spermatid that is to be sloughed off contains lipid droplets, mitochondria, ribosomes, endoplasmic reticulum, the caudally migrated Golgi apparatus, and numerous multivesicular and multigranular bodies. These membrane-limited bodies and the Golgi zone stain heavily for acid phophatase. Following extrusion, the residual bodies undergo a series of alterations: 1) disruption of the multigranular bodies with the release of free granules, 2) sequestration of granules, ribosomes, and reticulum inside double membrane-limited vacuoles derived from Golgi lamellae, 3) appearance of numerous single-membrane bound, cytoplasmic vacuoles, 4) fragmentation,

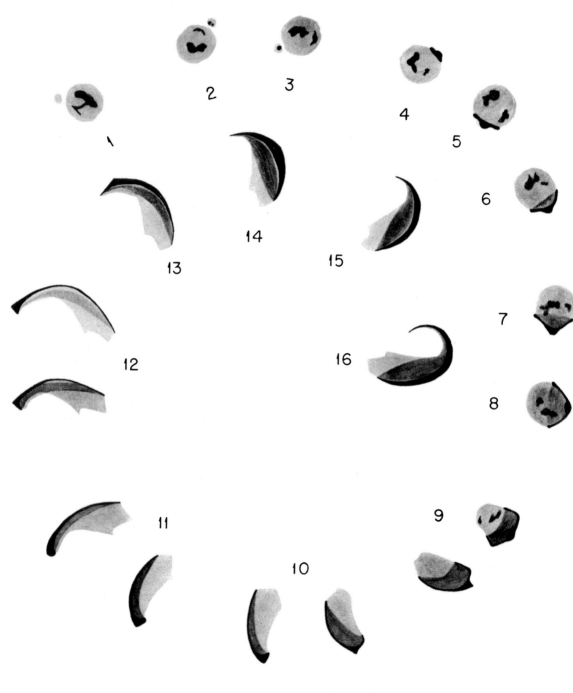

STAGES IN SPERMIOGENESIS

Spermiogenesis of the mouse as seen with periodic acid Schiff and hematoxylin staining of Zenker-formol fixed testis. Drawings are arranged in a spiral to demonstrate stages which overlap in a cycle of the seminiferous epithelium. Orientation of spermatids in relation to the basement membrane also is shown. Note especially the change between stages VII and VIII. 1–3 is the Golgi phase, 4–7 the cap phase, 8–12 the acrosome phase, and 13–16 the maturation phase.

STAGES IN SPERMIOGENESIS*

(The 16 stages below are based on changes in the acrosome and nucleus.)

Stage 1 — begins with the formation of a new group of spermatids, arising from the second meiotic division. The weakly staining idiosome may be seen close to the nucleus.

Stage 2 — this begins with the appearance of proacrosomic granules in the idiosome. The most frequent number of granules is two, with one being slightly larger than the other. The granules increase in size until they fuse.

Stage 3 — there is fusion of the proacrosomic granules into a single large granule adjacent to the nucleus.

Stage 4 — the granule formed in stage #3 becomes enlarged and flattens onto the nucleus, and begins its lateral extension over the nucleus, initiating the "cap" phase of spermiogenesis.

Stage 5 — there is extension of the cap so that a lateral view shows two projections from the acrosome granule.

Stage 6 — development of the cap has progressed so that its outer surface is visible in side view.

Stage 7 — there is further development of the head cap to cover from one-third to one-half of the nucleus.

Stage 8 — here the acrosomic phase of spermiogenesis is initiated. The young spermatids orient themselves with the acrosomic system toward the basement membrane, and elongation of the spermatid nucleus begins.

Stage 9 — there is now a definite change in the shape of the spermatid nucleus, with the caudal end appearing both narrower and angular. The nucleus begins to flatten, and the head cap begins its caudal progression over the dorsal or rounded side of the nucleus.

Stage 10 — migration of the acrosomic material has reached the dorsal, caudal angle of the nucleus. Both flattening and elongation continue, and the spermatid becomes narrower at its anterior end.

Stage 11 — there is now a transformation of the caudal angles of the spermatid from their previously rounded shape to sharp, acute angles. Flattening is more extreme, nuclear elongation continues, and the acrosomic material at the tip of the nucleus begins to shift caudally.

Stage 12 — the spermatid now reaches its greatest total length. The acrosome has a square anterior end, and appears as a wedge-shaped structure overlying the nucleus. The anterior extension of the nucleus begins to thin out, so that there remains only a thin thread of nuclear material underlying the anterior portion of the acrosome.

Stage 13 — there is now an abrupt shortening of the length of the spermatid by about 20%. The caudal angles assume the same shape as in the mature sperm. The acrosomic material migrates caudally, giving the appearance of a hook.

Stage 14 — the general shape of mature sperm is now attained. The acrosome appears as a crescent, and elongation at the tip results in a slender, sharp point.

Stage 15 — the nucleus now narrows, and the tip of the maturing sperm continues to develop. The acrosome is not wide, and resists staining.

Stage 16 — this is the condition of the spermatozoon at the time it is liberated from the seminiferous tubule. Both the acrosomic material and the nucleus extend to the extreme tip of the spermatozoon. A perforatiorium is probably present but not easily stained, or revealed.

* Based on work by E. P. Oakberg, 1956, Am. Jour. Anat. 99:391–413.

5) peripheral migration toward the tubular wall, and 6) phagocytosis of these migrating fragments by the Sertoli cells. The demonstration of acid phosphatase activity within free granules, the sequestration of Golgi lamellae, and both classes of vacuoles suggest that the initial body degradation occurs through liposomal cytoplasmic autophagy. (Pers. Com. S. E. Dietert) The nuclear material loses its identity and becomes lodged in a distinct head, covered by a bag-like acrosome, over the perforatorium. The acrosome carries the enzyme hyaluronidase or a zymogen-like precursor which, at fertilization, depolymerizes the hyaluronic acid jelly of the cumulus, and the exposed perforatorium carries a lysin to alter the chemical properties of the zona so that the sperm head can penetrate it. Enclosing the acrosome is the plasma membrane which extends around the entire spermatozoon. Over the distal half of the sperm head the plasma membrane carries a thickened inner aggregation of fine particles. Although the content of the sperm head is homogeneous, the DNA molecules of the haploid chromosomes are present in some form that preserves their gene sequences. A middle piece contains the mitochondria, the Golgi apparatus, and two centrioles; and the tail is made up of many flagella-like strands. The heads of many such spermatozoa lie embedded in the cytoplasm of a Sertoli cell, deriving nutrition therefrom until they are liberated during coitus.

acrosome

head

middle piece

tail

ELECTRON MICROGRAPH OF MOUSE SPERM HEAD

(Courtesy Dr. D. W. Fawcett)

MATURE MOUSE SPERMATOZOAN

The spermatozoon varies in length, width, and shape with the mouse strain. It generally has a hooked head about 0.0080 mm long, a short middle piece, and a very long tail, for an overall length averaging 0.1226 mm. Enzyme machinery controlled by the mitochondria in the middle piece facilitates the movement of the tail. The swimming of the spermatozoon depends on a waving or bending motion originating in the base of the tail, probably in the distal centriole of the middle piece, and is propagated distally.

As stated, in the process of transformation into a spermatozoon, a spermatid sloughs off a cytoplasmic tag but small vesicles appear within the retained Golgi apparatus, each containing a homogeneous granule. These coalesce to form an acrosome granule, which becomes attached to the surface of the nucleus. The two spermatid centrioles move to the plasma membrane to which the distal one becomes attached. A fine thread grows out from this distal centriole as a core for the developing tail. The two centrioles then align themselves in the direction of the nucleus, free from the surface and to the side opposite the acrosome. The base of the tail is thus drawn into the cytoplasm, along with the distal centriole, but the plasma membrane is also indented to form a sheath around it. Later the distal centriole forms a ring, still associated with the base of the tail, and finally lodges near the plasma membrane. Mitochrondria aid in development of the middle piece, and an outer ring of nine coarse fibers grows from the proximal centriole to form a sheath around the axial filament emanating from the distal centriole. All the cytoplasm of the spermatid is then shed except for a minute droplet that is lost in transit through the epididymis. After extrusion, the cytoplasmic masses undergo the alterations which are accompanied by fragmentation of the Sertoli site of indentation and phagocytosis of the fragments by the Sertoli cells.

DNA (deoxyribonucleotide) synthesis occurs only in spermatogonia and primary spermatocytes, since the succeeding stages do not involve mitotic divisions. The average life span of the spermatogonia is 27 to 30.5 hours, and DNA synthesis takes longer in type B than in type A. The nuclei of spermatids show a noticeable change in DNA pattern during the metamorphosis into spermatozoa. The DNA eventually becomes evenly distributed, and then resistant to desoxyribonuclease, so that the chromosomes undergo an orderly rearrangement. RNA may be found in the chromatoid body of a spermatid.

The process of spermatogenesis in the mouse is basically similar to that in any other mammal. One cycle of the seminiferous epithelium takes 207 ± 6 hours, and four such cycles occur between type A spermatogonium and mature spermatozoon. The production of mature spermatozoa from the original spermatogonial cell takes about five weeks in the mouse (and about twice as long in man). The testes, and particularly the mature spermatozoa, are the richest sources of animal hyaluronidase, and this enzyme effectively disperses the cumulus cells surrounding the mature ovum at the time of fertilization. Each individual spermatozoon carries enough of the enzyme to clear a path through the cumulus cells to the gel matrix of the ovum. The hyaluronic acid cement substance tends to bind together the granulosa cells of the cumulus, so that the sperm head must be supplied with abundant enzyme.

In summary the best evidence suggests the following major intervals:

Meiotic prophase - 12. 5 days

Spermatid stages - 9. 5 days

Last four stages of spermatogenesis - 5. 5 days and the maximum interval between type A spermatogonium and its release as mature spermatozoa from the seminiferous tubule is 35. 5 days.

C. DUCTS: Remnants of the embryonic excretory system that became functional parts of the male reproductive system are the rete testis, the efferent ducts, the tripartitite epididymis (caput, corpus, and cauda), and the ductus deferens. Each of these structures is paired. The rete testis is an anastomosing system of ducts into which the seminiferous tubules ultimately empty their contents. It is lined with low simple cuboidal epithelium. The rete opens into the collecting chamber (lacuna) located outside the tunica albuginea, which in turn opens into the three to seven connecting efferent ducts. Each duct has two parts, the first is short, convoluted, and surrounded by fatty tissue, and the second is more convoluted and surrounded by a connective tissue capsule continuous with that of the epididymis. The ducts join to form the first part of the catput of the epididymis. An efferent duct is lined with both low and high columnar epithelial cells, and hence its lumen is irregular in contour. Below the epithelial cells is a basement membrane, and outside this are some smooth muscle fibers. The caput of the epididymis is convoluted and divided into seven or eight segments (lobules), the second of which is lined with high columnar epithelial cells. These cells have no cilia, and their oval nuclei are variously placed. In the third segment of the caput the epithelium is lower, the nuclei are more uniformly situated, and the lumen is narrower. The lumen widens toward the cauda, and smooth muscle cells may be seen in the outer wall. The ductus deferens, which extends from the cauda of the epididymis is lined with high columnar epithelium, slightly stratified. The mucosal layer consists of deep longitudinal folds, whereas the lamina propria is made up of fibrous connective tissue. The outer longitudinal and inner circular smooth muscle fibers constitute a rather thick enclosing wall. The adventitia, or loose connective tissue, surrounds the whole structure. The ductus deferens enters an ampulla and then the urethra. The ampulla is lined with low columnar cells having little cytoplasm and large oval, darkly staining nuclei.

D. THE ACCESSORY GLANDS: The accessory glands do not contain or carry germ cells but are adjuncts to their proper functioning and transport. They include the seminal vesicles, (vesicular secretory glands), three pairs of prostate or coagulating glands, ampullary glands, bulbourethral glands, and preputial glands.

The paired seminal vesicles are long, lobulated glands curved at the lateral tips, and located next to the first pair of prostate glands. Each vesicle has a lumen with alveolar outpocketings and a lining of high columnar epithelium with oval nuclei near the base. The cytoplasm is basophilic and contains heavy, eosinophilic secretion granules. Smooth muscles and a connective tissue sheath surround the vesicle.

The coagulating glands, attached to the posterior margins of the seminal vesicles are the first of the three pairs of prostate glands. They secrete a substance that, when mixed with the secretions of the vesicular glands, forms a coagulum the presence of which in the vaginal orifice is considered proof of successful copulation. This usually means insemination and fertilization except in cases where the male has been vasectomized. Seen grossly, a coagulating gland appears to be homogeneous; yet it actually consists of a multi-folded mucous membrane with many projections into a central lumen. The lining is simple columnar epithelium with centrally placed nuclei and eosinophilic cytoplasm. The gland has two ducts, both lined with low cuboidal epithelium having slightly basophilic cytoplasm. As with most accessory glands, there are smooth muscles and an enclosing connective tissue sheath. The dorsal prostate glands are smaller and more rounded than the coagulating glands. Each has several ducts. Moreover, their mucous membrane is smoother, and depending upon the amount and accumulation of secretion, mucous folds may or may not be present in the ventral prostate glands. Their ducts are lined with low cuboidal epithelium, having slightly basophilic cytoplasm and deeply staining nuclei.

The paired ampullary glands are found around the base of the ductus deferens and open into the vestibule of the ampulla. They are lined with low cuboidal epithelium having large and oval nuclei and are thrown into delicate longitudinal folds, The tubules contain a dense, red-staining homogeneous secretion, which, on fixation, tends to coagulate toward the center of the lumen.

The paired bulbourethral glands (of Cowper) are very large and closely adherent to the penis, just outside the body wall. A duct from each gland appears to enter the anterior wall of the urethral diverticulum. The bulbourethral glands are both tubular and alveolar, surrounded by striated muscle, and lined with columnar epithelium having primarily basophilic cytoplasm. Their secretion has a staining reaction similar to that of mucin.

E. URETHRA: The connection of the bladder is lined with thinner and thinner layers of transitional epithelium, until it joins the urethra, which has a ventral lining of stratified squamous epithelium and a dorsal lining of low cuboidal epithelium. The major portion of the lining is the stratified squamous epithelium. The lamina propria is loose connective tissue, rich in blood vessels, and it is enveloped by a thick layer of striated muscle. Urethral glands (of Littré) consist of small alveoli lined with cells having predominantly granular and basophilic cytoplasm.

The preputial glands are large, flat, leaf shaped sebaceous-type glands. They may be homologues of the clitoral glands of the female. Their lining is polyhedral epithelium with pale staining nuclei. As the cells degenerate, they form a fatty secretion. Each preputial gland opens separately through a long duct at the tip of the prepuce (foreskin) and its function is probably lubrication.

F. PENIS: The penis consists of one thin corpus cavernosa and two thick corpora cavernosa. The thin corpus is an extension of the urethra, surrounded by a tunica albuginea (dense fibrous connective tissue) within which is a layer of circular smooth muscle fibers. The lumen of the urethra expands into the urethral bulb, forming paired lateral diverticuli. The two thick corpora cavernosa are also surrounded by the tunica, which separates proximally and allows the two cavernosa to join distally. The os penis, a small bone, may be found within the fibrous septum between the two thick corpora cavernosa and projects beyond the orifice of the penis. The glans penis (terminal) is covered with the prepuce or foreskin, consisting of stratified squamous epithelium that may contain a few scattered hair follicles. The root of the penis is attached to the pubic bone by the ends of the corpora cavernosa, associated with the ischio-cavernosus muscle.

THE ADULT FEMALE

The female reproductive system consists of paired ovaries and oviducts, a bicornuate uterus, a cervix, vagina, clitoral gland, and clitoris.

A. OVARIES: The ovaries are suspended by ligaments from the dorsal body wall, lateral to the kidneys. This bulge into the peritoneal cavity is covered by germinal epithelium instead of mesothelium. Each ovary is within a bursa from which liberated ova cannot escape. The suspensory ligaments are invested with smooth muscle fibers, which seem to extend into the ovarian coverings. Other ligaments connect each ovary to the anterior end of a uterine horn. These contain smooth muscles, which supply the uterine horn and the in-fundibulum of the oviduct. They also contain the epoöphoron, vestiges of the Wolffian bodies. The infundibular muscle connects with the hilus of the ovary. Each uterine horn is supported by a dorsal broad ligament (mesometrium) which contains some fat, as well as some smooth muscle fibers continuous with the muscles of the uterus itself.

An ovary is a small, pink structure with its surfaces covered by a thin, transparent, connective tissue membrane, the tunica albuginea or ovarian cap-sule. The whole is enveloped by mesothelium. The ovary of a mature mouse has an inner medullary portion (the zona vasculosa and stroma) and a more peripheral portion, the cortex, within which growing follicles may be seen.

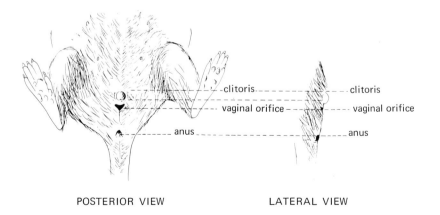

POSTERIOR VIEW   LATERAL VIEW

EXTERNAL GENITALIA OF ADULT FEMALE MOUSE

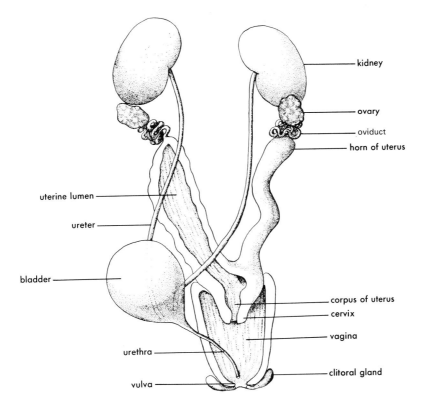

kidney

ovary

oviduct

horn of uterus

uterine lumen

ureter

bladder

corpus of uterus

cervix

vagina

urethra

clitoral gland

vulva

VENTRAL VIEW OF UROGENITAL SYSTEM OF FEMALE MOUSE
(From R. Rugh, "Vertebrate Embryology", Harcourt, Brace & World, Inc., New York, 1964.)

The small primary follicles, found just beneath the tunica, consist of oöcytes and their surrounding follicle cells, which appear at first to be squamous. The oöcyte nucleus is vesicular, with chromatin granules and a distinct nucleolus. As a follicle enlarges, the surrounding follicle cells become cuboidal, then layered, and finally are separated from the oöcyte by a clear, noncellular follicle cell secretion known as the zona pellucida. This plays an important role in the fertilization process. Enclosing each enlarging follicle are connective tissue fibers in the stroma, which together form the theca folliculi. As the follicle enlarges further, a distinction can be made between the theca interna, which has a rich blood supply and rather loosely arranged, constituent cells, and the outer and denser theca externa, in which the fibers are arranged concentrically.

The stroma is made up of dense fibrous connective tissue. Blood vessels penetrate the entire ovary, entering and leaving by the hilus. Outside the cortex is the single layer of cuboidal epithelium known as the germinal epithelium, and just outside this is the tunica.

The follicle continues to enlarge, since it is accessible to all the essential food elements from the blood (such as vitamins, and steroid hormones) but inaccessible to most enzymes, antigens, antibodies, and protein hormones. Small and irregular lacunae appear among the cells, and merge into an antrum, which becomes filled with follicular fluid, or liquor folliculi. The lining of the antrum

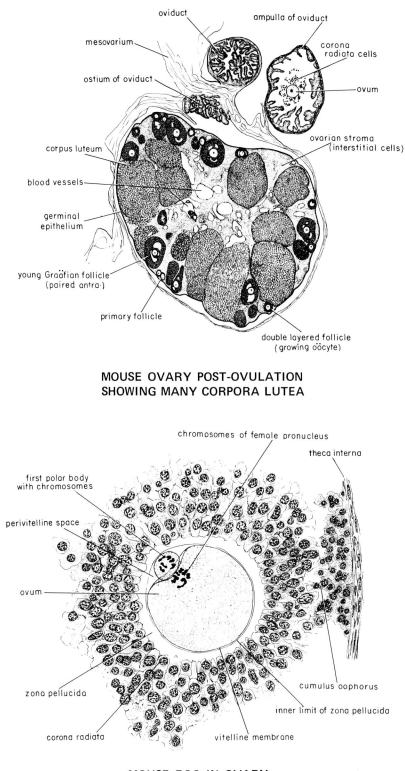

oviduct

ampulla of oviduct

mesovarium

corona radiata cells

ostium of oviduct

ovum

corpus luteum

ovarian stroma (interstitial cells)

blood vessels

germinal epithelium

young Graäfian follicle (paired antra·)

primary follicle

double layered follicle (growing oöcyte)

**MOUSE OVARY POST-OVULATION
SHOWING MANY CORPORA LUTEA**

chromosomes of female pronucleus

theca interna

first polar body with chromosomes

perivitelline space

ovum

cumulus oophorus

zona pellucida

inner limit of zona pellucida

corona radiata

vitelline membrane

**MOUSE EGG IN OVARY
(nearly ready for discharge)**

consists of stratified follicle cells that are conspicuously granular and hence are called granulosa cells. Granulosa cells surround the growing oöcyte and the zona pellucida to form the cumulus oöphorus, the matrix of which contains protein and hyaluronic acid, to be liquefied by the enzyme hyaluronidase from the spermatozoon. (The hyaluronidase probably comes from the acrosome, which is discarded at fertilization.) The zona pellucida accumulates around the growing ovum, tending to separate it more and more from the nutrition-bearing cumulus cells. The zona is a weakly acidic mucoprotein material. Since peroxide, trypsin, chymotrypsin, pronase or mold protease tend to dissolve it, it is believed that its main component is hyaluronic acid. The zona pellucida is a secondary membrane which is glossy, tough, resilient, elastic, and the product of follicle cells. While it is homogeneous, it has an irregular surface with microvilli from the surface of the vitelline membrane. Some cellular processes from the granulosa cell surfaces recede, due to the thickening of the zona, but maintain contact with the vitelline membrane. The protoplasmic extensions from the zona radiata penetrate the zona pellucida and channel yolk nutrition to the egg through its membrane. The zona is sloughed off during blastulation, probably owing to the expansion of the blastocyst as it acquires its blastocoel. Until then it may aid in the maintenance of a normal cleavage pattern and prevent the fusion of closely placed ova. It can be removed at any early stage of development by digestion with the enzyme pronase, which also disperses the cumulus cell.

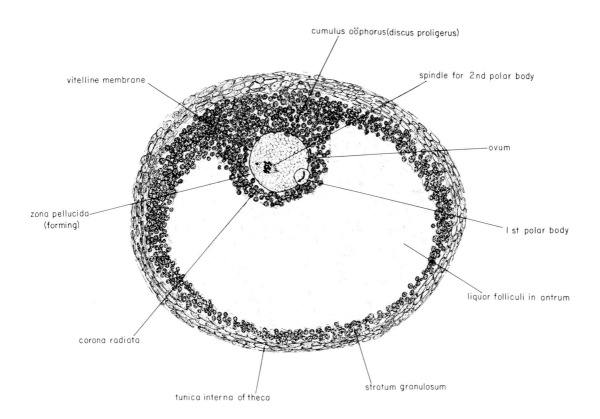

cumulus oöphorus(discus proligerus)

vitelline membrane

spindle for 2nd polar body

ovum

zona pellucida
(forming)

I st polar body

corona radiata

liquor folliculi in antrum

tunica interna of theca

stratum granulosum

MATURING MOUSE OVUM IN ENLARGING FOLLICLE

The cumulus cells immediately around the zona pellucida are radially arranged elongate cells, often attached to the ovum by delicate tubule-like cytoplasmic structures, and are collectively called the zona radiata. It is possible that the radially arranged cells aid in directing the spermatozoon to the ovum after its liberation at ovulation. They may also provide a sticky target to entrap the spermatozoon initially. An intact cumulus is believed essential to successful sperm penetration. As the antrum increases in size, the entire follicle enlarges and moves toward the surface of the ovary to project into the periovarial space. The developing ovum and its surrounding cells are then known as a Graäfian follicle, with an average diameter of over 500 microns. Often the mature ovum and its cumulus cells can be seen floating freely within the enlarged antrum, just prior to ovulation or liberation from the ovary. A Graäfian follicle can be seen with the naked eye as a bulbous projection from the surface of the ovary.

B. OÖGENESIS: At 8 days gestation the primordial germ cells of the female appear in the yolk sac splanchnopleure and are rich in alkaline phosphatase. They migrate by ameboid and undulating movement (as do the primordial germ cells in the males) along the dorsal mesentery to the mesenteric folds and arrive at the genital (germinal) ridges by 9 to 10 days gestation. By 11 days gestation the potential ovary shows its characteristic cortical proliferation in contrast with the medullary development of the potential testis. The controversy has not been resolved as to whether the migrating primordial germ cells are the lineal ancestors of all future ova or whether some ova are derived from the germinal epithelium of the genital ridge. The first theory is favored by most embryologists. These primordial germ cells are elongated, have definite clear cell membranes with globular mitochondria around the nucleus, nuclei of varying shapes apparently related to movement, and two or more nucleoli of high contrast. Beginning at about 13 days gestation meiotic prophase may be found, and this leads to ovum formation. Leptotene and zygotene stages may be seen in oöcytes at 14 and 15 days gestation and a high rate of oögonial division begins at this time. Numerous pachytene stages may be seen from 16 and 17 days to birth. Some continue through diplotene and diakenesis. DNA may be recognized in the meiotic prophases by 13 days and until 16 days gestation.

At about the time of birth some primordial ova acquire follicle cells, start on the maturation process, and shortly thereafter all have follicle cells. The total number of oöcytes at this has been estimated as up to 75,000, far in excess of the number that could ever mature. Many degenerate almost immediately, and others degenerate later or become follicle cells so that the stock is reduced to a fraction of the original number. The primary oöcytes may be distinguished from other cells of the ovarian cortex by their relatively large size, correspondingly large spherical nuclei, and by yolk particles in the cytoplasm. Although it has been suggested that some ova proliferate from the germinal epithelium of the ovary and that regeneration is possible after ovariectomy, this theory has not been confirmed. Young ova are large, have a spherical and heavily stained nucleus, and may occur in pairs. In the female mouse the germ cells pass through all of the oögonial divisions and reach the oöcyte stage while still in the embryo. About 3 days after birth the oöcytes reach a static state known as the dictyate stage, and this is maintained until a few hours (12) before ovulation. This interim may be as much as a year.

At 12 to 15 days after birth some antra appear, and at 17 days some follicles are almost maximum size. The ovum reaches its maximum size at the time that antrum formation begins. Follicular growth continues, and may even do so without the presence of an ovum. But as the ovarian egg increases, and as ovulation approaches, the cytoplasmic RNA (ribonucleicacid) is augumented sharply and many nucleoli can be seen. These early small nucleoli are low in RNA. Then the first ovulation occurs. The atretic follicles also increase in number, reaching a maximum at about 18.5 days. Once oögenesis has begun, it continues in regular $4\frac{1}{2}$ to 5 day cycles throughout the reproductive life to about 12 or 14 months of age. Ova degeneration occurs concomitantly. A fully formed and normal ovum consists of a few small Golgi granules, a thin layer of yolk, large Golgi granules, a nucleus, and abundant mitochondria. These mitochondria generally are concentrated in the periphery, and at fertilization they migrate to the region of the developing pronuclei and aggregate around them. During the growth of the oöcyte the DNA in its nucleus is reduced and a perinuclear band of ribonucleic acid makes its appearance in the cytoplasm. The mature ovum is slightly polarized in that on its dorsal side is a cortical zone of ribonucleic acid and on the ventral side appear some vacuoles.

According to some estimates at each estrus 1,000 or more ova are available for oögenesis, but only about 1% ever mature, for a 99% loss in potential. With estrus every $4\frac{1}{2}$ to 5 days, this means that in an average reproductive life of 2 to 12 months of age some 60,000 ova are available, of which only about 500 mature and only about 100 could possibly result in offspring.

Anomalies in oöcytes, aside from atretic ones, include giant oöcytes and polynuclear oöcytes. Polyovular follicles appear frequently, particularly in immature mice.

The first meiotic division of an oöcyte in a mature ovary begins 9 to 10 hours before ovulation, presumably stimulated by the luteinizing hormone from the previous ovulations. Its prophase is sometimes called the dictyate or leptoene stage. The first metaphase lasts roughly 6 hours. The second meiotic division in initiated immediately, but is arrested in metaphase where it remains until the ovum is fertilized. Thus the interval between the first and second meiotic divisions is entirely dependent on the time between ovulation and fertilization.

The DNA content of the female pronucleus is reduced to 26% of the primary oöcyte (which was tetraploid). The recently ovulated mouse egg has 5 to 10 times as much DNA as the ultimate normal diploid somatic cell. Some of the DNA is found in the mitochondria. Protein synthesis occurring in the developing mouse ovum prior to the RNA production by the nucleoli suggests ribosomes may be of maternal origin. The mature ovum has finely granular, homogeneous cytoplasm limited by a plasma membrane generally called the vitelline membrane (not to be confused with the membrane of the same name in invertebrates). The nucleoplasm is enclosed by a double membrane, perforated by pores. The mature ovum has a diameter of about 95 microns and a volume of about 200,000 cubic microns. Its fertilizable life span is 10 to 15 hours, and the loss of fertilizability is due to changes in the vitelline membrane, and the cortical cytoplasmic granules, occurring even before the zona pellucida loses its penetrability.

Ova are liberated from the adult female mouse ovary approximately every 5 days. In this interval two changes take place, maturation of the ovum and growth of the follicle. Correlating changes in the vaginal mucosa can be used to estimate the time of ovulation. Many of the stages, but not all, in the maturation of the ovum can be seen in any mature ovary and at any phase of the estrous cycle. A composite drawing of the ovary is given showing all the stages of oögenesis as well as a corpus luteum. The stages are as follows:

STAGES OF OÖGENESIS

Stage 1 — Oögonium, hardly distinguishable from other cortical cells of the ovary, and with no follicle cells. The amount of DNA is constant, but is diluted with the enlargement of the nucleus.

Stage 2 — Primary oöcyte, invested with a single loose layer of squamous epithelial (follicle) cells and having a nucleus slightly larger than those of the adjacent cells.

Stage 3 — Primary follicle, with a single layer of cuboidal follicle cells surrounding the oöcyte, the nucleus of which is enlarging.

Stage 4 — Double layer of follicle cells around the enlarging oöcyte.

Stage 5 — Many-layers of follicle cells around the enlarging oöcyte. Near the germinal vesicle is a yolk nucleus (Balbiani's body), and near it is a Golgi apparatus.

Stage 6 — Antral spaces are scattered among the follicle cells. Mitochondria form centers for yolk concentration; the diameter of the follicle is about 200 microns.

Stage 7 — Distinct antral spaces; the first polar body forms in the first maturation (meiotic) division, leaving a secondary oöcyte. Rodent polar bodies are characteristically large.

Stage 8 — Single fused antrum, with the oöcyte suspended in the cumulus oöphorus and the first polar body in the perivelline space.

Stage 9 — Antrum swollen with follicular fluid, and the ovum ready to erupt from the ovary, with its nucleus in metaphase of the second meiotic division; the diameter of the follicle about 500 microns. (See Peters and Borum 1961 for illustrations)

The first of the two maturation (meiotic) divisions of the ovum occurs at stage 7 and the second after invasion by the spermatozoon. Upon successful insemination and syngamy restoration of the diploid state of 40 chromosomes is accomplished.

With specific reference to the nucleus of the maturing ovum, another table of maturation may be presented. This relates to stages beginning at stage 7 in the preceding table.

1 — leptotene (sometimes called dictyate)
2 — late leptotene (or dictyate)
3 — chromatin mass (pachytene, diplotene)
4 — diakenesis
5 — first pre-metaphase I
6 — first meiotic metaphase I
7 — first anaphase I and telophase I, first polar body formation

8 — second pre-metaphase II
9 — second metaphase II (time of ovulation) and fertilization
10 — second polar body formation (penetration of the ovum by the spermatozoon and syngamy of the male and female pronuclei follow)

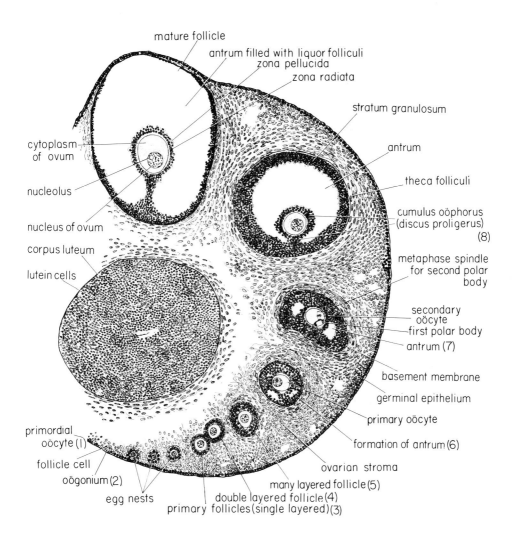

mature follicle
antrum filled with liquor folliculi
zona pellucida
zona radiata
stratum granulosum
cytoplasm of ovum
antrum
theca folliculi
nucleolus
cumulus oöphorus (discus proligerus) (8)
nucleus of ovum
metaphase spindle for second polar body
corpus luteum
lutein cells
secondary oöcyte
first polar body
antrum (7)
basement membrane
germinal epithelium
primary oöcyte
formation of antrum (6)
ovarian stroma
primordial oöcyte (1)
follicle cell
many layered follicle (5)
oögonium (2)
double layered follicle (4)
egg nests
primary follicles (single layered) (3)

DEVELOPMENT OF THE MOUSE OVUM AND OVARIAN FOLLICLE
(Parenthetical numbers refer to maturation stages — see text)

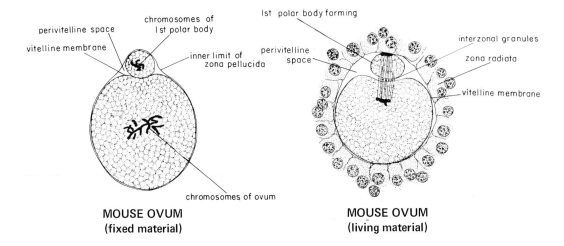

MOUSE OVUM
(fixed material)

MOUSE OVUM
(living material)

C. OVULATION: Ovulation is the liberation of the mature ovum from its follicle into the periovarial space. This process is not cataclysmic, as was once thought, but takes from 30 to 45 minutes. The ovary itself pulsates, most frequently and vigorously during estrus. The fluid within the enlarging antrum, the liquor folliculi, becomes more viscous as the time for rupture of the follicle approaches. There is some evidence that a second grade of liquor folliculi appears, first among cells of the cumulus oöphorus, causing the cumulus to become detached from its base and to float freely with its ovum in the antrum. During these preovulation changes the nuclear membrane of the ovum disappears, and the scattered chromatin granules aggregate as small, dense chromosomes. Unlike most animal cells an ovum often appears to have no centrioles or astral rays. Nevertheless, a spindle forms, and the newly assembled (tetraploid) chromosome line up in pairs along the equatorial plate. The chromosome pairs then divide and the two members of each pair move toward opposite spindle poles. One pole with its chromosomes becomes the first polar body, to be extruded and to lie within the vitelline membrane. (Stage 7). Since the first polar body is not always seen, it is believed to either escape the perivitelline space or become fragmented, or cytolyzed. Immediately the remaining chromosomes, which are diploid, anticipate the next division, which will give rise to the second polar body. The second polar body spindle is forming at the time of ovulation, and remains in metaphase until fertilization.

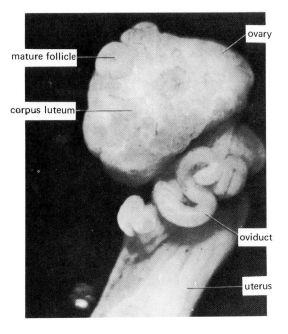

ACTIVELY OVULATING MOUSE OVARY

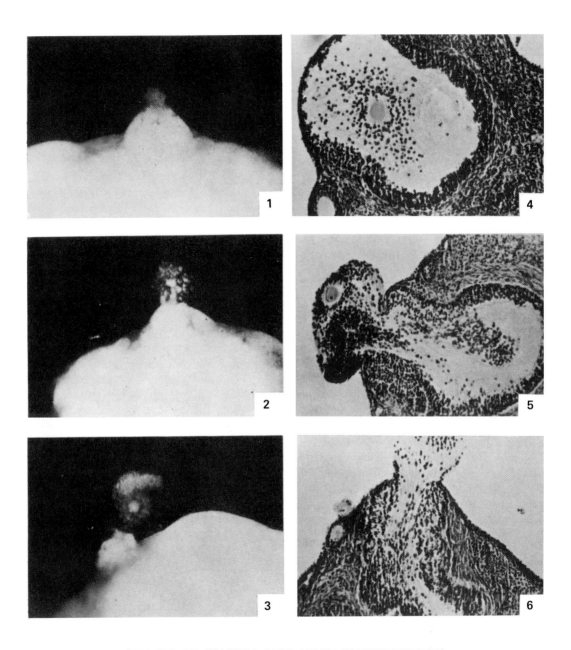

STUDIES ON *IN VITRO* OVULATION IN VERTEBRATES
S. NADAMITSU

(Jour. Sci. Hiroshima Univ., Ser. B. Div. 1, Vol. 20; 27-32, 1961)

Both in the vicinity of the matured follicle and in the medulla of the ovary will be seen large vessels engorged with blood. Each ovum must rupture through the granulosa cells, the stretched tunica albuginea, and the much flattened germinal epithelium. The periovarial space becomes congested with cells, fluid, and the ovum. Opposite to the area of rupture the theca interna and the granulosa cells may be thrown into folds, owing to the collapse of the follicle and the reduction in fluid pressure. Up to 10 ova may be liberated from a single ovary during a short period of time, to aggregate in the region of the infundibulum. The many cilia of the infundibulum quickly transport the mature ova into the ampulla of the oviduct, where fertilization can occur. Because of the slight stickiness of the liquor folliculi the ova tend to clump together.

The mean time for the onset of estrus in a normal female mouse is 4 to 6 hours after the onset of darkness in these nocturnal animals. Ovulation occurs from 2 to 3 hours after the onset of estrus. Both events are therefore correlated with the diurnal light cycle and can be shifted if that cycle is shifted. In general, ovulation occurs at about the mid-point of the dark period, sometime between 2 and 4 A.M. under normal conditions. The secondary oöcyte stays in metaphase for some 6 hours and then goes into anaphase, which takes 1.2 hours, and on into telophase, which takes only 0.2 hours. The separation of the second polar body depends upon the time of fertilization. The first sign of the approach of the final division is the condensation of the chromatin, which may be seen in the first hours of normal darkness on the night when ovulation is to occur.

Prepuberal mice can be induced to ovulate by treatment with hormones and generally respond by superovulating, producing up to 100 ova. Gonadotrophic hormones cause ovulation and render a female receptive to a male. A daily dose of 0.125 to 0.25 milligrams of progesterone ensures implantation and a dose of 1.0 to 2.0 milligrams maintains pregnancy. Relaxin facilitates delivery, administered at 13, 16, and 19 days gestation. A combination of progesterone and relaxin stimulates lactation. Since the uterus of a prepuberal female is spatially inadequate and ill prepared for such large numbers of ova, pre-implantation and fetal mortality are high and the fetuses that survive are best delivered by Caesarian section. In practice a female of 3 to 6 weeks of age is injected intraperitoneally with 10-15 international units (I.U.) of follicle stimulating pregnant mare serum (PMS) in 0.9% NaCl and 35 to 40 hours later with 10-15 I.U. of human gonadotropic hormone (HCG). In 12 to 16 hours she will be actively ovulating and may be exposed to sexually mature males to insure fertilization of the ovulated ova. This method is ideal for providing demonstration or teaching material of ova, or at about 24 to 32 hours after fertilization two cell embryos may be obtained from the ampulla of the oviduct. *

The rupture of the ovarian follicle is not associated with any appreciable hemorrhage in the mouse. There simply remains a small gap in the follicle wall. But the loss of tension following the escape of fluid allows the blood

*A basic culture medium for the 2 cell mouse embryo, as reported by Brinster (1965) follows. Ova should be cultured in small drops of medium under liquid paraffin oi.

	Gms/liter		Gms/liter
NaCl	6.97	$NaHCO_3$	2.106
KCl	0.356	Ca lactate	1.147
CaCl	0.189	Penicillin	100 u/ml
KH_2PO_4	0.162	Streptomycin	50 ug/ml
$MgSO_4$	0.294	Crystalline bovine	
		serum albumin	1.000

capillaries in the surrounding area to repair the damage in about 2 hours, after which the vestiges of the follicle are known as the corpus luteum. The cells of the theca interna project into the follicle, along with the remaining granulosa cells. They are soon invaded by capillaries to form radially arranged trabeculae which occupy the original antrum. The cells are later filled with a yellowish lutein substance. These changes require active cell proliferation in both the theca interna and the granulosa layers, as well as the growth of capillaries. The granulosa cells then hypertrophy to increase the volume of the corpus luteum. The transformation from granulosa cells to lutein cells is gradual. The granulosa cells are small with darkly staining nuclei and basophilic cytoplasm, while the lutein cells are large and polyhedral, with eosinophilic cytoplasm.

The corpus luteum marks the site of a follicle that has liberated its ovum. The presence of a number of such lutein masses in an ovary does not inhibit further ovulation, and they may be seen in an ovary with mature follicles. The presence of actively secreting corpora lutea is essential for the continued nutrition of the free blastocyst, and it prevents the degeneration of ova. However, as estrus approaches, the fat and lipoid content of the corpora lutea increases, and the lipid granules coarsen. On the other hand, after ovulation there is very little lipid found in the previous corpus luteum. If fertilization does not take place the corpora lutea begin to degenerate in 10 to 12 days, leaving only fibrous and fatty scar tissue. If fertilization does take place, lipids do not accumulate in the corpora until about halfway through gestation (8 or 9 days), and the granules are smaller than in the corpora of a non-pregnant animal. At 13 days of gestation a corpus has reached its maximum diameter of about 1 mm, and at 18 days it begins to shrink. It does not completely disappear for some days after parturition, remnants persisting during lactation and for 5 to 6 weeks. Evidence exists for corpora lutea of lactation, which are formed after the birth of a litter. These generally remain small, with their cells having small nuclei and cytoplasm relatively free of fat.

There are cyclic fluctuations in the gonadotropic hormone content of the pituitary gland during pregnancy, with maxima at about 12.5 and 15.5 days, and a gradual increase in the release of the hormone. Gonadotropic hormone is augmented during the last half of pregnancy by luteinizing hormone. Simultaneously the acidophilic granules of the pituitary tend to disappear, an event suggesting that they may be one source of the luteinizing hormone during pregnancy.

As the corpora lutea degenerate so do many follicles. Frequently degeneration begins in the ovum itself and proceeds to the surrounding cells and tissues. The ovum and its nucleus may exhibit bizarre configurations. If the zona pellucida has already formed, it remains as a halo around the disappearing follicle. The vitelline membrane arises from the ovum itself; the zona pellucida comes from the surrounding follicle cells; the cumulus oöphorus consists of follicle cells and an intercellular matrix of acidic mucopolysaccharide, hyaluronic acid, and protein. The corona radiata are the innermost layer of cumulus cells.

D. THE OVIDUCTS: The oviducts are tubes extending from the periovarial spaces to the uterine horns. They have been described as transport tunnels through which the fertilizing spermatozoon moves in one direction and the fertilized ovum in the other.   Each begins with a ciliated and fimbriated infundibulum (or ostium) within the periovarial space.   The ciliated epithelium of the infundibulum beats rapidly and forms a current in the basic (pH 8.05) oviduccal fluid that draws liberated ova into it and thence to the bulbous, and thin-walled ampulla.   The ampulla itself is not highly ciliated; it appears to be an expandable sac, dilating during estrus, in which the ova may accumulate to await fertilization.   The ampulla is continuous with a narrow looped tube of the oviduct lined with simple, low columnar non-ciliated epithelium.   The second loop of the oviduct exhibits peristaltic contractions some 12 to 16 seconds apart during ova transport.   This coiled duct joins a uterine horn eccentrically at its cephalic end, projecting into it.   The entire oviduct is invested with a coat of smooth muscle fibers, which aid in propelling the ova to the uterus.   There is probably a specific pattern of muscular contractions of the oviduct which causes the eggs to rotate as they progress downward.   There is also deposited onto the eggs, from the glandular cells of the oviduct, a mucous coating outside the zona pellucida, which may aid in the adhesion of the eggs to the uterine mucosa at implantation.   Adhesion may precede actual invasion by several hours.

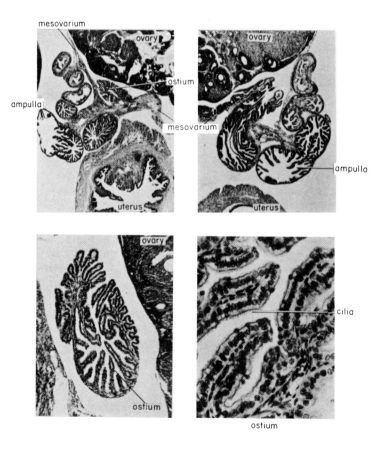

FIMBRIATED OSTIUM OF THE MOUSE

E. THE UTERUS: The uterus has two horns and a caudal section, corpus uteri, that is undivided.   In other words, it is shaped like a "Y" with a very short stem.   The bulk of its tissue is muscular; it has an outer layer of longitudinal smooth muscle fibers and an inner layer of circular smooth muscle fibers.   The lining of its folds is simple columnar epithelium, with numerous spiral tubular uterine glands.   The lamina propria contains small polyhedral cells with large, round nuclei as well as clusters of lymphocytes. The endometrium is really the mucosal layer in the nonpregnant animal, consisting of the lamina propria, the lining epithelium, the uterine glands, and many blood vessels.   The small polyhedral cells of the endometrium change into the large decidual cells of the placenta during pregnancy.   The fluid of the uterine lumen is slightly more alkaline than the surrounding peritoneal fluid.

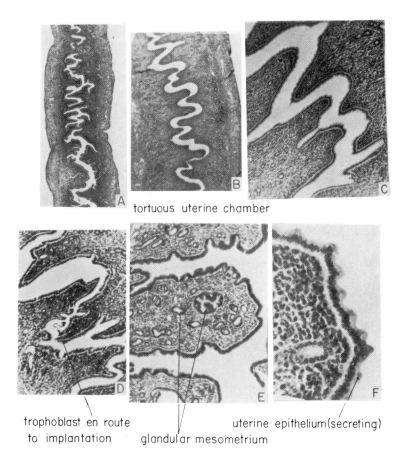

tortuous uterine chamber

trophoblast en route
to implantation   glandular mesometrium   uterine epithelium(secreting)

UTERUS OF THE MOUSE

The myometrium (peripheral to the endometrium) consists of the compact circular muscles, an enclosing layer of loose connective tissue with blood and lymph vessels, and finally the longitudinal muscles.   There the outermost enclosing layer of the uterus is a serious membrane which connects the uterine horns to the broad ligaments.

At their caudal ends the two horns are separated only by a septum consisting of longitudinal muscle and connective tissue. The tissue elements for implantation are absent from this region, which is lined with cuboidal cells. The corpus uteri projects into the short vagina, its mid-dorsal and mid-ventral wall being fused to the vaginal walls, to form a traplike space on either side (the fornix). The opening of the uterus into the vagina is through the cervix, which is lined with stratified squamous epithelium.

The maximum growth of the uterus occurs between 2.5 and 5 months of age, with the maximum RNA/DNA ratio at the latter time. The RNA content is correlated with the water content of the uterus and the amount of protein nitrogen synthesized. In a pregnant uterus the DNA content increases up to 15 days gestation and then levels off at 1400 to 1700 micrograms. The first pregnancy brings about the final maturation of the uterus. Electrical (EMG) techniques have been used to record the phasic activity in the non-pregnant as well as the early pregnant uterus, and changes in the intra-amniotic pressures of gravid mice. Spontaneous uterine contractions average one in 48 seconds on gestational day 10.5, and also at parturition, with a gradual increase in rate to days 19.5 and 20.5 when a mean interval of 32 seconds can be recorded. Pressures are greater in the earlier (11.5 day) gestation periods. Uterine contractions are autogenic in origin, and are not related to litter size in the pregnant mice. Strain specificity differences exist only during the early stages of pregnancy. Contraction rate does increase gradually as pregnancy progresses. (Pers. Com. M. L. Wood).

F. THE ESTROUS CYCLE:

The stages of maturation and estrus can be determined by analysis of vaginal smears. The changes are related to the diurnal light cycle and can be experimentally altered. The diurnal response may be controlled by the eyes, the central nervous system, and/or the anterior pituitary gland. The major normal stages and their characteristics are as follows:

PROESTRUS — anabolic, active growth in the genital tract; a swollen and congested uterus; an open vaginal orifice; largely nucleated, and some cornified, epithelium in the vaginal smear. The duration is 1 to 1.5 days.

ESTRUS — or heat; anabolic, active growth in the genital tract, a swollen and congested vulva, an open vaginal orifice, no leukocytes but both nucleated and cornified (not clumped) epithelia in the vaginal smear. Duration is 1 to 3 days.

METESTRUS 1 — catabolic, degenerative changes in the genital tract, clumped cornified epithelium exclusively in the vaginal smear.

METESTRUS 2 — catabolic, degenerative changes in the genital tract; nucleated and cornified epithelium and leukocytes in the vaginal smear. Both metestrus stages take from 1 to 5 days.

DIESTRUS — quiescent period of slow growth; nucleated epithelium, leukocytes, and some mucous in the vaginal smear. The duration is 2 to 4 days.

Vaginal smears are best obtained by means of an ordinary pipette, the tip of which has been flamed to a smooth, reduced aperture. A few drops of 0.9% sodium chloride solution are drawn into the pipette, introduced into the vagina and then retracted into the pipette. The fluid is transferred to a slide and

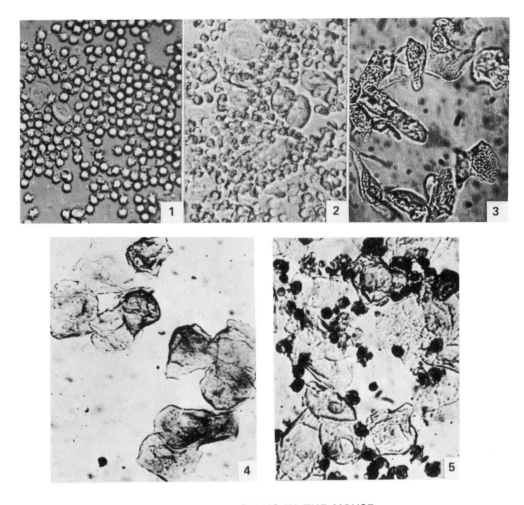

PHASES OF ESTRUS IN THE MOUSE

1 — Diestrus   — almost exclusively leukocytes, from vaginal smear.
2 — Pro-Estrus   — showing both leukocytes and nucleated epithelial cells in approximately
                              equal numbers.
3 — Early Estrus — showing clearly defined epithelial cells, some with distinct nuclei.
4 — Estrus   — large, squamous-type epithelial cells without nuclei.
5 — Post-Estrus — showing approximately equal numbers of leukocytes and epithelial cells,
                              but the latter are large, folded, and with translucent nuclei.

Mating generally occurs during stages 2 to 4 above, but these stages have relatively short duration. The major portion of the 5-day cycle is spent in diestrus.
       (From R. Rugh, "Experimental Embryology", Burgess Publishing Co., Minneapolis, 1962.)

mounted under a coverslip with a trace of methylene blue to add contrast and bring out the nuclei. Examination for cell types is carried out under low and then high power magnification, with reduced lighting. Cells may be nucleated or cornified (old, non-nucleated) epithelial cells or leukocytes. The nucleated epithelium is generally oval or polygonal, with obvious nuclei that stain readily. The cornified epithelium appears to be very thin and much folded. As estrus approaches, there may also be some mucous present, easily identified by the proper stain. The phases through which the vaginal cells go represent parallel changes in the entire reproductive tract and are hence diagnostic.

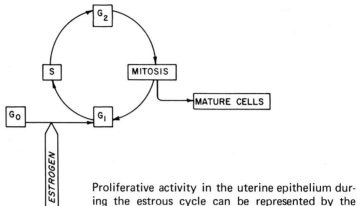

Proliferative activity in the uterine epithelium during the estrous cycle can be represented by the accompanying diagram. During diestrous, most of the cells are in what is called a G_0 phase — that is, cells that can synthesize DNA and subsequently undergo mitosis, but will not do so until the appropriate output of ovarian estrogen is attained. When this level of estrogen is reached there is a transition to a presynthetic or postmitotic interphase (G_1) — that is, to cells that will synthesize DNA (S phase) in the near future. These cells then enter the S phase and subsequently divide. Some of these cells progress through another generative cycle; others mature. The post synthetic or premitotic interphase is G_2. (See Perrotta '62).

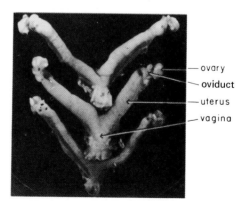

CHANGES IN THE UTERUS ATTENDANT UPON ESTRUS

Three entire genital structures of female mice.
Above — pro-estrus
Middle — estrus, note swelling of uteri
Lower — post-estrus, note shrinkage of uteri

G. VAGINA, CLITORIS, AND CLITORAL GLANDS: As noted the vagina is lined with stratified squamous epithelium which undergoes cyclic changes. The mucous membrane of the vagina displays no glands; the lamina propria is vascular and fibrous; the thin muscular coat contains both circular and longitudinal fibers; and the outer wall consists of loose connective tissue. The vagina is dorsoventrally flattened. Its opening to the exterior is the vulva, anterior to which is the clitoris, homologous to the penis of the male but lacking in erectile tissue. There is a small clitoral pouch (fossa) into which open the urethra dorsally and the two clitoral glands laterally. The clitoral glands are comparable to the preputial glands of the male, except that each contains a single hair follicle and elaborates a sebaceous-like secretion.

THE CERVIX

H. MAMMARY GLANDS: The mouse has five pairs of mammary glands, three thoracic and two inguino-abdominal. In the female they undergo changes associated with estrus, pregnancy, and lactation. The glands are rudimentary in the male.

The nipple of each mammary gland consists of three epidermal layers; the stratum germanitivum, the stratum granulosum, and the stratum corneum, covered by elastic skin. A single major duct in the stratum germanitivum leads into subcutaneous fat pads, where it forms a framework of ducts, none of which is connected with any other mammary gland. Each duct is lined with cuboidal epithelium and surrounded by circular connective tissue fibers. Just before puberty (at 6 weeks of age) and just before estrus in the mature mouse the subcutaneous ducts extend and expand, and their lining epithelium becomes mitotically active. A "spreading factor" from the end bulbs of the growing ducts helps break down some component of the surrounding connective tissue, possibly the ground substance.

The amount of "spreading factor" in the mammary gland during pregnancy closely parallels growth of the gland, reaching a maximum at two-thirds of the way through the gestation period and declining during the last third of the period, when the cells of the gland begin their secretion. Practically speaking the elaboration or activation of the factor ceases as secretion begins.

At about 11 days gestation the epithelial elements of the mammary gland start increasing, to form alveoli, and at 14 days an alveolar system is well established, in anticipation of lactation. Secretory activity begins in the alveoli near the nipple and progresses distally, until at 17 days gestation the entire gland is involved. Vascularization of the gland is concomitant with alveolar

TABLE 4

CHANGES IN THE REPRODUCTIVE TRACT CORRELATED WITH ESTRUS

STAGE	VAGINAL SMEAR	OVARY & OVIDUCT	UTERUS	VAGINA
PROESTRUS	Epithelium, cornified cells and leukocytes	Follicles large (380μ) with liquor folliculi. Few mitoses.	Some hyperemia and hydration. Mitosis, few leukocytes. Glandular	Cell proliferation, many mitoses, few leukocytes. Gaping vulva. Vaginal weight maximal.
ESTRUS	More cornified than nucleated epithelial cells.	Follicles 550μ in diameter. Ovulation: Oviduct enlarged (distended). Mitoses in germinal epithelium and follicular cells. Plasma progesterone maximum.	No leukocytes: maximum mitosis and hydration . Glands active.	Outer nucleated layer lost, replaced by cornified cells. Deep cellular proliferation. Gaping vulva.
METESTRUS I	Exclusively and abundant cornified cells.	Corpora lutea; ova in oviduct; some follicular atresia.	Hydration and distension decreased; leukocytic invasion.	Cornified cells desquamated; leukocytes begin to appear.
METESTRUS II	Abundant cornified and nucleated epithelial cells; and leukocytes begin to appear.	Corpora lutea enlarging; ova moving toward uterus. Less activity in germinal epithelium.	Leukoctyic invasion; mitosis rare; uterine wall and epithelium degenerating. Glands least active.	Many leukocytes and layers of true epithelial cells.
DIESTRUS	Both cornified and nucleated epithelial cells, plus mucous	Follicles begin rapid growth for next ovulation.	Mucous, glands and walls collapsed; many leukocytes, anemic. No gland activity, beginning of regeneration.	Leukocytes and epithelial cells; begin active proliferation. Vaginal weight minimal.

development, and the fat pads tend to disappear. During lactation, which reaches its peak at 14 days after parturition, the ducts and alveoli are dilated with milk. Suckling generally continues for 3 weeks. Any unused glands (and all the glands at the end of suckling) regress. In a resting gland fat droplets accumulate in the cytoplasm of the fibroblasts, the fat pads reform, the lumina of the ducts narrow, and the surrounding connective tissue sheaths thicken. The gland remains at rest until the next pregnancy. When senility sets in, it begins a gradual involution.

Estradiol - 17β alone promotes proliferation of the mammary epithelium but causes little alveolar differentiation. Progesterone alone promotes both epithelial proliferation and alveolar differentiation. The two hormones together exercise an additive action on cell proliferation and greatly augment alveolar differentiation. Testosterone administered to a female at 12 days gestation inhibits development of the mammary glands. It has no effect on the mammary glands of a male. Curiously, the second (or thoracic) and fourth (or inguinal) pairs of glands are the most affected.

Chapter  3 NORMAL DEVELOPMENT OF THE MOUSE

GESTATION DAYS	STAGE
1	1 to 2 cell stage in ampulla of oviduct
2	2 to 16 cell, in transit through oviduct to uterus
3	Morula, in upper uterus
4	Free blastocysts in uterus, shedding of zona pellucida
4.5	Implantation beginning, inner cell mass, trophoblastic cone
5	Pendant inner cell mass, endoderm, proamniotic cavity, primitive streak
6	Implantation complete, extraembryonic parts developing, uterine reaction
7	Ectoplacental cone, amniotic folds, primitive streak, mesenchyme, heart, pericardium forming, head process
7.5	Early neurula, neural plate, chorioamniotic stalk, embryonic lordosis, allantoic stalk beginning, pendant inner cell mass with 3 cavities, exocoelom, amniotic cavity, somites beginning to differentiate, foregut
8	Somites 1 to 4, visceral arch I, Reichert's membrane, pre-germ cells in yolk sac endoderm, embryonic lordosis, ectochorionic cyst fused with ectoplacenta and also with allantoic stalk; early regression of yolk sac; heart primordia, thyroid, optic sulcus, 1st aortic arch.
9	Somites 5 to 12, visceral arch II, disc and yolk sac placenta, embryo begins to reverse lordosis to curve ventrally with germ layers in proper relation, otic invagination, liver, 2 pharyngeal pouches, nephrogenic cord.
9.5	Somites 21 to 25, yolk stalk closes, primary germ cells migrating via mesentery to final site, primitive streak gone, tail, limb and lung buds forming, mesonephric tubules, 3 pharyngeal pouches.
10	Somites 26 to 28, visceral arch III, lung buds, pronephros reaches cloaca, posterior limb bud and lens forming, GI tract developing, sense organs differentiating, aortic arches I, II, and III.
11	Somites 29 to 42, total length 6.2 mm., thyroid, umbilical hernia, meso- and metanephros in early stages of formation; ventral pancreatic rudiment, uretric bud and mesonephric ducts to U.G. sinus, endocardial cushion fused, epiphysis and hypophysis evaginate, subcardinals formed, olfactory fibers to the brain, pigment in retina, lens cells elongate, aortic arches IV and V.

GESTATION DAYS

12	Somites 43 to 48, bronchi, aortic arch V gone, and VI reduced, vitreous humor, pancreatic rudiments, mammary welts, ribs and centrum chondrify, secondary bronchi, posterior cardinals degenerate, epithelial cords in testis, choroid fissure closed, superior vena cava enters heart.
13	Somites 49 to 60, cephalization resulting in brain differentiation, spinal ganglia, histological sex differentiation, active and complete circulation, atrioventricular valve, interventricular septum complete, nerves in optic stalk, otic capsule pre-cartilaginous, esophageal sub-mucosa thickens, neuroblasts in retina, enucleate red cells 1%.
14	Somites 61 to 63, total length 9.6 mm., digital development, umbilical hernia receding, mesonephros degenerating and metanephros becoming functional, diaphragm completed, gonad primordia becoming vascular, intestinal villi, ossification of frontal and zygomatic, saccule and utricle separated, enucleate red cells 25%, aortic pulmonary semilunars.
15	Somites 64 and 65, snout protruding from face and lifts off chest, hair follicles developing, digits clear, body contour more rounded, cartilage in humerus, centrum and ribs ossifying, nucleated red cells 5%, stratum granulosum, cerebellum fused at mid-line.
16	All somites formed, differentiating from anterior to posterior, fetal stage, eyelids, pinnae cover ears, umbilical hernia withdrawn, ossification proceeding, 16 mm. length, corpus callosum formed, centrum ossified, nucleated red cells down to 1%, proliferation of gastric glands.
17	Eyelids sealed, extra embryonic membranes reach maximum development, tail alone now 10 mm. length.
19 to 20	Birth

Thus far we have described the genital anatomy of male and female mice, and given a brief chronology of development. Now we may proceed with the details of development from mating and fertilization, through cleavage, blastulation, gastrulation, implantation, and organogenesis, to birth of the fully formed, relatively independent mouse.

A. MATING: The female mouse, like most other female mammals (except the anthropoids and man), copulates only during estrus when ova are or become ready for fertilization. Since estrus usually begins around midnight, mating is most common during the night hours generally about 2 A.M. However, it can occur in the early morning or late evening, during short exposures of female to male. Thus the time variable in estimating embryonic or post-fertilization age can be reduced. Evening matings resulting in fertilization probably depend upon the survival of the spermatozoa in the female genital tract until the subsequent ovulation, and morning matings resulting in fertilization upon the presence of ova in the ampulla of the oviduct. Our practice is to mate from 8 to 8:45 A.M., to catch ripe eggs ovulated 6 to 8 hours earlier. Overnight matings give a time range of some 16 hours, instead of two or less.

Generally a male is placed in a box containing five or six females and removed 45 minutes or more later. Evidence of successful mating is a vaginal plug, a coagulum of fluid from the vesicular and coagulating glands of the male that occludes the vaginal orifice. The plug can be produced even by a vasectomized male. It usually hardens to such a degree that mechanical removal can injure the vaginal mucosa and the uterine ligaments. Our experience with this short period of mating has shown that we get an average of 6.98% vaginal plugs, and 92.4% of these result in pregnancies. When sexually experienced males are available an average of 8.5% plugs may be achieved in a randomly exposed group of females having various stages of estrous. When neither male or female is sexually experienced, but sexually mature (10 + weeks of age), the average percentage of plugs may be as low as 3.5% Theoretically only about 10% of a large group of sexually mature females might be found in estrous at the time of mating, hence receptive of the male. Spermatozoa introduced into the vagina may be found in the ampulla of the oviduct within minutes, but they retain their motility and fertilizing power so that as much as 8 hours may elapse between copulation and fertilization of the ova.

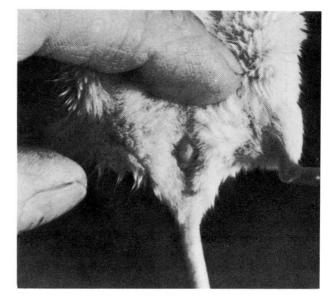

**VAGINAL ORIFICE OF MOUSE,
SHOWING VAGINAL PLUG**

Mating involves active inspection of the female genitalia by the male and lack of resistance by the female, with a tendency toward lordosis. Courtship continues for 15 to 20 minutes or less if the male is vigorous and sexually experienced. Mounting is attempted and, if successful, repeated a number of times until intromission has been accomplished and semen has been deposited in the vagina. It is estimated that an average of 60,000,000 sperm may be deposited in an ejaculate. A brief period of quiescence, particularly on the part of the male follows, after which normal activities are resumed. Other males present may attempt to mount the female subsequently but usually fail. Litters have been produced fathered by two males. Some males can mate with two (in 45 minutes) or even three (in 2 hours) susceptible females in succession making all such females pregnant. Under normal conditions coitus precedes ovulation by sometimes as much as 5 hours but may follow ovulation by as much as 8 hours. Strain differences in frequency of mating or time for recovery before a second mating can be achieved by a male vary from 1 hour to 4 days. Hybrid mice are most vigorous and recover most rapidly.

Some semen passes out through the cervix, but since the vaginal plug may persist for several days, more remains in the uterus. Leukocytes rapidly infiltrate the lumen when it is filled with semen and during the first day after mating most of the spermatozoa disappear owing to the phagocytic action of the scavenging white cells. Apparently only 100 or so spermatozoa survive to reach the ampulla of the oviduct to fertilize the available ova.

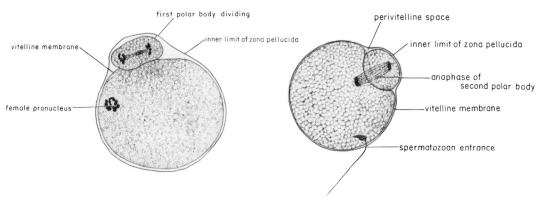

PREFERTILIZATION — MATURE OVUM   FERTILIZATION — ½ HOUR AFTER MATING

The progress of spermatozoa to the ampulla is aided by the peristaltic action of the muscles of the oviduct wall. Sperm can pass from cervix to ampulla in 15 minutes.

B. FERTILIZATION:

Fertilization consists in the activation of the ovum by a spermatozoon and the union of the male and female pronuclei, a process known as syngamy. It is a means of preventing the natural death of the ovum. The entire operation occurs in the ampulla of the oviduct, where the fertilizable life of the ovum has been estimated at 15 hours. Activation stimulates the ovum to complete its interrupted second meiotic division, by which it becomes haploid, and syngamy restores the normal diploid number of chromosomes. Second polar body formation may take as long as 2 hours.

Since ova are surrounded by a slightly sticky secretion, they tend to clump together but the clumping does not seem to interfere with fertilization. However, the cumulus oöphorus cells and the zona pellucida must undergo certain changes before fertilization can take place. The zona pellucida is essentially a homogeneous but layered matrix with irregular areas of varying density, and both inner and outer surfaces ragged. * The cumulus (corona) cells have extensions embedded in the zona which break after a few hours in the ampulla, possibly as a result of dispersal by the enzyme hyaluronidase. The corona cells retract their cytoplasmic extensions from the zona canaliculi.

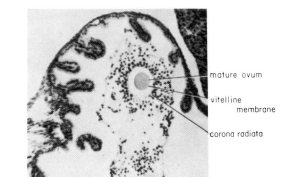

AMPULLA OF MOUSE OVIDUCT

Mature mouse ovum in ampulla awaiting fertilization by mature spermatozoon which must penetrate the surrounding zone radiata (cumulus oöphorus cells). This is aided by the elaboration of the enzyme hyaluronidase from the head of the sperm.

* The zona pellucida can be preserved best by glutaraldehyde-osmium or permanganate.

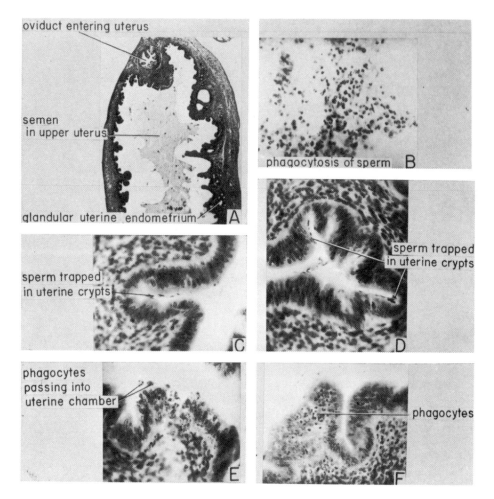

MOUSE UTERUS 1 HOUR AFTER MATING

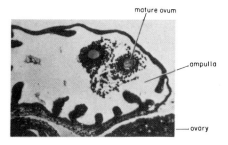

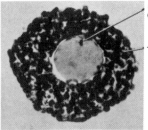

MOUSE OVA IN AMPULLA
AWAITING FERTILIZATION

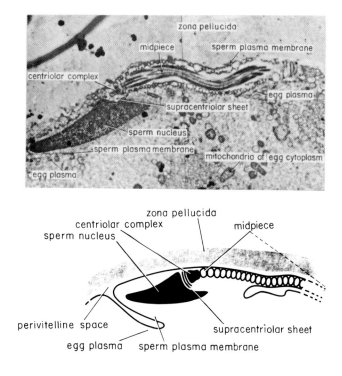

Upper figure - Rat sperm in process of penetrating egg, shown in electron microscopy.

Lower figure - A diagrammatic representation of sperm seen above, illustrating the behaviour of the egg and sperm plasma membranes during sperm entrance.

(Courtesy of D. G. Szollosi and H. Ris, 1961)

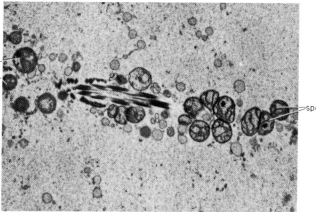

Electron micrograph of mouse egg one day after fertilization showing connection between mid-piece and principal piece of sperm tail. Sperm mitochondria to right and egg mitochondria to left.

(Courtesy of Dr. D. Szollosi)

Originally these extensions or processes have knoblike connections with the egg surface, and are presumed to convey nutriments to the egg during its development. Also, egg cortex microvilli pass into the zona. Both the extensions from the follicle cells and the microvilli are retracted upon explusion of the first polar body and the appearance of the perivitelline space. Remnants of the microvilli are seen to be most numerous at the point of polar body explusion, and at the site of sperm head entrance into the vitellus. About an hour elapses as sperm pass through the cumulus and zona pellucida. There is some evidence that fertilizin is present in the zona pellucida, and there may be some species specificity although hybrid crosses are easily accomplished. If present, it is probably similar to the fertilizin in other forms, consisting of a glyco-protein of high (82,000+) molecular weight. It presumably reacts with an antifertilizin substance identified as an acidic protein in a spermatozoon, its major function would be to attach a homologous spermatozoon to the coating of the ovum. If such a sperm agglutinin exists, it is in the zona. It is possible that the cortical granular response of the ovum releases an agent into the perivitelline space that reduces the penetrability of the zona pellucida to additional spermatozoa. It is believed that with the emission (abstriction) of the first polar body some perivitelline fluid is formed, and then when the sperm enters the vitellus causing the emission of the second polar body, further perivitelline fluid accumulates with the contraction of the vitellus. Such a "zona reaction", believed to take from 10 minutes to several hours, would be entirely independent of the second polar body spindle and would tend to block polyspermy although it is not completely effective. Some 20% of mouse eggs are found to have supplementary

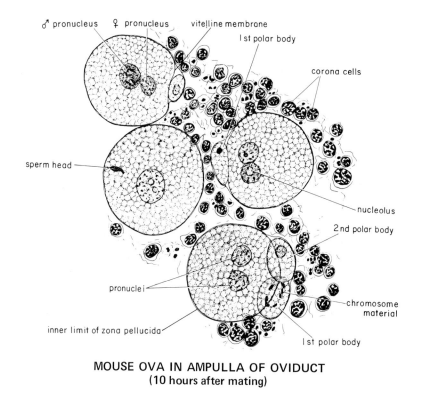

MOUSE OVA IN AMPULLA OF OVIDUCT
(10 hours after mating)

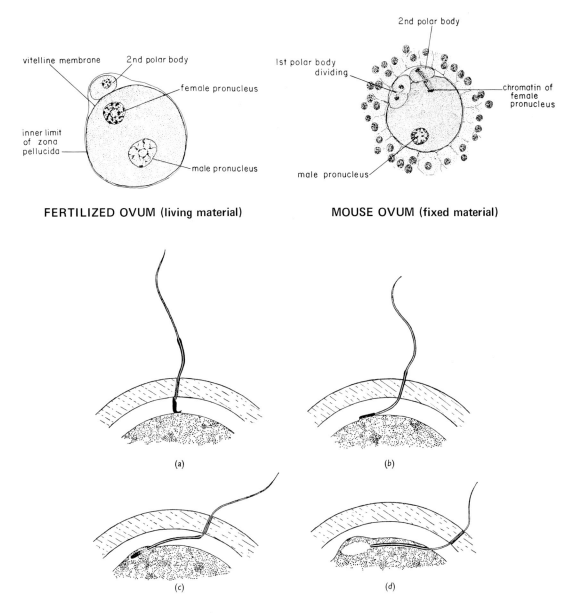

FERTILIZED OVUM (living material)   MOUSE OVUM (fixed material)

The figures illustrate four stages in the penetration of a rodent egg by a spermatozoon. (a) The spermatozoon head has just passed through the zona pellucida and made contact with the vitellus. (b) The spermatozoon head is lying flat upon the vitelline surface to which it is now attached. (c) The whole of the spermatozoon mid-piece has entered the egg and the head has passed through the surface of the vitellus. The spermatozoon head shows an early phase in its transformation to a male pronucleus—the posterior end of the head is becoming indistinguishable from the egg cytoplasm. (d) The head and mid-piece of the spermatozoon have now entered the vitellus. Transformation of the head has proceeded to the stage immediately before the appearance of nucleoli. The cytoplasmic elevation over the spermatozoon head, just evident in (c) has now become much larger.

(From Austin & Braden '56, J. Exp. Biol. 33:358-365)

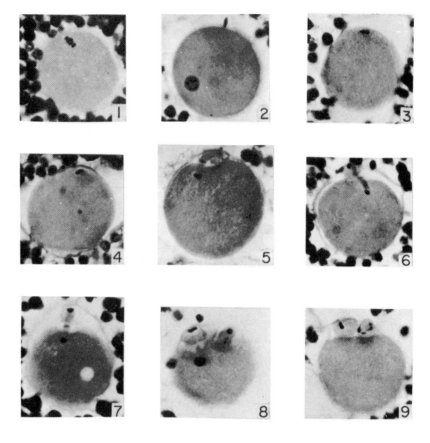

SPERM ENTRANCE AND SECOND
POLAR BODY FORMATION

Fig. 1 — Metaphase chromosomes of second maturation spindle at periphery of ovum found in ampulla surrounded by cumulus oöphorus cells.

Fig. 2 — Sperm head making contact at egg surface and female pronucleus seen at opposite pole, following second maturation division.

Fig. 3 — Sperm head (alone) just beneath surface of egg.

Fig. 4 — Sperm head moving into egg from surface.

Fig. 5 — Second maturation (metaphase) spindle seen tangentially.

Fig. 6 — Second maturation spindle in anaphase showing second polar body formation.

Fig. 7 — Same as Fig. 6 but direct view showing interzonal granules on spindle, and everted second polar body.

Fig. 8 — Two polar bodies in perivitelline space and mature egg nucleus within cytoplasm.

Fig. 9 — Clear view of two polar bodies surrounded by vitelline membrane.

spermatozoa. Another device for preventing polyspermy is the cumulus oöphorus cells which disappear after invasion by a spermatozoon and seem to alter the penetrability of the vitelline membrane.

The spindle for the formation of the second polar body, which has been lying parallel to the surface, turns 90 degrees so that one pole of the spindle is near the surface and the other deep within the ovum. The more peripheral pole attracts half of the chromosome material into the second polar body. The polar body has little cytoplasm. A nucleus occasionally reconstitutes within it, distinguishing it from the first polar body and its division products. The two polar bodies are rarely seen together in a mouse egg because of the plane of the section.

A female may be artificially inseminated by injection of semen through a 28 gauge needle into the periovarial space, the ampulla, the oviduct, or the uterus 1.5 to 4.5 hours after mating with a vasectomized male. Mid-ventral laparotomy is required unless insemination of only one side is proposed.

An active spermatozoon in the ampulla of the oviduct retains its capacity for fertilization for 8 hours, and generally all of the eggs of a single female are fertilized by 6 hours after mating. The delay is sometimes due to the necessity for the maturation of the cumulus cells. Some sperm-egg collisions occur within 15 minutes of mating. Cooled sperm can be used for fertilization up to 24 hours.

There is no evidence for the necessity of sperm capacitation in the mouse, as there seems to be in the rat. The mature spermatozoon passes rapidly through the zona pellucida and loses its acrosome, then comes into immediate contact with the vitelline membrane. There may be some evidence of a solid sperm penetration filament (SPF) which extends from the apex of the sperm head as it progresses through the zona pellucida, causing it to take a curved course. The presence of the SPF in the mouse has not been conclusively demonstrated as it has in the rabbit, pig, and sheep. The sperm head largely consists of deoxyribonucleoprotein, and does not penetrate the vitelline membrane at right angles but tangentially. The vitelline membrane may outpush somewhat, but the major effort is made by the spermatozoon. The acrosome probably aids in penetration. The fertilizing spermatozoon often remains attached to the surface of the egg for 30-40 minutes before proceeding. Penetration has been described as (1) a sort of specific phagocytic engulfing of the spermatozoon by the egg membrane   (2) a brief rupture in the membrane permitting entrance of the spermatozoon, followed by closure of the membrane over the spermatozoon. The latter method appears to be the more likely. Penetration requires about 30 minutes. Usually the head separates from the middle piece and tail, and the tail does not necessarily enter the ovum. The tail undulates, propelling the sperm head forward and that part of the tail which penetrates the egg stops this movement, but the portion still in the perivitelline space continues to undulate. The forward motion averages 10 to 20 microns per lash of the tail. Also, the sperm plasma membrane remains intact around that portion of the spermatozoon still outside the vitellus, but at the surface of the ovum it is continuous with the egg plasma membrane. The sperm head which penetrates the vitellus lacks both the nuclear and plasma membranes.

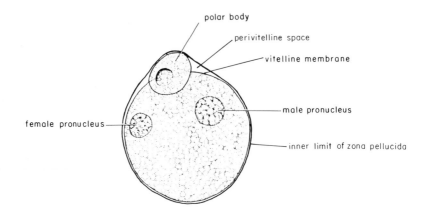

FORMATION OF MALE AND FEMALE PRONUCLEI

The activation of the ovum by the fertilizing spermatozoon starts before the enlargement of the head of the spermatozoon which occurs from 1 to 2 hours after penetration. The chromatin elements of the head disappear for a short while during this process. Swelling of the head begins at the posterior end and progresses anteriorly. The volume of the resulting pronucleus is several hundred times the original head volume, and usually two or three times the volume of the female pronucleus. The pronuclear volume depends upon material drawn from the cytoplasm, which limits the ultimate sizes. Both pronuclei appear to have double walls, and uniformly fine granules. The interval between penetration and formation of the male pronucleus is short in the mouse as compared with other rodents, but the interval between penetration and the first cleavage is long.

After fertilization, the zona pellucida expands, to give an overall diameter to the zona and ovum of about 114 microns. Since the ovum tends to shrink about 15%, a space appears, known as the perivitelline space, between it and the zona pellucida. The first polar body is occasionally visible in this space. The second polar body is discharged into it within 2 or 3 hours after fertilization. Then begins a boiling motion of the egg cytoplasm.

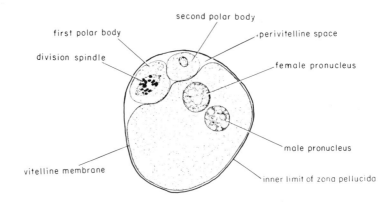

COMPLETION OF POLAR BODY FORMATION

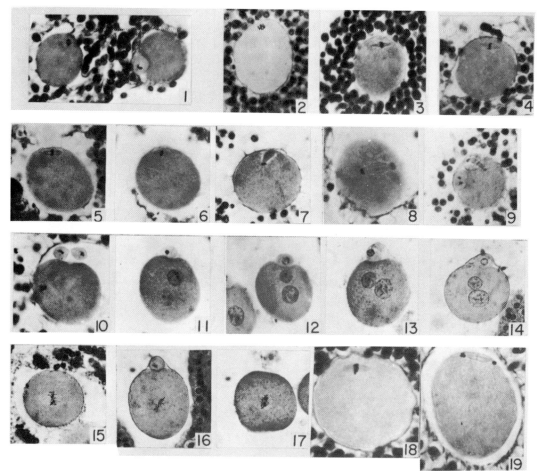

SECOND MATURATION DIVISION AND FIRST CLEAVAGE IN AMPULLA OF MOUSE

Fig. 1 — Two mouse ova surrounded by cumulus cells. One to left shows metaphase plate of second maturation spindle, hence is as yet unfertilized and one to right shows prominent first polar body within vitelline membrane.

Fig. 2 — Distinct chromosomes in metaphase at periphery anticipating first maturation division following fertilization.

Fig. 3 to 6 — Lateral views of metaphase spindle for second maturation division; condition of all ova in ampulla, awaiting fertilization.

Fig. 7 — Polar view of second maturation spindle.

Fig. 8 — View of entire spindle in anaphase of second maturation division, at surface of egg.

Fig. 9 — Tangential view of second maturation spindle in process of everting second polar body alongside first polar body, both within perivitelline space.

Fig. 10— First and second polar bodies free within the perivitelline space.

Fig. 11— Section showing male and female pronuclei (female nearest polar body) and everted second polar body.

Fig. 12 to 14 —Pronuclei approaching each other, and persistent polar body between the ovum and the perivitelline membrane (zona pellucida).

Fig. 15 — Metaphase spindle for first cleavage division of mouse egg about 14 hours after fertilization.

Fig. 16 — Same as Fig. 15 but showing polar body.

Fig. 17 — Beginning of cleavage furrow and early anaphase of chromosomes.

Fig. 18 — Enlarged view of mouse egg at time of fertilization, when the chromosomes are in metaphase prior to the second maturation division which occurs following the stimulation of fertilization (as in Figs. 3 to 6).

Fig. 20 — Lateral view of polar body formation showing entire spindle highly enlarged, with interzonal granules (as in Fig. 5).

Both male and female pronuclei may be seen as early as 6 hours after mating. The nucleus to cytoplasmic volume ratio of 1:30 is consistent for all mouse eggs, and is large compared with those of other mammals. The nucleus has a prominent nucleolus. The two pronuclei synthesize DNA before their fusion and each, with many distinctive acidophilic nucleoli and devoid of RNA, move toward each other, meeting near the center of the ovum. Occasionally the nucleoli may be found indented into the nuclear membrane. As they do so, they are considerably reduced in size. There is some question whether the primary mechanism in the cytoplasmic union of the gametes is a fusion of their membranes or a form of specific phagocytosis. Fertilization is considered to be complete when the pronuclei are almost contiguous and the first cleavage spindle is forming. Before the first cleavage the membranes of the pronuclei disappear, and the respective chromosomes form and split longitudinally. At the first cleavage two members of each chromosome tetrad move into each of the blastomeres so that a distinction between male and female chromatin elements is no longer possible. In other words representatives of each chromosome pass into the resulting blastomeres.

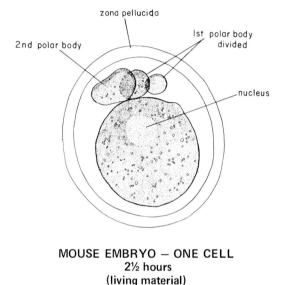

MOUSE EMBRYO — ONE CELL
2½ hours
(living material)

Anomalies associated with fertilization include anengamy (an abnormal number of pronuclei); *polyandry* (extra male pronuclei); *polygony* (extra female pronuclei); *dimegaly* (two sizes of spermatozoa) or *polymegaly* (many sizes of spermatozoa).

C. EARLY DEVELOPMENT:

a. *Cleavage:* Cleavage is total (holobastic) but slightly unequal in this meiolecithal egg. Its plane is believed to pass through the positions occupied by the centers of the two pronuclei in syngamy. But even in the earliest stages of development, patterns may be discerned that are peculiar to the species, and these may be related to cytological characteristics later to develop. Visible changes occur in the cristae of mitochondria of the early embryonic cells. Thus, there is no experimental evidence of the regulative capacity (totipotency) of the early mouse blastomere except that later blastocysts can be mechanically fused to derive an oversized blastocyst. The first of the cleavages begins in the ampulla of the oviduct about 24 hours after fertilization and continue for 2 to 3 days, even as the embryo moves into the uterus. Factors influencing its form are entirely maternal but the rate of cleavage is the inherent property of the zygote. The spermatozoon supplies the kinetic center in its proximal centriole and may affect the time to the second cleavage. The time interval to the first cleavage from fertilization ranges from 17 to 57 hours in different strains of mice. Difficulties in getting in-vitro zygotes to pass from the 1 to the 2 cell

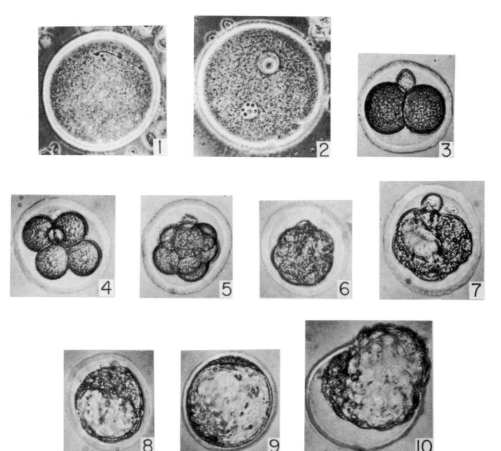

EARLY DEVELOPMENT OF THE MOUSE
(In vitro culture)

Fig. 1 — Fertilized egg 6 hours after ovulation, slightly compressed for better phase contrast photography. Note spermatozoon recently penetrated, lying beneath the vitelline membrane.

Fig. 2 — Zygote at 12 hours after ovulation showing male and female pronuclei, a polar body, remains of sperm middle piece and tail, and dispersing cumulus oöphorous.

Fig. 3 — Two-cell stage at 36 hours after ovulation, with polar body, within zona pellucida.

Fig. 4 — Four-cell stage at 58 hours, with polar body and zona pellucida.

Fig. 5 — Eight-cell stage, with polar body at 67 hours after ovulation, still in zona pellucida.

Fig. 6 — Morula stage at 82 hours after ovulation.

Fig. 7 — Early blastocyst stage at 93 hours with small blastocoel, and persistent polar body.

Fig. 8 — Blastocyst showing inner cell mass at 105 hours.

Fig. 9 — Blastocyst expanding with enlarging blastocoel to fill the zona pellucida and stretch it.

Fig. 10 — Blastocyst emerging from the zona pellucida in a hatching reaction which causes the rupture of the zona.

(Courtesy A. H. Gates, 1965, Ciba Foundation Symposium on Preimplantation Stages of Pregnancy, p. 270–288, eds. G. F. W. Wolstenholme and M. O'Connor; Pub. J. & A. Churchill, London.)

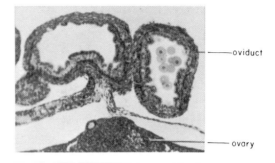

TWO CELL MOUSE EMBRYOS
32 hours after conception
(en route to the uterus)

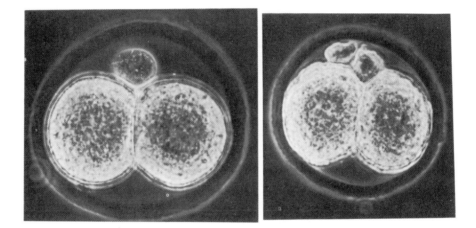

**TWO-CELL MOUSE EMBRYOS PHOTOGRAPHED BY PHASE,
SHOWING SINGLE & DOUBLE POLAR BODIES**

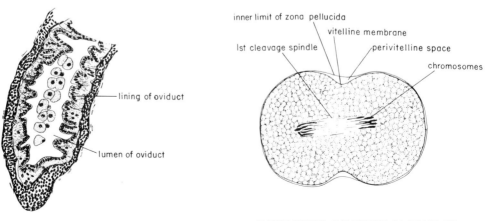

MOUSE OVA IN OVIDUCT
(2 cells — 24 hours)

BEGINNING OF FIRST CLEAVAGE

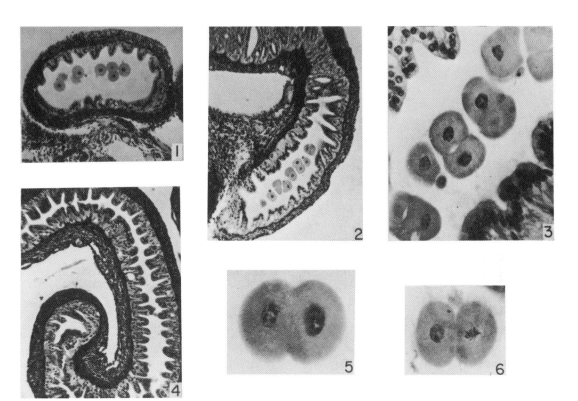

TWO CELL MOUSE EMBRYO
24 hours after fertilization

Fig. 1 — Ampulla of oviduct of mouse showing presence of five fertilized eggs in 1 and 2 cell stages.

Fig. 2 — Longitudinal section of oviduct showing mostly 2-cell mouse ova enroute to the uterus.

Fig. 3 — Enlarged view of two-cell mouse eggs showing distinct nuclei and cleavage furrow (space) between blastomeres. Note persistent polar body.

Fig. 4 — Longitudinal section of major portion of mouse oviduct showing its glandular but non-ciliated epithelium.

Fig. 5 — Enlarged view of two-cell mouse embryo showing clear nuclei. Zona pellucida was present but did not stain.

Fig. 6 — Two-cell stage with one blastomere anticipating the second cleavage division by metaphase spindle formation.

stage suggest that there is something in the oviducal environment requisite to this initial cleavage. It may take as long as $3\frac{1}{2}$ hours for all of the mature ova from a single female to be fertilized. This may contribute to the range of development within the members of the preimplantation ova of any female, but it is also possible that development could be impeded or arrested and the rate of cleavage altered by other means. The initial elongation of the fertilized ovum, anticipating cleavage, is probably the result of the arrangement of the spindle and asters although the latter structures are not readily demonstrable in the mouse zygote. The spindle fibers appear to be fine straight tubules while the chromosomes are ill-defined, electron-dense masses. There is some question as to whether the centrioles responsible for the first cleavage spindle arise de novo in the egg

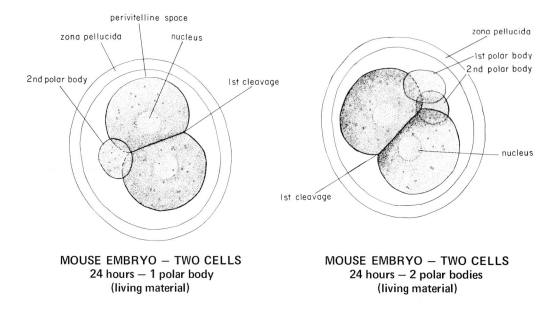

MOUSE EMBRYO — TWO CELLS
24 hours — 1 polar body
(living material)

MOUSE EMBRYO — TWO CELLS
24 hours — 2 polar bodies
(living material)

cytoplasm, but under some influence from the sperm nucleus. Before the first cleavage, the chromatin contents of the two pronuclei are not mixed, but thereafter, nuclear contents combine. The amount of DNA in each nucleus doubles before each division and the nucleoli become increasingly active centers of RNA synthesis. The nuclei also contain a considerable amount of protein, some of which is synthesized during early cleavage.

Succeeding cleavages occur after shorter and shorter intervals, soon becoming asynchronous. The second cleavage (to four cells) occurs at about 37 hours after fertilization; the third (to eight cells) at about 47 hours after fertilization, etc. The blastomeres of the 8 cell stage are labile and shift with considerable motility and yet they are sufficiently adhesive to retain their contigual relations with each other.* There is some evidence that by this cleavage the cells are committed to either embryonic or trophoblastic development. The fourth cleavage, at about 60 hours after fertilization, results in a solid ball of cells known as the morula. This is seen enroute to the uterus, or in the upper part of the uterus at 3 to 4 days after fertilization. The second cleavage takes the longest time, about 6 hours. Later divisions are reduced to about 10 minutes. An estimated early cleavage time table follows:

Coitus - 0
First Cleavage spindle - 21 to 28 hrs.
Two cell stage - 21 to 43 hrs.
Four cell stage - 38 to 50 hrs.
Eight cell stage - 50 to 64 hrs.

Sixteen cell stage - 60 to 70 hrs.
Blastocyst stage - 66 to 82 hrs.
Transport to uterus - 66 to 72 hrs.
Implantation - 4 to 5 days

*Bovine serum albumen (BSA) alone will support development of the 8-cell mouse embryo. It may also survive on glucose or malate alone, without albumen. It seems that there is an ability to utilize different substrates during various stages of development of the mouse embryo. At the 2-cell stage both pyruvate and BSA are necessary. Probably the best medium for the development of the 2-cell mouse embryo is 2.5 to 5.0 x 10^{-4} M pyruvate, plus 2.5 to 5.0 x 10^{-2} M$_2$ lactate. Waymouth's medium (MB 752/1) allowed 2 and 4 cell stages to develop to the blastocyst stage. Glucose is not necessary for the early cleavages.

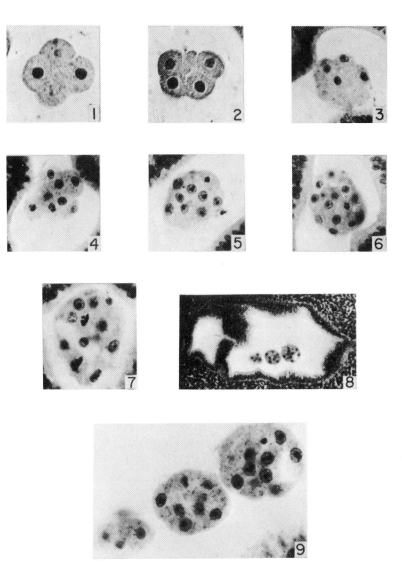

EARLY CLEAVAGE MOUSE EMBRYO

Fig. 1 — Four cell mouse embryo, two cells of which show their nuclei in this section.

Fig. 2 — Four plus cells with persistent polar body. Probably 6 or 8 cells since this was a section.

Fig. 3 — Third cleavage, eight or more cells.

Fig. 4 — Later than third cleavage, 8 nuclei showing in this section.

Fig. 5 — Probably 16 cell stage, nine nuclei showing plus the polar body.

Fig. 6 — Morula stage showing earliest indication of blastocoel formation (beneath topmost nucleus). This stage is en route to the uterus.

Fig. 7 — Earliest blastula (trophoblast) stage. Note blastocoel, and blastomeres in mitotic division.

Fig. 8 — Low power view of sections of three morula-blastula stages in upper uterus. Sections are at different levels but two at left were probably at stage represented by median section of stage at right.

Fig. 9 — Enlarged view of three morula-blastula stages in uterus at 3.5 days.

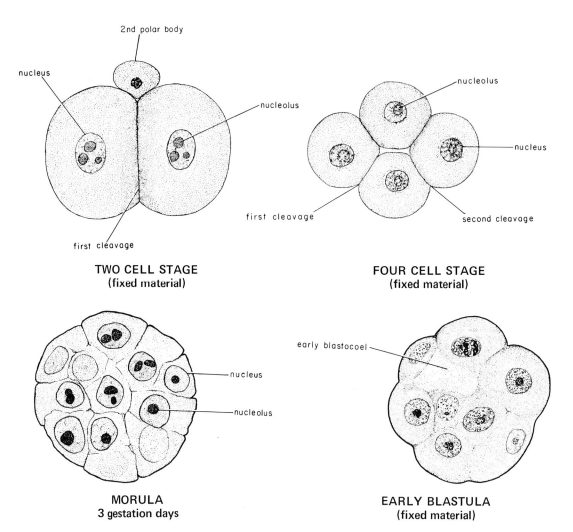

TWO CELL STAGE
(fixed material)

FOUR CELL STAGE
(fixed material)

MORULA
3 gestation days

EARLY BLASTULA
(fixed material)

Eggs do not all cleave at the same time intervals, nor are all the cleavages of a single egg complete, so that many early embryos possess uneven numbers of cells. There may be as many as two cleavages difference between the most advanced and most retarded eggs by the time they reach the uterus of any female. Some of this delay is due to variations in sperm penetration time, and may occasionally be so out of synchrony with the uterine endometrium as to complicate the process of implantation. As division proceeds, the cells become progressively smaller. The nuclei decrease in volume, the nucleoli diminish in size and number and disappear, and the chromosomes condense before each cleavage. Material associated with the nucleoli is hardly visible in the two-cell stage, but becomes increasingly prominent until by the 16 cell stage it almost obscures the nucleolus. Extranucleolar nuclear RNA production continues and the nucleoli become very active centers of RNA synthesis. Both nucleoli and cytoplasm synthesize protein.

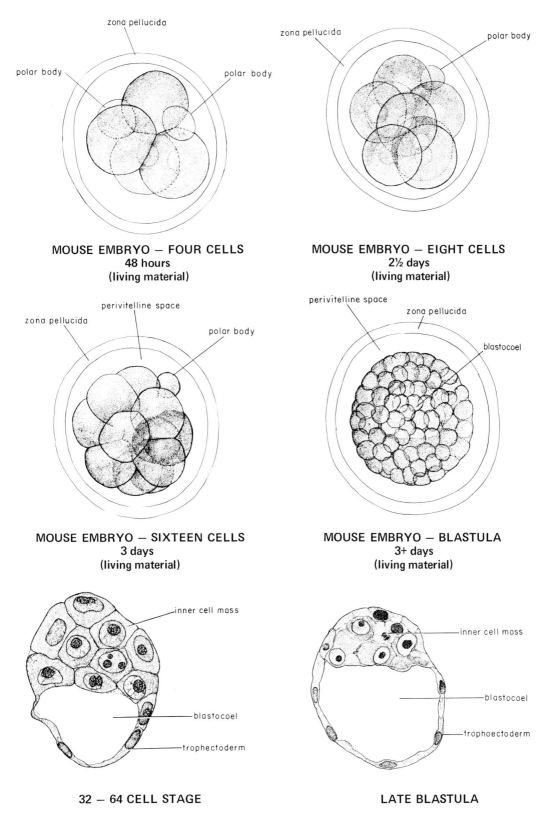

MOUSE EMBRYO — FOUR CELLS
48 hours
(living material)

MOUSE EMBRYO — EIGHT CELLS
2½ days
(living material)

MOUSE EMBRYO — SIXTEEN CELLS
3 days
(living material)

MOUSE EMBRYO — BLASTULA
3+ days
(living material)

32 — 64 CELL STAGE

LATE BLASTULA

   b. *Blastulation:* In the 32 to 64 cell stage the morula (in the uterus) acquires an eccentrically placed, slit-like cavity filled with fluid. This is the
beginning of the blastocoel and of the process of blastulation, deriving the
blastocyst stage which undergoes cycles of expansion and contraction. The
blastocoel enlarges but the total protoplasm does not increase, since the embryo
is unable to absorb any nutriment. Cytoplasmic structure alters a great deal
during blastulation. A single layer of lining cells, the trophoblast, encloses

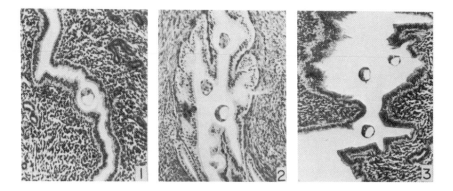

MOUSE MORULAE AND BLASTULAE IN UTERUS — 3.5 g.d.

the blastocoel except at one side, where a knob of cells, the inner cell mass,
forms. It has been suggested that the rapidly dividing cells of the blastocyst
may be the precursors of the trophoblast while the more slowly dividing cells
are the precursors of the inner cell mass, or the embryo proper. In any case,
all of the blastocysts possess short, relatively uniform microvilli on their
outermost surfaces, and all trophoblast cells have ribosomes in clusters and
rosettes. Glycogen, lipid granules and crystalloid inclusions can be found in
the early mouse blastocyst cells. Blastocoelic fluid first arises from the blastomeres at the end of the fifth cleavage.

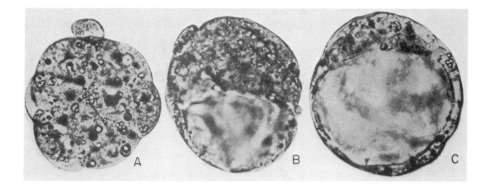

Fig. A — Morula stage, note persistent polar body.
Fig. B — Blastocyst at 76 hours; note inner cell mass and early blastocoel.
Fig. C — Large blastocyst at about 5 days, note trophoblast surrounding
             inner cell mass.   q

(Photographs from Carnegie Institute of Washington Publication Embryology # 148,
1935; W. H. Lewis and E. S. Wright.)

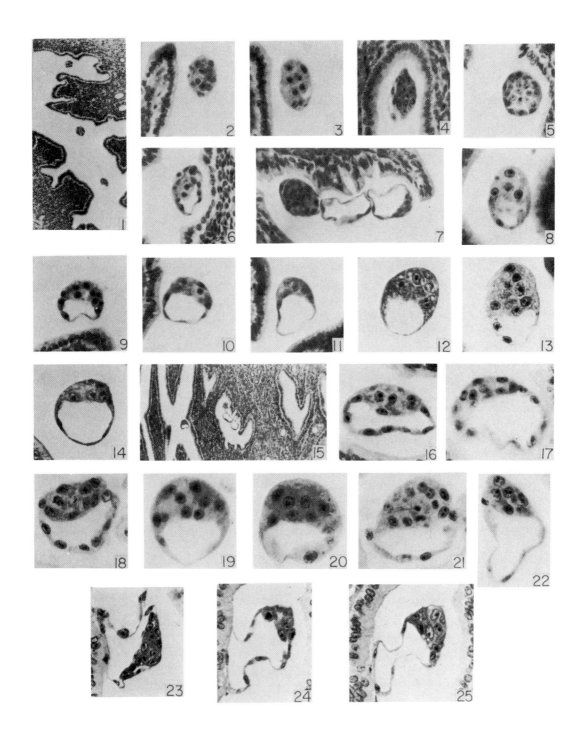

MOUSE MORULA TO TROPHOBLAST
3.5 to 4.5 days

The embryo is free within the lumen of the uterus at $4\frac{1}{2}$ days after fertilization but the inner cell mass, blastocoel, and trophoblast are undergoing changes in preparation for implantation. These changes may be triggered by a secretion from the ovary; they do not take place in an ovariectomized mouse or in one subcutaneously injected with 2.5 milligrams of progesterone in 0.1 cubic centimeter of corn oil about 4 days after mating. The abembryonic portion of the blastocyst is a syncitium. The blastocyst (blastula) exhibits a high rate of respiratory metabolism, which must consume much of its nutrient reserve, and starts a "hatching" action by which it sheds the zona pellucida. This action includes both an enlargement and a rhythmic undulating movement. It is not caused by any secretion or activity of the uterus. The initial volume changes of the blastocyst are noted at 4 days post-fertilization. Single large contractions of the blastocyst almost result in its disappearance, and these are interspersed with several partial contractions. The frequency of the large contractions is every 6 to 8 hours, and of the smaller contractions every 20 to 100 minutes. Some contractions are slow (5 to 6 minutes), while others are

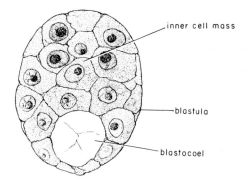

LATE BLASTULA

relatively fast (15 to 20 seconds). Such cycles are seen in liberated or denuded blastocysts, and it is believed that hatching cannot occur without these pulsations. (Cole and Paul '63). It is preceded by the appearance of slender and sticky protoplasmic processes (alkaline secreted) extending through the zona pellucida from the abembryonal pole cells of the blastocyst. At the time of hatching the blastocyst is no longer spherical but slightly ovoid, and it varies considerably in size, measuring an average of 96 microns in length, or about 108 microns with the intact zona pellucida. The denuded blastocysts are very sticky. The uterus is slightly more acid than the oviducts and this may contribute to the dissolution of the zona pellucida.

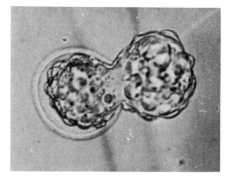 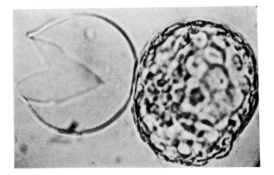

LATE MOUSE BLASTOCYST ESCAPING FROM ITS ZONA PELLUCIDA

(Two magnifications: Exp. Cell Res. 32:205-208, 1963. Courtesy Dr. R. L. Brinster)

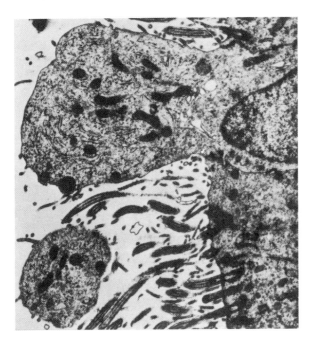

Infundibulum of mouse oviduct at time of egg transport. Apical part of ciliary and secretory cell, the latter having a smooth surface membrane with a few microvilli, and some dense protein-like granules. The endoplasmic reticulum has parallel membranes with ribosomes and well developed Golgi apparatus. x 15,000.

(Courtesy S. Reinius, originally published in Proc. Vth World Congress on Fertility and Sterility, Stockholm 1966, Excerpta Medica Foundation, Amsterdam.)

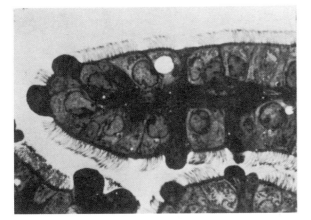

Infundibulum of mouse oviduct. Ciliate cells dominate, secretory cells among them ("peg cells"). x 1,100.

(Courtesy S. Reinius, originally published in Proc. Vth World Congress on Fertility and Sterility, Stockholm 1966, Excerpta Medica Foundation, Amsterdam.)

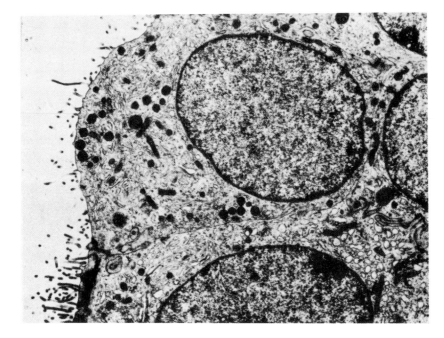

Secretory cell in ampulla of mouse oviduct at time of egg transport. Secretory granules, granular and agranular endoplasmic reticulum as well as Golgi apparatus are well developed.

(Courtesy S. Reinius, originally published in Proc. Vth World Congress on Fertility and Sterility, Stockholm 1966, Excerpta Medica Foundation, Amsterdam.)

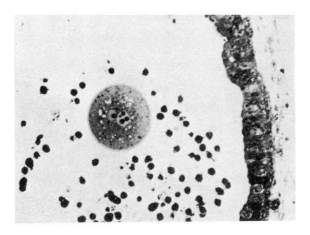

Fertilized egg in pronuclear stage, surrounded by corona radiata cells and found in ampulla of mouse oviduct. Toluidine blue. x 320.

(Courtesy S. Reinius, originally published in Proc. Vth World Congress on Fertility and Sterility, Stockholm 1966, Excerpta Medica Foundation, Amsterdam.)

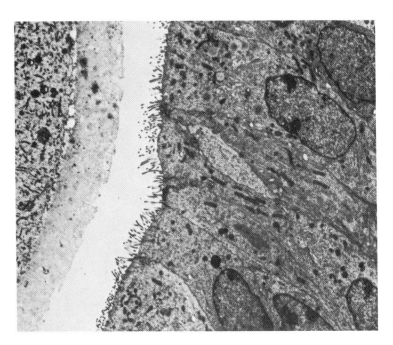

Portion of two-cell mouse egg with zona pellucida in relation to the epithelium in the isthmus of oviduct. Microvilli are long and regular. The endoplasmic reticulum is granular and forms parallel membranes aggregated in groups. x 8,960.

(Courtesy S. Reinius, originally published in Proc. Vth Congress on Fertility and Sterility, Stockholm 1966, Excerpta Medica Foundation, Amsterdam.)

Two cell egg in isthmus of mouse oviduct. Toluidine blue. x 320.

(Courtesy S. Reinius, originally published in Proc. Vth Congress on Fertility and Sterility, Stockholm 1966, Excerpta Medica Foundation, Amsterdam.)

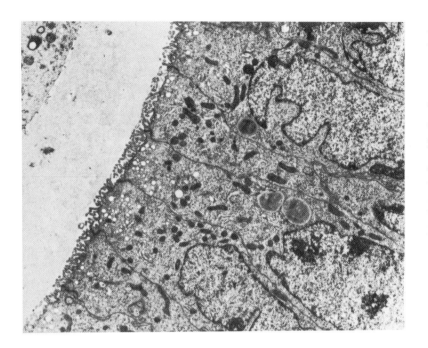

Mouse egg — epithelial relationship in the utero-tubal junction. Microvilli are low and regular, apical vesicles are present. × 11,000.

(Courtesy S. Reinius, Excerpta Medica, in press.)

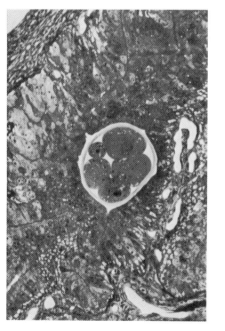

Mouse egg in utero-tubal junction. Note firm contact with surrounding epithelium. Toluidine blue. × 320.

(Courtesy S. Reinius, Excerpta Medica, in press.)

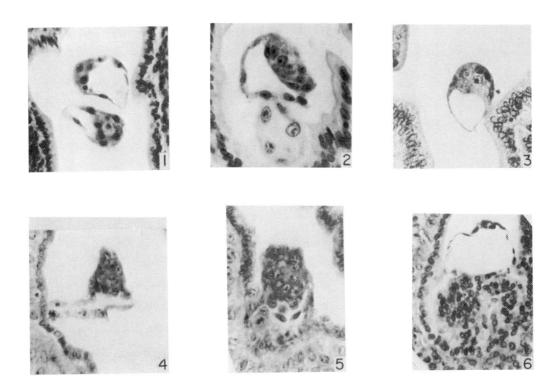

INITIAL CONTACT OF MOUSE EMBRYO WITH UTERINE WALL
(3.5 gestation days)

Fig. 1 — Two blastocysts in uterine cavity prior to implantation contact.

Fig. 2 — Later blastocyst and beneath it three cells of a morula, both in uterine cavity.

Fig. 3 — Blastocyst showing contact of lateral trophectoderm with uterine epithelium, prior to any obvious erosion.

Fig. 4 — Distal contact of trophectoderm of later blastocyst, indicating point of implantation.

Fig. 5 — Proliferating inner cell mass and invasion of trophectoderm into uterine epithelium.

Fig. 6 — Erosion and invasion of uterine epithelium and mucosa by trophoblast.

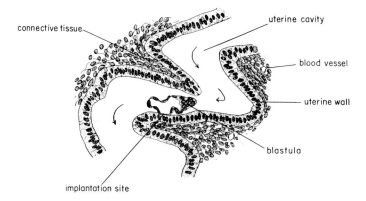

IMPLANTATION OF BLASTULA IN MOUSE
4 g.d. (after ovulation)

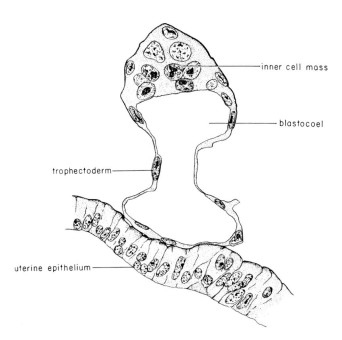

LATE BLASTULA
(site of implantation)

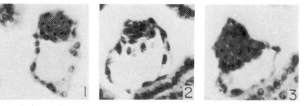

uterine cavity   blastocoel   inner cell mass

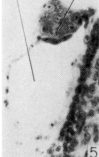

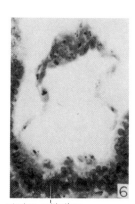

embryonic ectoderm

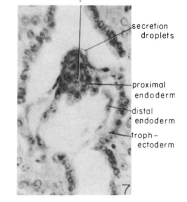

secretion droplets

proximal endoderm

distal endoderm

troph-ectoderm

uterine epithelium

MOUSE IMPLANTATION
4.5 to 5.0 days

Fig. 1 — Blastocyst free within the uterine lumen showing distinct layer of endoderm cells beneath inner cell mass.

Fig. 2 — Trophectoderm making first contact with uterine epithelium. Note endoderm cells which appear to be free within the blastocoel early distal endoderm.

Fig. 3 — Inner cell mass increasing by mitosis.

Fig. 4 — Larger blastocyst, still free within the uterine lumen, but endoderm layer distinct.

Fig. 5 — Lateral contact of trophectoderm with uterine epithelium over wide area.

Fig. 6 — Blastocyst trapped in an epithelial crypt, with implantation contact at pole opposite inner cell mass — the most frequent situation.

Fig. 7 — Implantation in crypt, with contact on two sides of trophectoderm. Note extensive proliferation of inner cell mass, and distal endoderm.

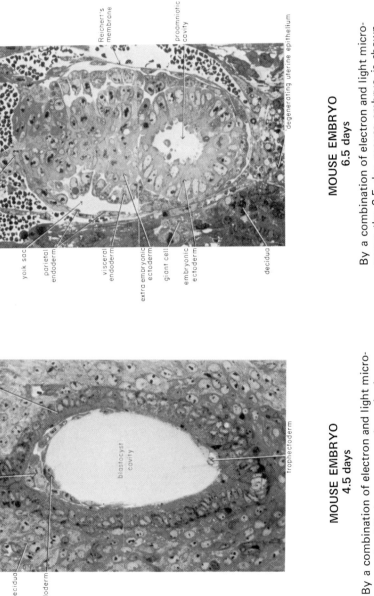

MOUSE EMBRYO
6.5 days

By a combination of electron and light micro-scopy the 6.5 day mouse embryo is shown.

(Courtesy S. Reinius, Zeit. Zellforsch. 68:711-723, 1965.)

MOUSE EMBRYO
4.5 days

By a combination of electron and light micro-scopy the 4.5 day mouse embryo is shown.

(Courtesy S. Reinius, Zeit. Zellforsch. 68:711-723, 1965.)

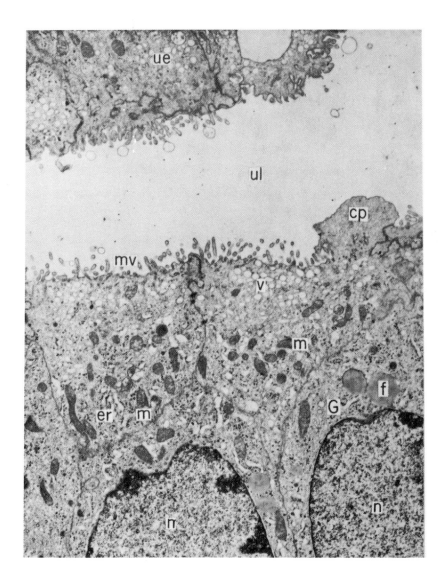

The surface of the uterine epithelium in pre-attachment stage gestation day 4½; cp — cytoplasmic protrusion; er — endoplasmic reticulum; f — fat granule; G — Golgi apparatus; m — mitochondria; mv — microvilli; n — nucleus; ue — uterine epithelium; ul — uterine lumen; v — apical vesicles. x 10,700.

(Courtesy S. Reinius, Z. Zellforsch. 77:257-266, 1967.)

The trophoblastic cells which surround the blastocoel of the blastocyst are transformed by the giant cells which appear in the contact and invasion area. This usually begins at the abembryonic pole, and in any litter will be seen in various stages of progress in its individual members. Blastocysts at this time may measure as much as 186 microns average, and the decidual swellings may be seen. Nucleolar activity of the embryonic cells is accelerated, deriving cytoplasmic (ribosomal) RNA which is of the high guanine-cytosine type. It seems that this giant cell transformation of the trophoblasts and changes in the size and form of the inner cell mass are initiated by a secretion from the ovary, because ovariectomized mice do not show these reactions. This concept is supported by the reaction to the subcutaneous injection of progesterone about 4 days after insemination.

Blastocysts obtained from old mice, from 13 to 24 months of age, fail to implant and develop when transferred to still older uterine environments. However, when transferred to uteri of young mice 3 to 7 months of age, just as many blastocysts develop as from reproductively active females. Reduction in litter size with increasing maternal age is therefore probably due to an increasingly less favorable uterine environment rather than to a decline in viability of the ova.

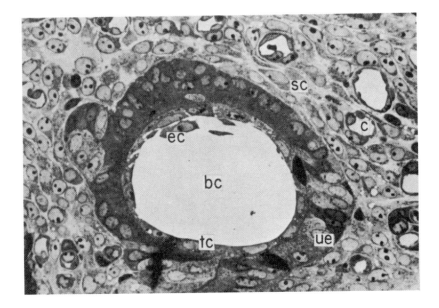

An attached blastocyst, gestation day 5, showing no decidual reaction. bc — blastocyst cavity; c — capillary; ec — endoderm cell; sc — stromal cells; tc — trophoblast cell; ue — uterine epithelium. Toluidine blue staining. x 3,700.

(Courtesy S. Reinius, Z. Zellforsch. 77:257-266, 1967.)

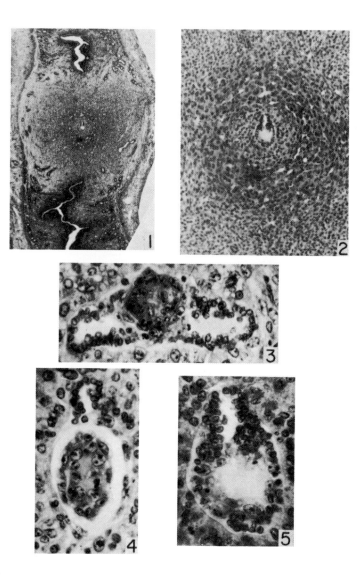

MOUSE EMBRYO
5 days

Fig. 1 — Low power view of mouse embryo at 5 days gestation, showing extensive decidual reaction of the uterus. Note lumen to either side of implantation site.

Fig. 2 — Higher power view of 5 day embryo and surrounding decidua.

Fig. 3 — Transverse section through cells of embryonic ectoderm and on either side portions of the blastocoel or yolk cavity.

Fig. 4 — Tangential section through 5 day embryo showing yolk cavity, and split between the inner embryonic ectoderm and outer endoderm.

Fig. 5 — Five day egg cylinder well embedded into uterine epithelium, and early development of central proamniotic cavity.

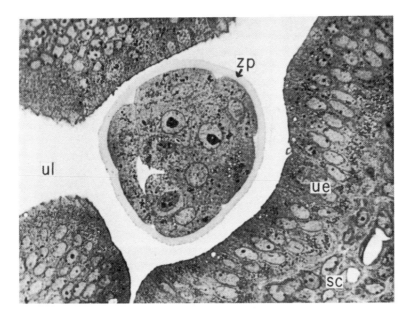

An early blastocyst on gestation day 4, free in uterine lumen. sc —
stromal cells; ue — uterine epithelium; ul — uterine lumen; zp — zona
pellucida. Toluidine blue stain. x 590.

(Courtesy S. Reinius, Z. Zellforsch. in press.)

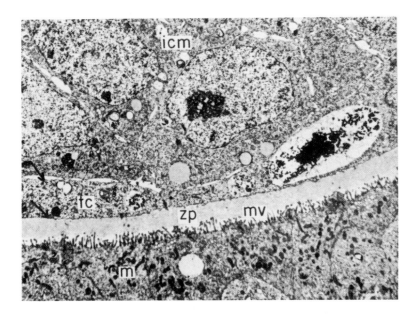

Part of the inner cell mass of a blastocyst, gestation day 5, still surrounded by a thin zona pellucida. Note relation to the uterine epithelial cell surface. icm — inner cell mass; m — mitochondria; mv —
microvilli; tc — trophoblast cell; zp — zona pellucida. Arrows indicate
outer and inner cell membrane of the trophoblast cell. x 3,700.

(Courtesy S. Reinius, Z. Zellforsch. 77:257-266, 1967.)

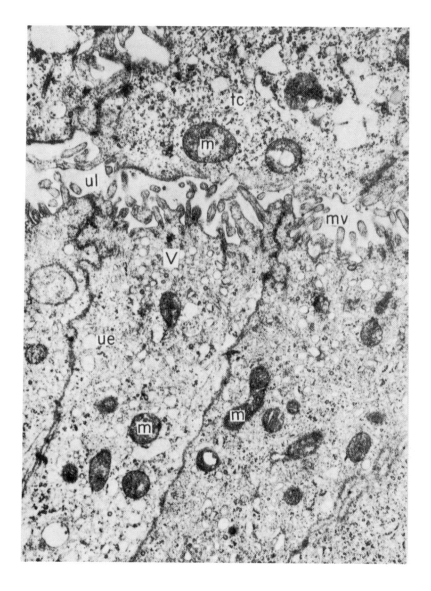

Blastocyst — uterine cell contact just prior to attachment, gestation day 5 when the zona pellucida is lost. m — mitochondria; mv — microvilli; tc — trophoblast cell; ue — uterine epithelium; ul — uterine lumen; v — apical vesicles. x 19,000.

(Courtesy S. Reinius, Z. Zellforsch. 77:257-266, 1967.)

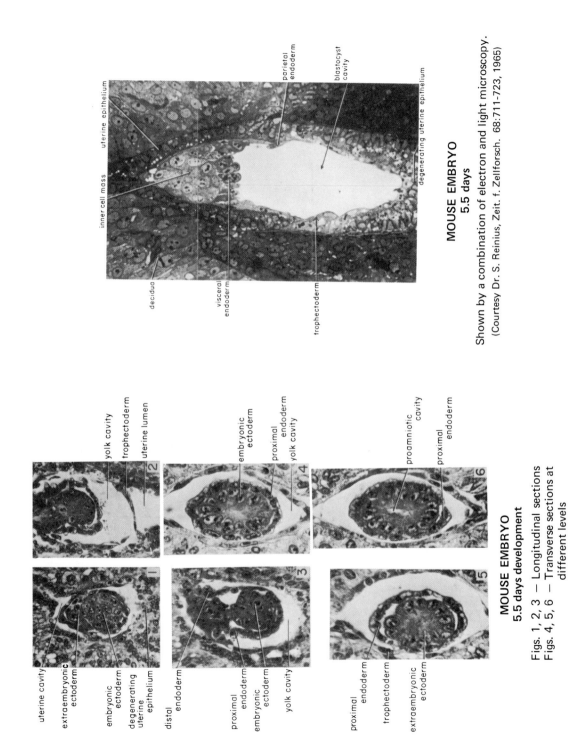

MOUSE EMBRYO
5.5 days

Shown by a combination of electron and light microscopy.
(Courtesy Dr. S. Reinius, Zeit. f. Zellforsch. 68:711-723, 1965)

MOUSE EMBRYO
5.5 days development

Figs. 1, 2, 3 — Longitudinal sections
Figs. 4, 5, 6 — Transverse sections at
                different levels

MOUSE DECIDUAL REACTION
5.5 days

Fig. 1 — Low power view of entire cross section of uterus containing 5.5 day mouse embryo. Note extent of decidual reaction within the distended uterus.

Fig. 2 — Enlarged view of embryo from Fig. 1. Note uterine lumen above and degenerating uterine epithelium below the embryo.

Fig. 3 — Transverse section through the extra embryonic level showing no proamniotic cavity, but distinct proximal endoderm.

Fig. 4 — Longitudinal section of embryo showing ectoplacental cone (above) and split of proamniotic cavity extending toward the cone. Note slight indentation at mid-embryo level between embryonic and extra embryonic regions.

Fig. 5 — Transverse section at extra-embryonic level showing separated proximal endoderm, and spacious yolk cavity.

Fig. 6 — Tangential section showing the difference between the bubbly proximal endoderm and the more squamous-like embryonic endoderm.

Fig. 7 — Slightly younger embryo showing proximal and distal endoderm, embryonic and extra-embryonic ectoderm, and degenerating uterine epithelium (below), and maternal vascular bed above.

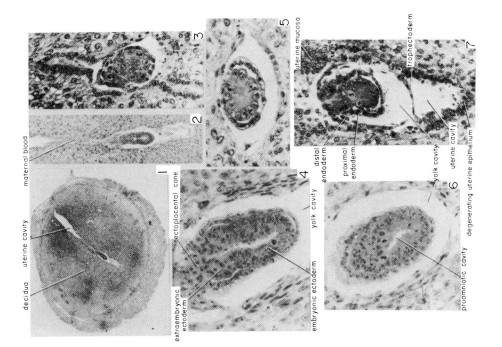

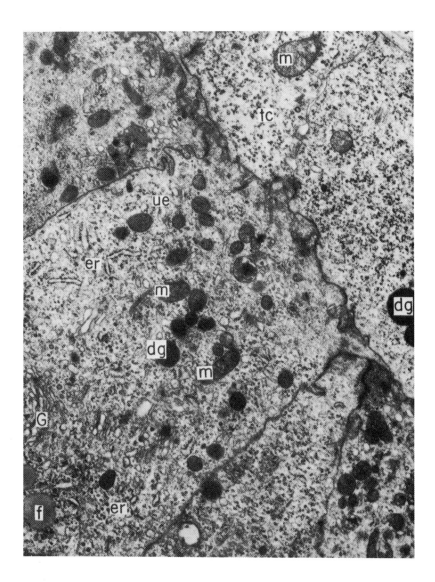

Blastocyst attachment, gestation day 5½. er — endoplasmic reticulum; dg — dense granule; f — fat granule; G — Golgi apparatus; m — mitochondria; tc — trophoblast cell; ue — uterine epithelium; x 15,400.

(Courtesy S. Reinius, Z. Zellforsch. 77:257-266, 1967.)

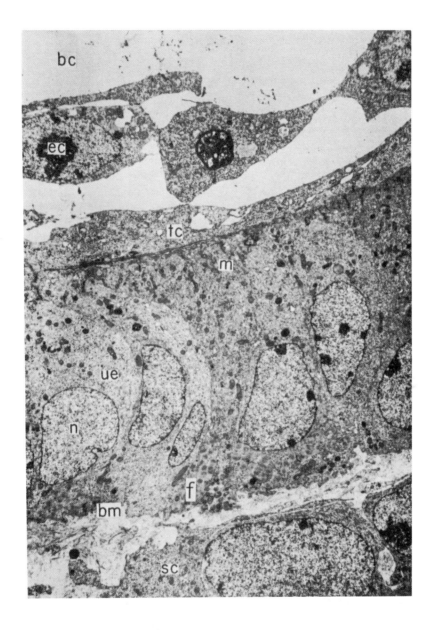

Blastocyst attachment, gestation day 5½. bc — blastocyst cavity; bm — basal membrane; ec — endoderm cell; f — fat granule; m — mitochondria; n — nucleus; sc — stromal cell; tc — trophoblast cell; ue — uterine epithelium. x 4,500.

(Courtesy S. Reinius, Z. Zellforsch. 77:257-266, 1967.)

c. *Gastrulation:* Gastrulation, or the formation of two primary germ layers occurs as the embryo is being implanted. The part of the inner cell mass nearest to the blastocoel splits off to form the endoderm. The remaining cells make up the ectoderm.

The developing embryo comes to lie within a crypt in the uterine mucosa on the ventral or anti-mesometrial side of the lumen. It is the uterus and not the blastocyst that determines the ventral or anti-mesometrial site for implantation. The mucosa is thrown into numerous folds, with those of the opposite sides closely interdigitated, so that every embryo will be trapped in a crypt. The factors that control the spacing of embryos within the uterus are not known, but with a normal litter of about 10 it seems to be most efficient. In experimental super-ovulation many of the embryos die at the time of attempted implantation. The blastocyst implanting in a endocrinologically prepared uterus, causes the nearby endometrium to develop a refractory zone in which no other blastocysts can implant. The onset of degenerative changes in nearby ova is coincidental with the rupture of blood vessels in the area of the implanting blastocyst. The initial stages of implantation appear to be the entrapment of and a secretion from the trophoblast, which adheres to the mucosa. Leukocytes of the uterus rush to the site of contact, and mucosal cells loosen and seem to degenerate (or be digested away). The embryo, minus its zona, its engulfed by the mucosa, to begin its parasitic-type relation with the maternal tissues.

Growth of the embryo as a whole is virtually impossible until implantation because no nutrition is available to its cells. Hence the total embryonic mass has been stabilized, even though cleavage has resulted in an increasing number

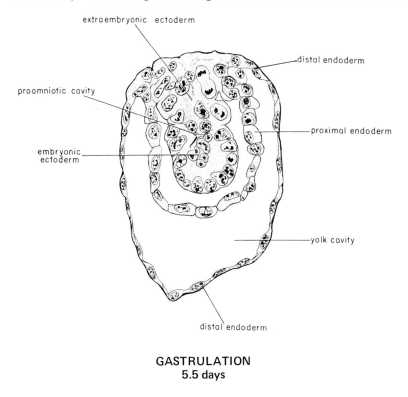

GASTRULATION
5.5 days

of smaller and smaller cells. Nevertheless, these important changes have been in progress; the blastocoel now known as the yolk cavity, has further enlarged; the inner cell mass, now known as the egg cylinder, has divided transversely into an outer layer of small extraembryonic ectoderm, an underlying layer of large embryonic ectoderm cells, and a central core of proximal endoderm cells, from which a tongue of distal endoderm extends into the blastocoel. There is some evidence of nutritional uterine milk elaborated at $4\frac{1}{2}$ days at the mesometrial pole of the implanting embryo. Reichert's membrane, the yolk cavity and the yolk sac effectively separate the embryo from the surrounding maternal blood. Lactic dehydrogenase activity in the mouse embryo decreases tenfold during the first five days of development.

d. *Implantation:* Implantation depends upon a delicately balanced, synergistic action of estrogen and progesterone on the uterine endometrium to provide the optimal conditions for attachment and embedding of the embryo. There appears to be an "estrogenic surge" at about $4\frac{1}{2}$ days, just before the time for implantation, initiated by secretion of the lutenizing hormone. Ova will not implant when transplanted into uteri that are not thus prepared. Actually there is only a very short period (24 hours) during which a transplanted blastocyst can survive in the mouse uterus, this organ being generally inimical to implantation. On postconceptive days 4 or 6 the uterus will not accept the blastocysts. Best results are achieved when the donors' post-conception time is slightly in advance of that of the recipient. Progesterone from the corpora lutea prepares the endometrium for implantation, while estrogens and progesterone are balanced for maintenance of the implantation. Implantation may consist of five steps: adhesion of the trophoblast to the uterine mucosa, penetration of the mucosa by the trophoblast, invasion and spread of the embryonic components in the mucosa, active response of the maternal tissues, arrest of proliferation of the maternal tissues when optimum conditions have been reached and active corpora lutea secure the implantation. The zona pellucida is dissolved gradually around the blastocyst prior to its attachment, so that the zona-free blastocyst exists for only a very short time. The blastocyst attachment to the uterine epithelium is established when the trophoblast and uterine cell membranes lie only 150 Angstroms apart. The epithelium of the uterus does not immediately show any degenerative response, and the invasion precedes the decidual reaction.

The uterine mucosa reacts quickly to invasion by an embryo at $4\frac{1}{2}$ to 5 days gestation by growth of the decidua which is rich in nutrient glycogen. The decidual reaction, which can be invoked by inanimate objects and does not involve any destruction of mucosa, is distinct from the active invasion of the endometrium by the trophoblast by a considerable interval of time. Possible functions of the decidua are suggested.

a. Parturitional aid
b. Nutritional source
c. Protective (most likely)
d. Hormonal (secretes prolactin)

Those who favor the protective function suggest the decidua protects the fetus against immunological attack by the mother, or defends the mother against the invasive activities of the trophoblast. At $5\frac{1}{2}$ days the embryo measures about 250 microns in diameter, and the egg cylinder about 70 microns in diameter.

MOUSE EMBRYO
6 days gestation

Fig. 1 — Transverse section of embryo at extra-embryonic level, showing surrounding decidual reaction of the uterus.

Fig. 2, 3, and 4 — Longitudinal sections of the same embryo at several levels. Note extent of maternal hemorrhage (above) and uterine degeneration (below).

Fig. 5 — Slightly enlarged view of tangential section of embryo showing extra-embryonic ectoderm and proximal endoderm. Note surrounding leukocyte activity.

Fig. 6 — Enlarged view above ectoplacental cone showing maternal vascular bed and a few secondary giant cells.

Fig. 7. — Transverse section just at level of proamniotic cavity and proximal endoderm.

Fig. 8 — Longitudinal section of typical 6 day embryo which is beginning to show the line of demarcation between the extra-embryonic and embryonic regions.

Fig. 9 — The egg cylinder showing some extra-embryonic ectoderm (above), bubbly proximal endoderm, compact embryonic ectoderm within, and faint proamniotic cavity.

Fig. 10 — Enlarged view of transverse section through extra-embryonic level showing the nature of the proximal endoderm, and the mitotic activity of the extra-embryonic ectoderm.

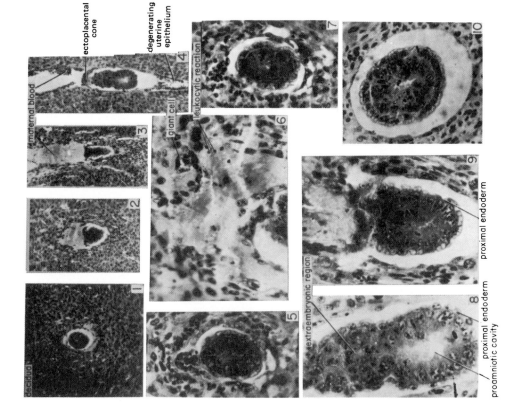

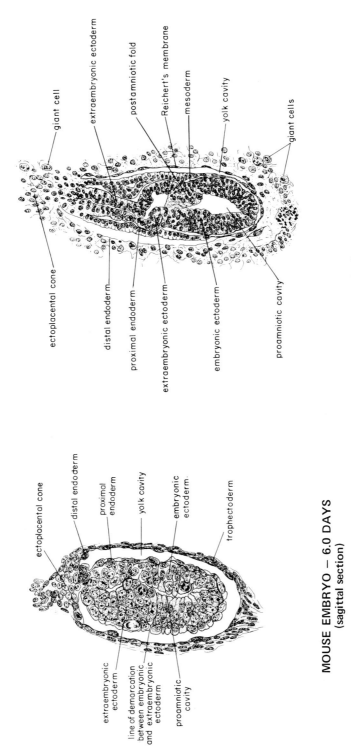

giant cell

extraembryonic ectoderm

postamniotic fold

Reichert's membrane

mesoderm

yolk cavity

giant cells

ectoplacental cone

distal endoderm

proximal endoderm

extraembryonic ectoderm

embryonic ectoderm

proamniotic cavity

MOUSE EMBRYO – 6.5 DAYS
(sagittal section)

ectoplacental cone

distal endoderm

proximal
endoderm

yolk cavity

embryonic
ectoderm

trophectoderm

extraembryonic
ectoderm

line of demarcation
between embryonic
and extraembryonic
ectoderm

proamniotic
cavity

MOUSE EMBRYO – 6.0 DAYS
(sagittal section)

But considerable variation in development is seen in mouse blastocysts as they reach the uterus. The decidua is darkly staining with some lipid granules visible. At $6\frac{1}{2}$ days the embryo measures about 300 microns in diameter. Except for some degenerating cells anti-mesometrially, the uterine epithelium in this region has disappeared, leaving a hyperemia which may encourage implantation.

The uterine mucosa dorsal to the embryo is the site of the discoid hemochorial placenta, which divides the uterine lumen in two. By 8 days gestation a new connection is established between the two resulting cavities, but it is now ventral to the embryo. Active proliferation of the uterine mucosa cuts off the decidua from the uterine muscle ventrally, so that the embryo is suspended in a new portion of the lumen.

The decidua capsularis is ventral to the embryo, on the antimesometrial side, and contains many large, bi-, tri-, and even tetranucleate cells. When first formed, it is relatively thick, but, as the embryo grows, it is stretched and becomes thin. It eventually partitions the embryo from the uterine lumen. The decidua basalis, which looks like mucosa, is dorsal (mesometrial) to the embryo, and gives rise to much of the placenta. At 5 days gestation a vascular area can be identified between the decidual zones, in which blood is loose in lacunae or sinusoids, bathing the embryo almost directly. At 7 days gestation the decidua encloses the embryo and separates it completely from the lumen of the uterus. The impression is that many maternal capillaries have been ruptured by the invasion of the embryo, but the two blood systems never unite. They are effectively separated by Reichert's and yolk sac membranes. Some of the more peripheral sinusoids become so engorged that they shed their blood into the uterine lumen at about 8 days gestation, causing bleeding which reaches the vagina at 12 or 13 days gestation. This bleeding is natural and a certain sign of pregnancy.

During implantation the embryo may derive some nourishment from degenerating mucosal cells at the implantation site. There is a widespread occurrence of glucose, glycogen, lipids, phosphatases, iron, calcium, and many other nutritional necessities such as vitamins and enzymes in the endometrium which may provide the nutritional requirements of the early stages of blastocyst implantation. Both decidua and trophoblast contribute glycogen, with the peak at about 15 days gestation. The decidua is rich in glycogen and the enzyme glucose-6-phosphatase. The glycogen content of the trophoblast increases from the time of implantation to about 17 days gestation. The yolk sac is rudimentary, there being little or no yolk, but it is involved in the embryonic circulation, absorbing nutrition from the maternal blood bath. It also provides glycogen, with the peak at about 18 days gestation. The cytoplasm of the embryonic cells is basophilic, and their nucleoli have concentrations of RNA as implantation approaches, suggesting that protein synthesis, although demonstrated in unovulated ova and blastocysts, is not quantitatively significant.

The placenta is formed by the fusion of the decidua basalis, the ectoplacental cone, the chorion, and parts of the allantois, all so interfused that they can no longer be distinguished. Estrogens have not been detected in the mouse placenta, perhaps owing to technical difficulties or too low concentrations.

Placental fusion occurs very rarely in the mouse. This is probably due to the same factors which control the efficient spacing of the implantations. Even closely approximated blastocysts in the same uterine crypt can form their independent placentae. Foreign material in the uterus prevents implantation because it inhibits the decidual reaction. It is however a most effective contraceptive when the uterus is preparing for implantation. There is some evidence that females with large, heavy pituitary glands have a greater number of successful implantations than other females, but the reverse relation does not hold.

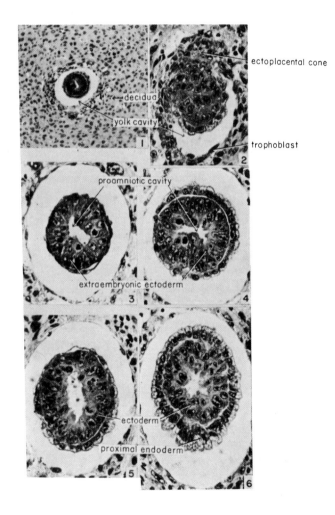

SECTIONS OF MOUSE EMBRYO
6.5 days gestation

Fig. 1 — Low power showing decidua surrounding embryo.

Fig. 2 — Longitudinal section showing yolk cavity and ectoplacental cone.

Fig. 3 — 6 — Transverse sections at various levels of the embryo at 6.5 days showing the early formation of the proamniotic cavity.

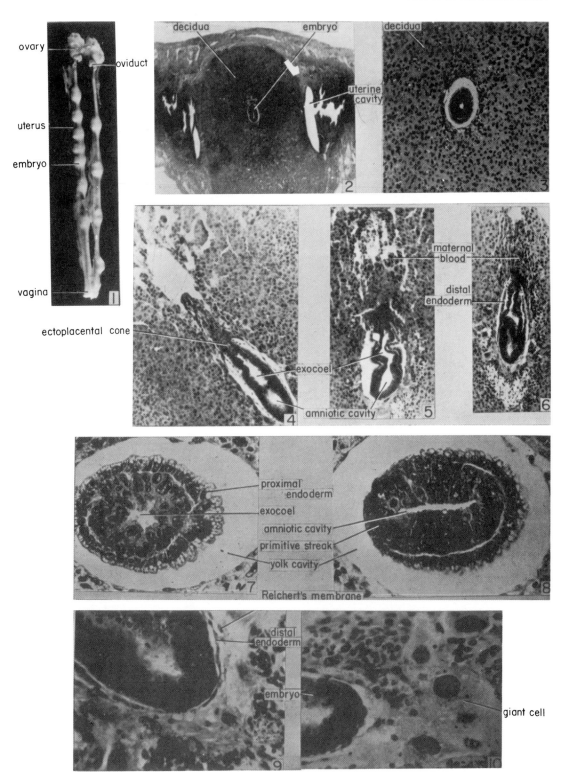

ovary

oviduct

uterus

embryo

vagina

1

decidua embryo decidua

uterine cavity

2 3

ectoplacental cone

maternal blood

distal endoderm

exocoel

amniotic cavity

4 5 6

proximal endoderm

exocoel

amniotic cavity

primitive streak

yolk cavity

7 Reichert's membrane 8

distal endoderm

embryo

giant cell

9 10

MOUSE EMBRYO — 6.5 days

The egg cylinder is divided into two regions, one dorsal with darkly staining high columnar cells having elongated nuclei and one ventral with lightly staining, smaller cells with rounded nuclei. The dorsal region is extra-embryonic ectoderm, which gives rise to the extra-embryonic structures (such as membranes), and the ventral region is the embryonic ectoderm, which gives rise to various structures of the embryo proper. Even at $5\frac{1}{2}$ days gestation, when the staining differences disappear, there is a sharp line of demarcation between the two regions. A proamniotic cavity appears in the center of the embryonic ectoderm, followed by a similar cavity in the extra-embryonic ectoderm. The two cavities join to form an elongated narrow lumen. The growing ectodermal mass of the egg cylinder remains separated from the yolk cavity by proximal endoderm. Hence the relations of the cell layers of the embryo in rodents are just the reverse of those in most other vertebrate embryos; the thick inner mass is ectoderm, which gives rise primarily to neural structures; and the thinner outer layer is endoderm, which gives rise to the gastointestinal tract lining. Both project deeper and deeper into the yolk cavity to obliterate it. In rare instances a slitlike extension of the proamniotic cavity through the ecto-placental cone to the outside is seen suggesting that the inversion of the germ layers is due to limitations on growth ventrally into the yolk sac, which causes invagination. As the proamniotic cavity enlarges, the embryo gradually fills the yolk cavity.

Cells proliferated from the extra-embryonic ectoderm dorsalward make up the ectoplacental cone, which approaches the lumen of the uterus and by $6\frac{1}{2}$ days gestation comprises about half the length of the embryo. The cone invades the maternal tissues as it grows, rupturing blood vessels in its path and producing

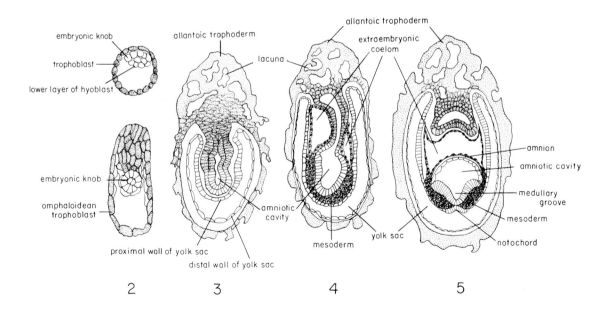

FORMATION OF THE AMNION IN THE MOUSE

(Redrawn after Jenkinson in "Vertebrate Embryology", Oxford & London, 1913.)

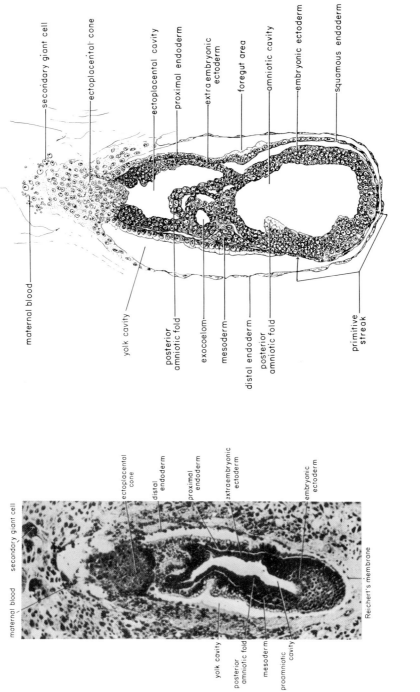

MOUSE EMBRYO – 7.5 days
(sagittal section)

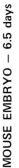

MOUSE EMBRYO – 6.5 days

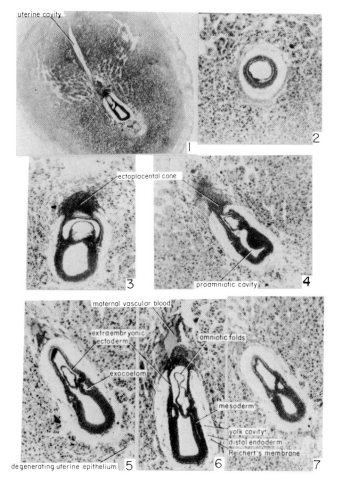

MOUSE EMBRYO
7 days gestation

Fig. 1 — Transverse section through entire uterus showing embedded 7 day mouse embryo. Above it is a portion of the uterine lumen; near the ectoplacental cone is the maternal vascular bed; and below the embryo is the degenerating uterine epithelium.

Fig. 2 — Transverse section through the extra embryonic level showing large proamniotic cavity and relatively thin extra-embryonic ectoderm. The distal endoderm is seen as an enclosing membrane-like structure in the yolk cavity.

Fig. 3, 4 — Longitudinal sections at different regions showing (3) exocoelom and mesoderm interrupting the proamniotic cavity and this cavity as continuous (4) in both embryonic and extra-embryonic levels.

Figs. 5, 6, and 7 — Note clear distinction between embryonic and extra-embryonic levels, enlarged exocoelom and the amniotic folds (anterior and posterior). Reichert's membrane is becoming clearer.

some scattered giant cells. It later contributes to the placenta. The primitive
streak begins at the posterior margin of the egg cylinder as a thickening of
embryonic ectoderm. It establishes the primary axis of the embryo. Possibly
the streak is homologous with the dorsal lip of the blastopore in amphibians as
organizational center. It is a center of rapid growth and a derivative of all the
primary germ layers; and there is no remnant of it in a later embryo. During implantation the embryo is oriented in a crypt with its long axis at right
angles to the long axis of the uterus. Since the embryo develops ventrally to the
egg cylinder, its dorsoventral axis is at right angles (perpendicular) to the long
axis of the uterus, and its anterior-posterior axis is at right angles to the mesometrium. This orientation persists until organo genesis starts at 8 days gestation.

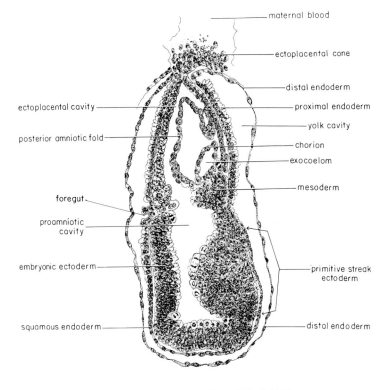

MOUSE EMBRYO — 7.0 DAYS
(sagittal section)

At $6\frac{1}{2}$ days gestation the third germ layer, the mesoderm appears as scattered mesenchyme cells between the primitive streak and the proximal endoderm. The cells move between the ectoderm and endoderm, even toward the
extraembryonic region, where they acquire a cavity known as the exocoelom.
They also form the yolk sac membrane, which envelops later the embryo and is
discarded at birth. By 7 days they have all but separated the two other primary
germ layers except near the primitive streak. Dorsal to the exocoelom is a
diverticulum in the extraembryonic ectoderm, resulting from the proliferation
of mesoderm, known as the proamniotic fold. This is continuous with lateral
and posterior amniotic folds, all products of mesodermal activity. The proamniotic fold constricts the middle of the egg cylinder.

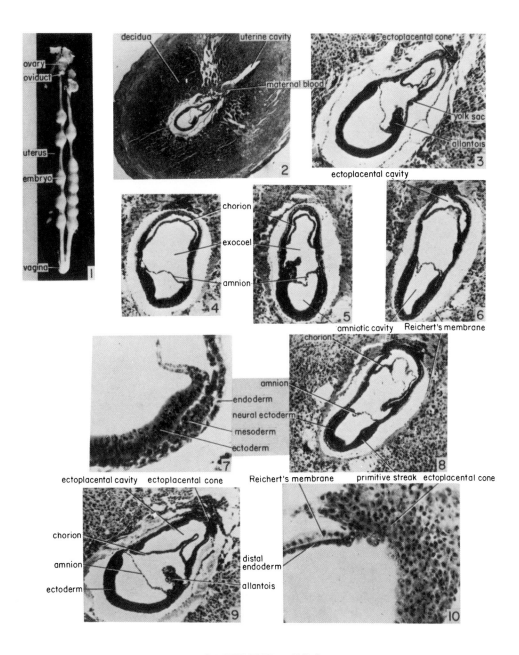

MOUSE EMBRYO — 7.5 days

As stated, the proamniotic fold arises posteriorly within the proamniotic cavity but shortly becomes continuous across the cavity. As it does so, it acquires small lumina (lacunae), which coalesce to form the exocoelom lined with mesoderm and surrounded by ectoderm. The closed off portion of the proamniotic cavity is the amniotic cavity, lined with embryonic ectoderm surrounded by mesoderm. A third cavity persists dorsally as the ectoplacental cavity. The dual membrane between the exocoelom and the amniotic cavity is the amnion, or amniotic membrane. It separates extra-embryonic from embryonic regions, and, with the later complete reversal of the embryo, comes with its cavity, to enclose the embryo from outside. The chorion separates the exocoelom from the ectoplacental cavity and is nonfunctional. Both amnion and chorion are composed of ectoderm and mesoderm but not in the same arrangement.

e. *Head Process Formation:* Mesoderm originating from the primitive streak moves laterally and posteriorly between the other germ layers, never anteriorly. At 7 days gestation, however, the embryonic ectoderm at the ventral tip of the egg cylinder thickens to begin forming the head process, cells of which are continuous with those of the primitive streak but move anteriorly away from the streak and laterally almost around the anterior half of the egg cylinder. The endoderm over and anterior to the head process changes gradually from squamous type to almost columnar type and indents near the point separating the embryonic and extraembryonic regions.   The indentation is the earliest indication of the foregut.

The head process, therefore, constitutes a distinct head-like projection. The development of the cephalic structures is due largely to the formation of the foregut, which tends to lift the anterior end of the embryo upward above the heart area. As the head process grows anteriorly, it extends downward and away from the amniotic cavity, forming a head fold and trapping the cephalic end of the foregut as a blind pocket. A similar process, but in the opposite direction, forms a tail fold and traps the hindgut at the posterior end of the embryo.

f. *Early Organogenesis:* Beginning at about $7\frac{1}{2}$ days gestation the mouse embryo shows signs of organ differentiation. Between the thickened ectoderm of the head process and the related endoderm appears the notochord probably derived from endoderm, from which it is indistinguishable in some regions. Mesenchyme spreads fanlike outward from the sides of the notochord, filling the spaces between the other two primary germ layers. At 7 days the ectoderm anterior to the primitive streak and above the notochord thickens to form the neural or medullary plate and at 8 days it acquires a depression known as the neural groove. The primitive streak, head process, neural groove and notochord establish the anterior-posterior axis of the embryo.

The head process actively extends itself anteriorly, carrying with it the related and underlying endoderm, which will constitute the lining of the foregut. Head process ectoderm, foregut endoderm, and notochord are not always clearly delineated in the mouse so that some investigators believe that foregut endoderm is at least in part derived from head process ectoderm and notochord.

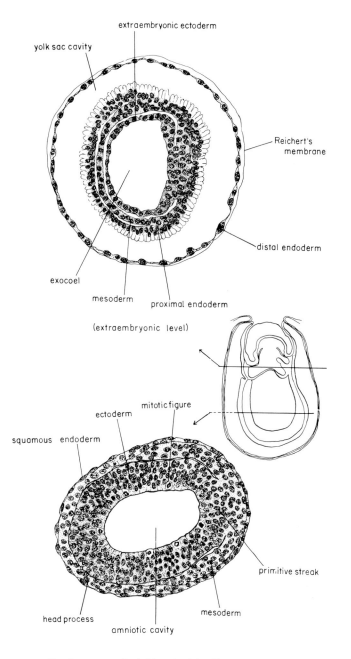

yolk sac cavity

extraembryonic ectoderm

Reichert's membrane

distal endoderm

exocoel

mesoderm

proximal endoderm

(extraembryonic level)

mitotic figure

ectoderm

squamous endoderm

primitive streak

head process

amniotic cavity

mesoderm

(head process and primitive streak level)

MOUSE EMBRYO — 7.5 days

Mesenchyme arising from the posterior limit of the primitive streak grows anteriorly in the exocoelom to form the allantois, which contains numerous small lacunae. Thus the mouse allantois lacks the endodermal lining characteristic of the allantois in most vertebrates. The allantois expands until it fills the exocoelom, and by 8 days it is fused with the chorion. It joins the primitive streak to the ectoplacental cone, whose blood vessels link the embryo to the uterus.

The foregut, first seen at 7 days, is a pocket beneath the cephalic neural primordia at 7 days. Directly beneath the foregut is loose mesenchyme which forms the heart primordia, at this time. The hindgut differentiates, by 7 3/4 days. Thus the two ends are the first portions to appear. Their openings into the intervening region are the anterior and posterior intestinal portals. Eventually the anterior ectoderm breaks through the stomodeum to the foregut, and the posterior ectoderm breaks through the proctodeum to the hindgut. The midgut forms through a downward growth of the endoderm on either side, connecting the foregut and hindgut regions in a zipper-like action. The entire embryo exhibits a lordosis curve, to be rectified when the three germ layers assume their proper relationships, with the ectoderm outermost, the mesoderm next, and the endoderm innermost. This is a reversal of the situation in the embryo at 7 days. The reversal starts at the foregut.

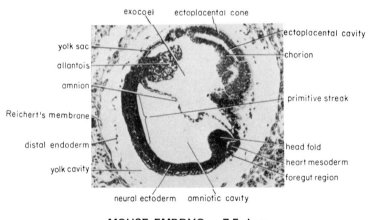

MOUSE EMBRYO — 7.5 days
(sagittal section)

Almost as soon as the mesenchyme aggregates adjacent to the notochord (the paraxial mesoderm), it is organized into paired, segmental (metameric) somites. The first pair to appear at 7 3/4 days is the most anterior, just anterior to the primitive streak and just posterior to the future hindbrain. The second pair forms immediately behind the first but is cleanly separated from it and lags behind it slightly in differentiation. By 8 days there are at least 4 pairs of somites. Eventually there are 65 pairs, each sharply delineated from its neighbors. The chronological and linear appearance of the somites can be taken as an accurate guide to the embryonic age.

At 8 days the intermediate cell mass (nephrotome) arises as the precursor of the excretory system. Laterally it is continuous with the parietal (or lateral plate) mesoderm, which splits into two layers, forming an intervening

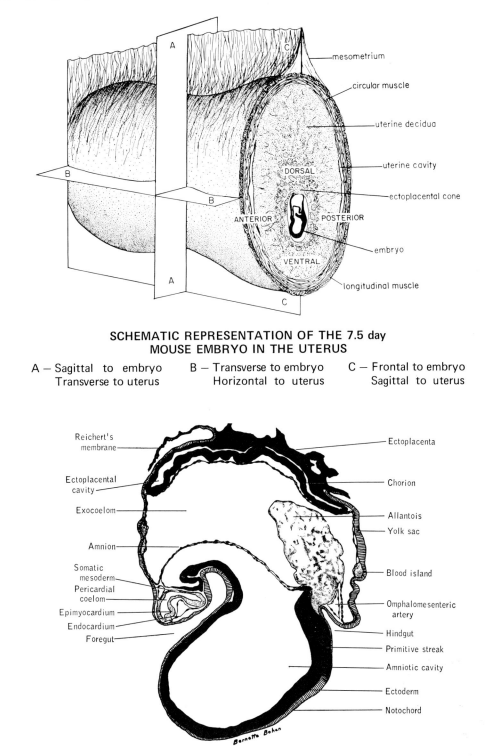

mesometrium

circular muscle

uterine decidua

uterine cavity

DORSAL

ectoplacental cone

ANTERIOR   POSTERIOR

embryo

VENTRAL

longitudinal muscle

**SCHEMATIC REPRESENTATION OF THE 7.5 day
MOUSE EMBRYO IN THE UTERUS**

A — Sagittal to embryo   B — Transverse to embryo   C — Frontal to embryo
Transverse to uterus   Horizontal to uterus   Sagittal to uterus

Reichert's membrane

Ectoplacental cavity

Exocoelom

Amnion

Somatic mesoderm

Pericardial coelom

Epimyocardium

Endocardium

Foregut

Ectoplacenta

Chorion

Allantois

Yolk sac

Blood island

Omphalomesenteric artery

Hindgut

Primitive streak

Amniotic cavity

Ectoderm

Notochord

Bernette Bohan

PARTLY DIAGRAMMATIC SAGITTAL SECTION OF EMBRYO of 7 days 18 hours (X100)

(From "The Biology of the Laboratory Mouse," 2nd Ed., Jackson Laboratory, Editor, Earl L. Green. Copyright
© 1966. McGraw-Hill, Inc. Used by permission of McGraw-Hill Book Co.)

coelom. The outermost layer is ectoderm associated with somatic mesoderm, and the two together are called somatopleure. The somatopleure is continuous with the amnion. The innermost layer is endoderm, associated with splanchnic mesoderm, and the two together are called splanchnopleure. The intervening coelom is continuous with the exocoelom. The coelom is not divided into paired cavities in the mouse but extends beneath the foregut and head fold as the single pericardial cavity. The original shape of the coelom is that of an inverted U(∩), but, as the anterior intestinal portal moves posteriorly with the downgrowth of gut endoderm, the shape changes to an inverted V(Λ) and finally to an inverted Y (Λ), with a single anteriorly directed tubular cavity as the stem of the inverted Y(Λ).

The circulatory system derives from blood islands, aggregations of mesenchymal cells in the mesoderm of the splanchnopleure. Among these aggregations appear lacunae which join together to form a continuous blood vascular system. The contained cells give rise to some blood cells as early as 7 days gestation. At this time the heart develops, the splanchnic mesoderm of the pericardial cavity forming the epicardium (outer heart membrane) and the myocardium (muscles) of the heart wall. Between the splanchnic mesoderm and the endodermal floor of the foregut loose mesenchymal cells, probably of endodermal origin, form the endocardium (mesothelial lining of the heart). The endocardium forms at first a straight tube, but, as it grows within the confines of the pericardial cavity, it is coiled upon itself to form the various chambers of the heart. The heart is soon joined by the aortic arches which pass through the visceral arches. Anterior extensions of these vessels become the paired ventral and dorsal aortae. The ventral aortae proceed anteriorly into the head as the external carotid arteries and the dorsal aortae merge posteriorly into the single large vitelline or omphalomesenteric artery, leading to the yolk sac. Blood from the vitelline artery returns to the embryonic heart, (after discharging its waste and acquiring nutrition from the yolk sac) by way of paired vitelline or omphalomesenteric veins. Thus function begins as developmental (structural) changes continue. The rudimentary yolk sac is intimately associated with the placenta to absorb nutrition from the mother, to provide a channel for excretion, and to afford a membrane for protection. It eventually envelops the amnion and the exocoelom.

At $7\frac{1}{2}$ days gestation the amnion covers the dorsal surface of the embryo. Owing to the inversion of the germ layers, it eventually surrounds the embryo except at the umbilical cord. The embryo finally seems to float freely within the amniotic fluid, anchored only by the cord.

g. *Reichert's Membrane:* Besides the amnion and yolk sac, the mouse embryo has a third extraembryonic protective membrane called Reichert's membrane. This is found only in rodents. At $5\frac{1}{2}$ days the trophoblast surrounds the embryo except at the region of the ectoplacental cone and is in close contact with the uterine mucosa. On the inner surface of the trophoblast are a few scattered endoderm cells, which at $6\frac{1}{2}$ days compose an almost continuous sheet. Between the sheet and the trophoblast develops the thin, tough, homogeneous, noncellular, acidophilic Reichert's membrane. It is usually first seen at the ventral extremity of the egg cylinder. Since it expands with the enlargement of the embryonic mass during blastulation, it must be elastic. It has been suggested that it is a secretion of the related endoderm, which could explain its growth and expansion. It is definitely a protective covering for the embryo.

h. *Giant Cells:* From 6 to about 14 days gestation excessively large cells, called giant cells, lie between Reichert's membrane and the decidua. Neither their origin nor their function is clear cut; some investigators believe that they support Reichert's membrane, and others attribute steroid-hormone production to them. There is also recent evidence that they aid in the transformation of the trophoblast and are, therefore, necessary for proper implantation.

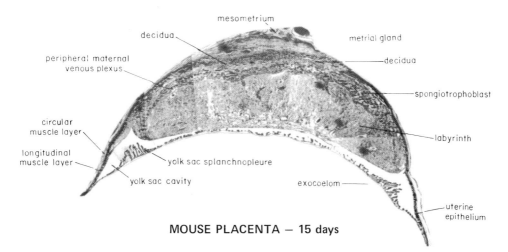

MOUSE PLACENTA — 15 days

It seems that they are under the control of the ovary, since ovariectomized or superovulated mice have difficulty or fail in the natural growth of the inner cell mass, the expansion of the blastocyst, and the transformation of the trophoblast requisite to implantation. Isoantigens appear in significant quantity in the trophoblast by 4 days gestation and the maternal tolerance of an embryonic homograft may be due to an inert barrier of giant cells between the mother and the developing embryo. They may therefore provide protection in the nature of an immune reaction. They first appear between the ventral side of the embryo and the decidua and continue to enlarge for several days and move into the uterine cavity ventral to the embryo. Other and more numerous giant cells appear at the sides of the ectoplacental cone, from which they are derived, and move posteriorly. By 8 days these have protoplasmic processes which traverse the vascular area between the decidua and Reichert's membrane. Still other giant cells, which are multinucleate, may be found in the decidua at 7 days. These are known as synplasia. There are also secondary giant cells, which, although conspicuous for their size, are not as large as the others.

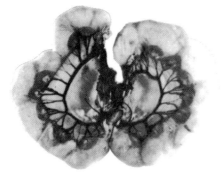

Paired gravid uteri of mouse at 18 days gestation showing the distribution of india ink injected into the uterine arteries and stopping at the maternal placenta. Since none of the ink particles are found in the embryo or embryonic placenta, there is no direct connection of the two circulatory systems.

Chapter  CHRONOLOGY OF DEVELOPMENT: 8-16 days

A. THE 8 DAY EMBRYO: The 8 day embryo can be identified by its 4 pairs of somites. The body is much coiled, like an extended letter S with the two ends curled in the same direction, and a deeply concave mid-region. Optic evaginations and otic (auditory) invaginations are present. Remnants of the primitive streak persist. In transverse section the neural plate and under-lying endoderm are flattened, in anticipation of the reversal of the germ layers. The neural groove is still open throughout its length, but the anterior neural ectoderm is beginning to thicken and the anterior end of the groove (neuropore) to close. The foregut has thyroid and liver primordia. The midgut remains wide open, the allantois is fused with the chorion, and the yolk sac has begun to develop blood islands.

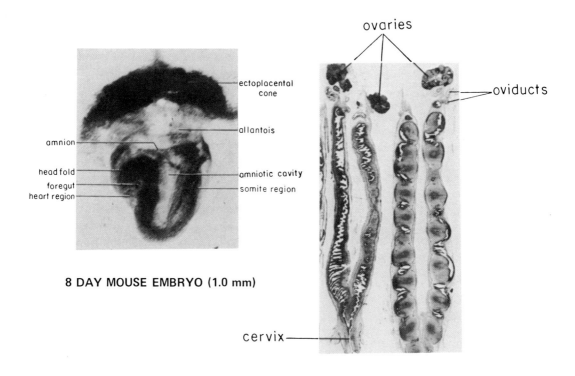

8 DAY MOUSE EMBRYO (1.0 mm)

NON-GRAVID AND GRAVID
UTERI OF MOUSE (8 g.d.)

Growth of the foregut and hindgut initiates the reversal of the anterior-posterior relationships so that the embryo eventually acquires the C shape characteristic of most vertebrate embryos, with the open part of the C (the midgut) ventrally directed. Transverse sections show that the head end is turning clockwise while posterior regions are turning counterclockwise. The midgut remains connected with the yolk cavity but also rotates rapidly before the completion of the terminal twisting. It rolls to one side, taking with it the associated splanchnopleure, which fuses ventrally to almost close to the midgut by 8 3/4 days. The result is much as if one pushed upward against the convex midgut, making it concave. The embryo then turns its left side ventrally, leaving the placenta dorsal.

The coelom is partially closed off ventrally to the head process as the pericardial cavity, containing the tubular heart. Epi-, myo-, and endocardium are all present, as is the dorsal mesocardium. However, the mesocardium soon disappears. Folds in the endocardial tube, necessitated by the active proliferation of heart tissue, aid in the formation of the four-chambered heart. The first pair of aortic arches is also visible. Early forming blood islands and lacunae leading to vessels may be seen in the yolk sac.

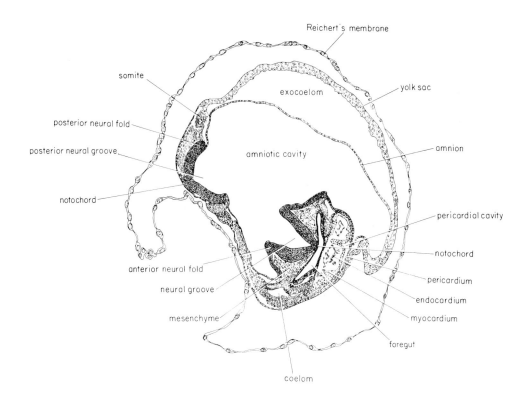

8 DAY MOUSE EMBRYO
(transverse section)

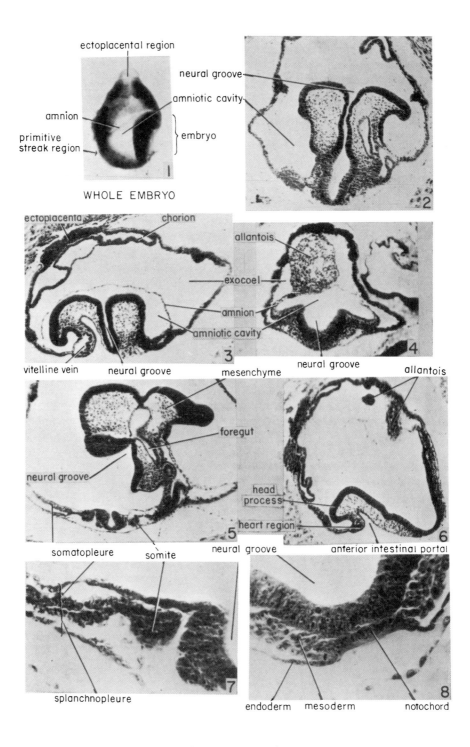

WHOLE EMBRYO

MOUSE EMBRYO — 8 days
(Photographs)

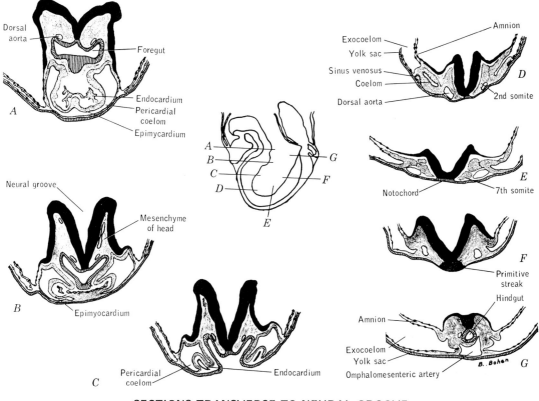

Dorsal aorta

Foregut

Endocardium

Pericardial coelom

Epimycardium

A

Exocoelom

Yolk sac

Sinus venosus

Coelom

Dorsal aorta

Amnion

D

2nd somite

Neural groove

Mesenchyme of head

B

Epimyocardium

Notochord

7th somite

E

Primitive streak

F

Hindgut

Amnion

Exocoelom

Yolk sac

Omphalomesenteric artery

B. Bohen

G

Pericardial coelom

Endocardium

C

SECTIONS TRANSVERSE TO NEURAL GROOVE OF 8-DAY 1-HOUR, 7-SOMITE EMBRYO

The location of each section is indicated on the key diagram.

(From "The Biology of The Laboratory Mouse", 2nd Ed., Jackson Laboratory, Editor, Earl L. Green. Copyright © 1966. McGraw-Hill, Inc. Used by permission of McGraw-Hill Book Co.)

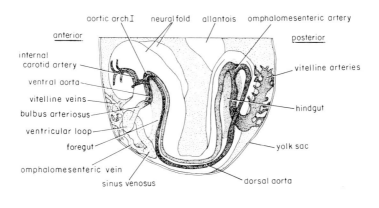

aortic arch I   neural fold   allantois   omphalomesenteric artery

anterior

posterior

internal carotid artery

ventral aorta

vitelline veins

bulbus arteriosus

ventricular loop

foregut

omphalomesenteric vein

sinus venosus

vitelline arteries

hindgut

yolk sac

dorsal aorta

CIRCULATORY SYSTEM OF 8 DAY MOUSE EMBRYO

(Redrawn from Green '66, "Biology of the Laboratory Mouse, McGraw-Hill, Inc.)

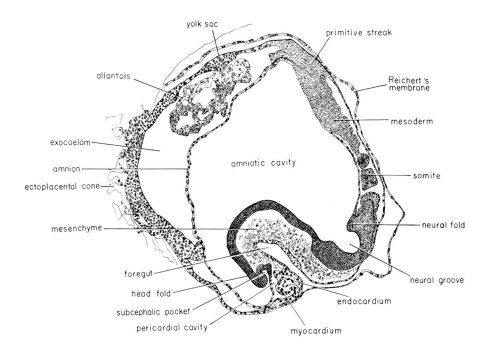

8 DAY MOUSE EMBRYO — HEAD PROCESS
(sagittal section)

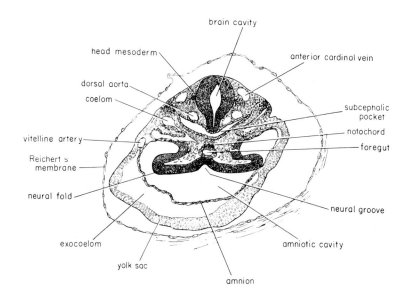

8 DAY MOUSE EMBRYO
(transverse section)

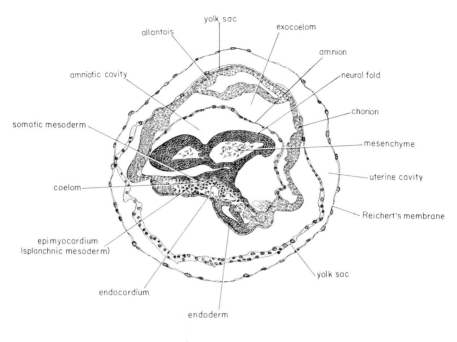

8 DAY MOUSE EMBRYO
(transverse section)

B. THE 8½ DAY EMBRYO: At $8\frac{1}{2}$ days gestation two major activities are under way in the embryo. One is the development of somites and of neuromeres, which indicate the metamerism or segmentation of the muscles, skeleton, brain and spinal cord. The other is the beginning of placentation. There is abundant mesenchyme, some of which has been organized into 8 to 12 somites; some early heart parts can be identified; blood islands are plentiful in the yolk sac, blood vessels begin to appear; the nephrogenic cord can be seen; the stomodeum is invaginating in the oral region; the first two pairs of visceral pouches are visible; the mid-gut is still slightly open but the intestinal portals are converging toward the mid-body region, thus closing off part of it. The amnion, which covers the embryo dorsally, forms lateral limiting sulci on either side of the body folds. The amniotic cavity is enlarging. Reichert's membrane is distinct and intact, completely separating the embryo and its extra-embryonic membranes from the maternal tissues. The untwisting of the embryo is not yet complete. It is like a corkscrew, with cephalization enlarging the anterior end and the posterior end coiled in the opposite direction.

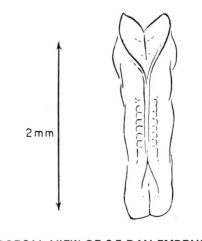

2 mm

DORSAL VIEW OF 8.5 DAY EMBRYO

(Theiler and Stevens 1960)

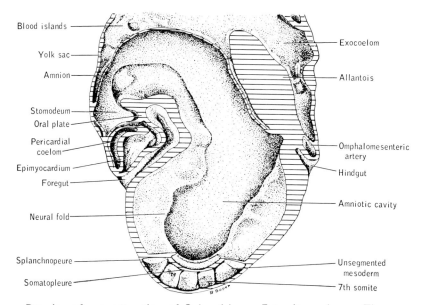

Drawing of reconstruction of 8-day 1-hour, 7-somite embryo. The recon-
construction is cut in the midsagittal plane and only the right half shown
except at the ventral extremity where the last 4 somites and part of the undivided mesoderm of the left side are included. Cut areas are shown by
horizontal shading.

(From "The Biology of The Laboratory Mouse," 2nd Ed., Jackson Laboratory, Editor,
Earl L. Green. Copyright © 1966. McGraw-Hill, Inc. Used by permission of McGraw-
Hill Book Co.)

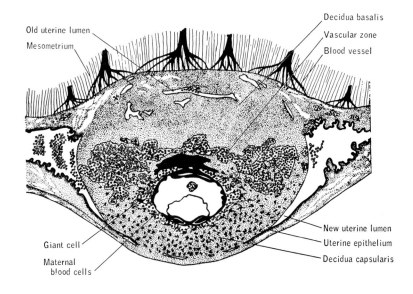

Longitudinal section (partly diagrammatic) of uterus at site of
implantation of 8 day 6 hours, 5-somite embryo. Cut parallel
to mesometrium.

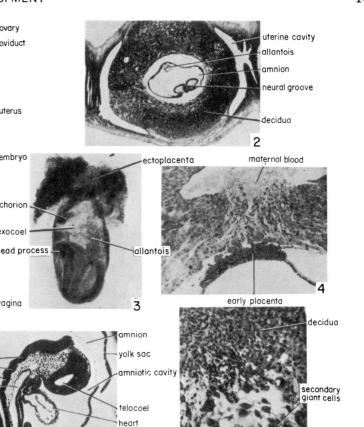

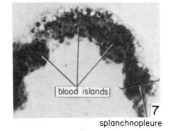

MOUSE EMBRYO — 8.5 days

Fig. 1 — Photograph of entire uterus of gravid female at 8.5 days gestation showing 14 sites of implantation of embryos.

Fig. 2 — Transverse section through entire decidua of 8.5 day embryo.

Fig. 3 — Embryo dissected out of the uterus with membranes intact.

Fig. 4 — Section through early placentation, showing both embryonic and maternal tissues.

Fig. 5 — Sagittal section showing cephalic flexures, head process, anterior intestinal portal, and transverse section of neural groove indicating a twist in
the body at this level.

Fig. 6 — High power view of implantation site and area of invasion of maternal
tissues, with secondary giant cells.

Fig. 7 — Blood islands in splanchnopleure.

Fig. 8 — Giant cells of trophoblast which probably are functional in invasion of
the maternal tissues.

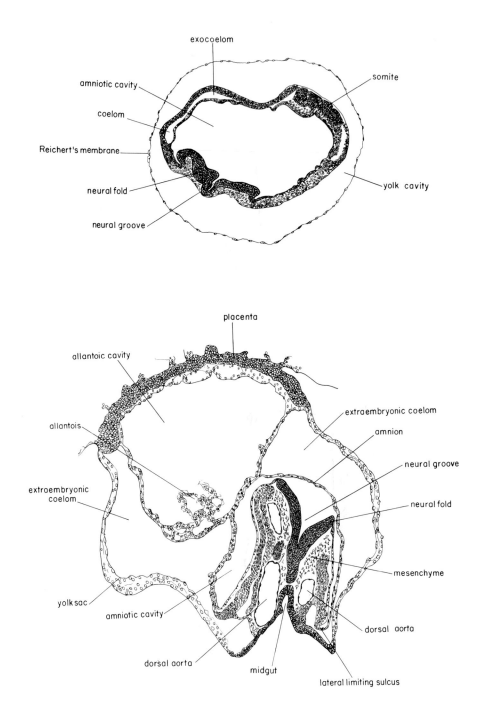

8.5 DAY MOUSE EMBRYO
(transverse section)

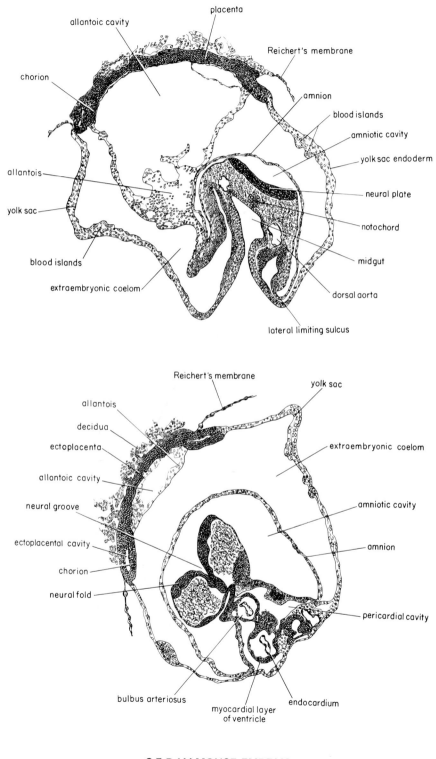

8.5 DAY MOUSE EMBRYO
(transverse section)

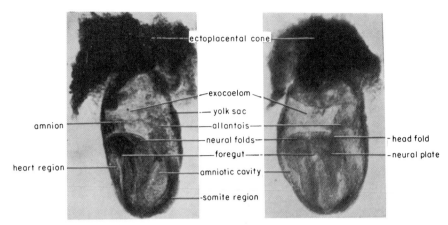

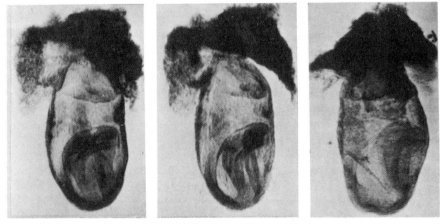

8.5 DAY MOUSE EMBRYO (1.25 mm)

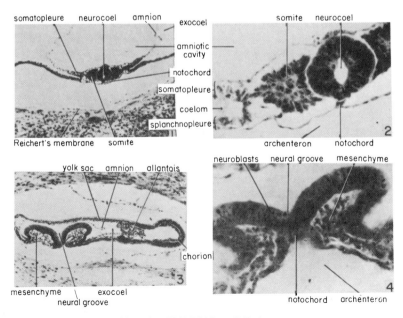

MOUSE EMBRYO — 8.5 days
(transverse sections)

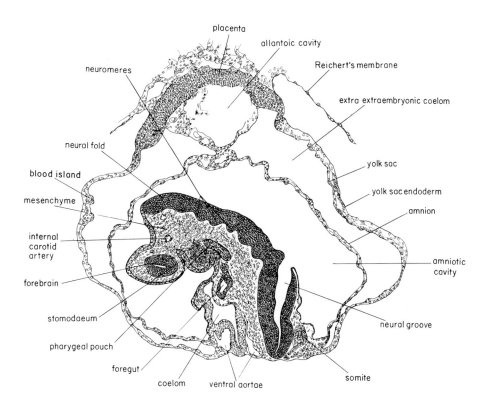

8.5 DAY MOUSE EMBRYO
(sagittal section)

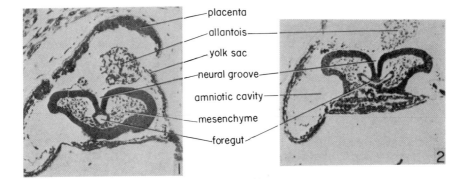

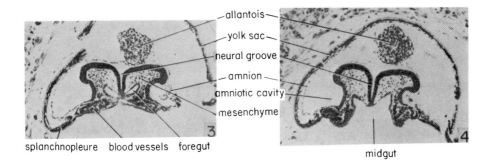

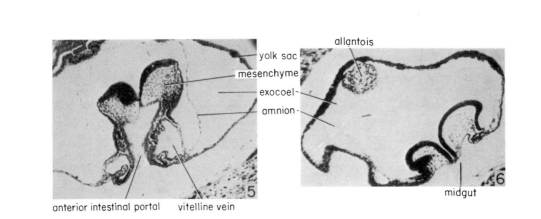

NEURULATION IN MOUSE EMBRYO — 8.5 days
(Photographs)

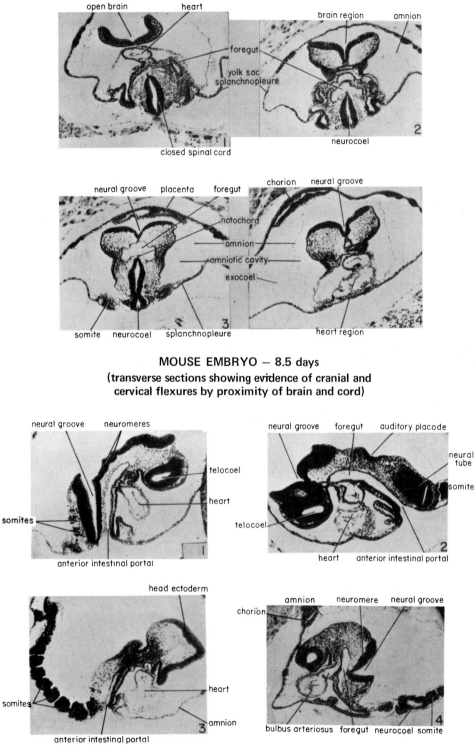

MOUSE EMBRYO — 8.5 days
(transverse sections showing evidence of cranial and
cervical flexures by proximity of brain and cord)

MOUSE EMBRYO — 8.5 days
(sagittal sections)

C. THE 9 DAY EMBRYO: Probably the most significant and involved changes in the embryo occur between $8\frac{1}{2}$ and 9 days, when organogenesis is greatly accelerated and the contour of the embryo is being altered simultaneously.   The most devastating congenital anomalies can be produced in embryos subjected at this time to various types of trauma.   These include stunting, edema, hydrocephaly and microcephaly, microphthalmia, degeneration of part of the neural tube, and related skeletal deformities.   On the other hand, this is about the earliest stage at which the embryo can be dissected out without injury. It is clearly a vertebrate and can be confused with a variety of other vertebrate embryos.   It has a distinct head, body, and tail, and the allantois protrudes from near the base of the hindgut.   It is almost completely enclosed in the amnion.

The 9 day embryo has 13 to 20 somites.   The neural tube is closed except for the posterior neuropore and the roof of the myelencephalon, and the entire nervous system is undergoing cellular differentiation.   Visceral arches I (mandibular) and II (hyoid) are present, as is Rathke's pocket; the large paired sacs may be seen on either side of the myelencephalon; the protruding allantois and heart can be identified and the forelimb buds are forming.

In sagittal section the 9 day mouse embryo resembles the 72 hour chick embryo, with the brain cavities indicated, all sense organ primordia (optic, otic and olfactory) apparent, the four chambered heart formed, the stomodeum broken through and Rathke's pocket formed.   The four chambered heart is now pulsating regularly.   The sinus venosus and the bulbus arteriosus can both be located; the paired dorsal aortae have fused into a single vessel, which extends into the tail fold after giving off the vitelline artery; segmental arteries

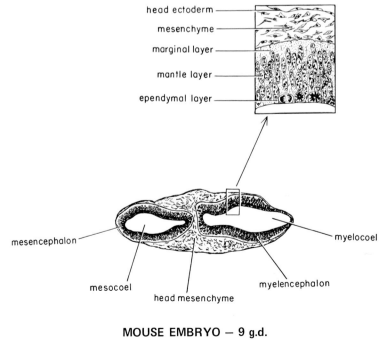

head ectoderm

mesenchyme

marginal layer

mantle layer

ependymal layer

mesencephalon

myelocoel

mesocoel

myelencephalon

head mesenchyme

MOUSE EMBRYO — 9 g.d.
(transverse section)

may be seen to issue from all along the aorta, between the somites; and the external and internal carotids and cerebral arteries proceed anteriorly. The anterior and posterior cardinal veins are forming.

The embryo is still so twisted, particularly at the posterior end, that a transverse section may cut it at three levels. The placenta is developing rapidly with the decidua capsularis separating the embryo from the uterine lumen and the decidua basalis forming a widening junction between embryo and maternal tissues. Blood islands and capillaries are numerous in the yolk sac.

Brain and spinal cord sections show marginal, mantle and ependymal layers, with the ependymal layer particularly exhibiting many mitoses. The acousticofacial ganglion (cranial nerves II and VIII) is developing just anterior to the otic vesicle.

The heart and adnexa deserve special mention at this time, when the circulatory system begins to play a major role in development. Metabolic wastes must be removed, and nutritional elements must be collected and delivered to every part of the rapidly growing embryo. The heart is well developed and functional, with four chambers each lined with endocardium, consisting of myocardium, and surrounded by pericardium. These chambers form as function is initiated. The sinus venosus enters the right atrium, and at the same level the right ventricle opens into the bulbus arteriosus. Anteriorly, near the thyroid primordium in the floor of the pharynx, are the paired ventral aortae, which feed the aortic arches and then become the external carotid arteries. Lateral to the external carotid arteries are the anterior cardinal veins. As usual, the veins are thin-walled in comparison with the arteries. The ventricles are more posterior. Between the ventricles are the interventricular foramen, septum, and sulcus.

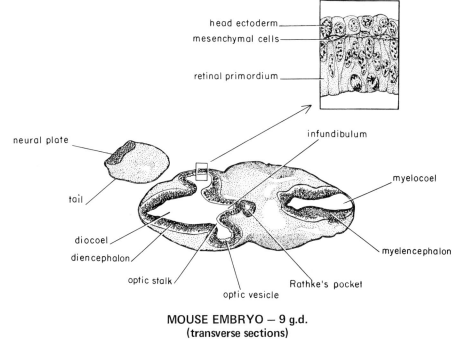

head ectoderm

mesenchymal cells

retinal primordium

neural plate

infundibulum

tail

myelocoel

diocoel

diencephalon

myelencephalon

optic stalk

Rathke's pocket

optic vesicle

MOUSE EMBRYO — 9 g.d.
(transverse sections)

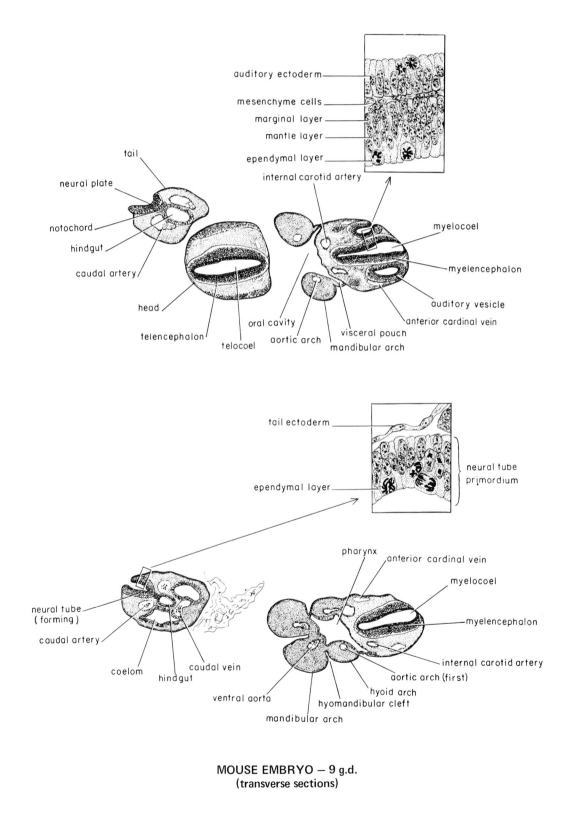

MOUSE EMBRYO — 9 g.d.
(transverse sections)

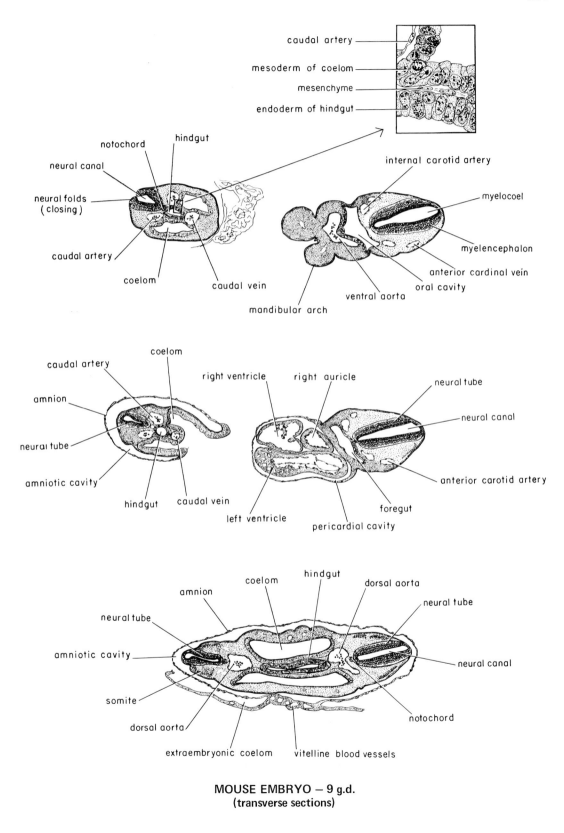

caudal artery

mesoderm of coelom

mesenchyme

endoderm of hindgut

notochord

hindgut

neural canal

internal carotid artery

neural folds
(closing)

myelocoel

myelencephalon

caudal artery

anterior cardinal vein

coelom

oral cavity

caudal vein

ventral aorta

mandibular arch

caudal artery

coelom

right ventricle

right auricle

neural tube

amnion

neural canal

neural tube

amniotic cavity

anterior carotid artery

hindgut

caudal vein

foregut

left ventricle

pericardial cavity

coelom

hindgut

dorsal aorta

amnion

neural tube

neural tube

amniotic cavity

neural canal

somite

dorsal aorta

notochord

extraembryonic coelom

vitelline blood vessels

MOUSE EMBRYO — 9 g.d.
(transverse sections)

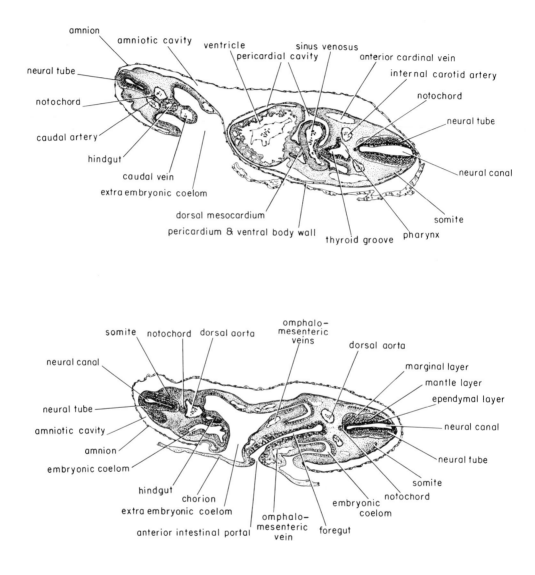

MOUSE EMBRYO — 9 g.d.
(transverse sections)

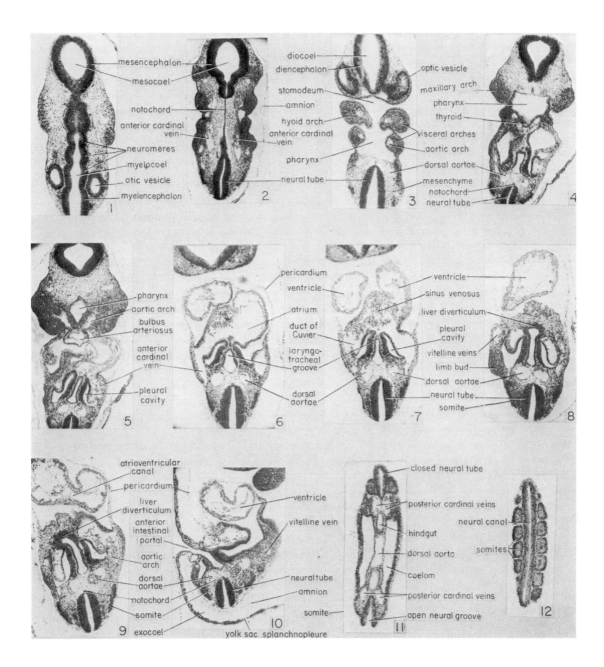

MOUSE EMBRYO — 9 days
(Photographs)

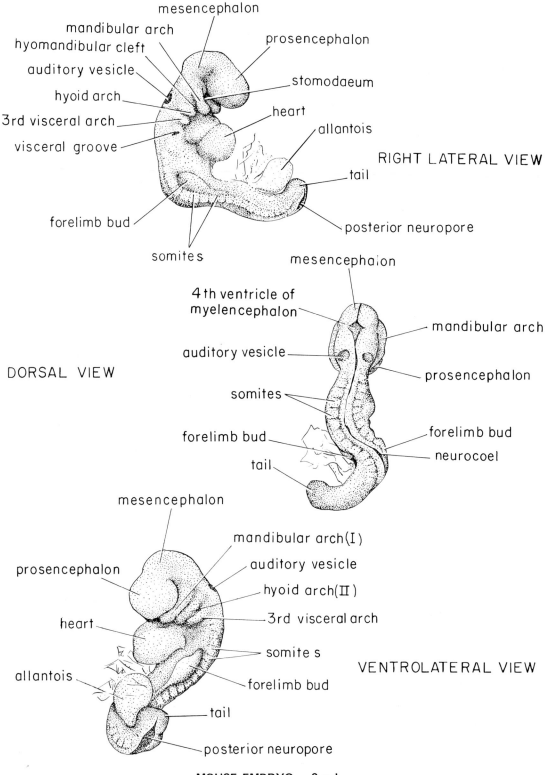

RIGHT LATERAL VIEW

mesencephalon

mandibular arch
hyomandibular cleft
auditory vesicle
hyoid arch
3rd visceral arch
visceral groove

prosencephalon

stomodaeum

heart

allantois

tail

posterior neuropore

forelimb bud

somites

DORSAL VIEW

mesencephalon

4th ventricle of
myelencephalon

auditory vesicle

somites

forelimb bud

tail

mandibular arch

prosencephalon

forelimb bud
neurocoel

mesencephalon

prosencephalon

heart

allantois

mandibular arch(I)
auditory vesicle
hyoid arch(II)
3rd visceral arch

somites

forelimb bud

tail

posterior neuropore

VENTROLATERAL VIEW

MOUSE EMBRYO — 9 g.d.

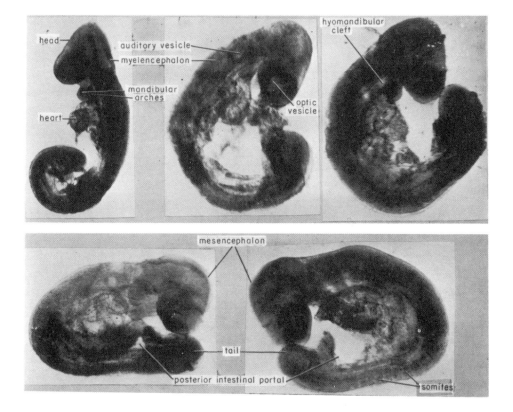

9.0 DAY MOUSE EMBRYO (2.0 mm)
(Photographs)

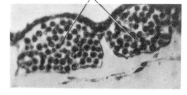

MOUSE EMBRYO — 9 days

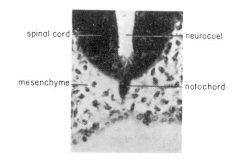

MOUSE EMBRYO — 9 days

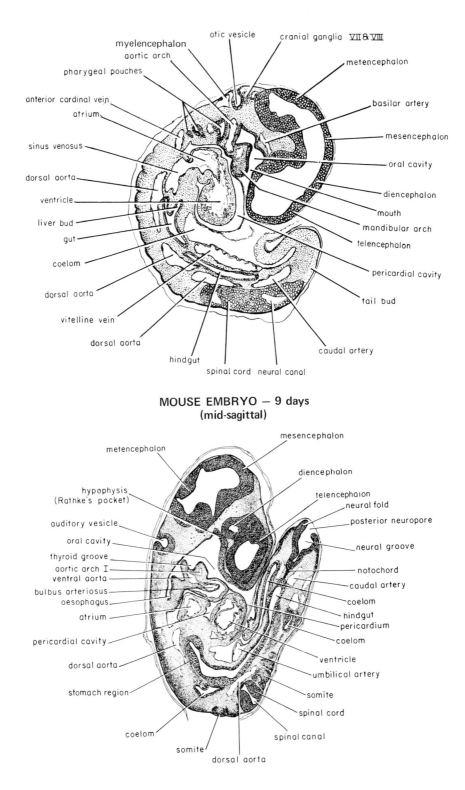

MOUSE EMBRYO — 9 days
(mid-sagittal)

PARASAGITTAL SECTION OF 9 DAY MOUSE EMBRYO

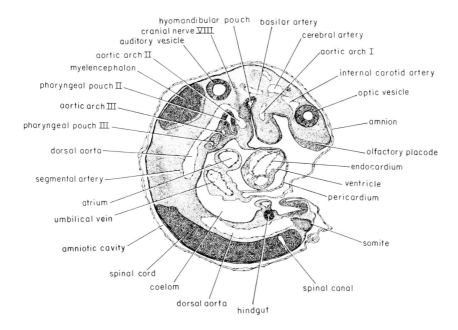

PARASAGITTAL SECTION OF 9 DAY MOUSE EMBRYO

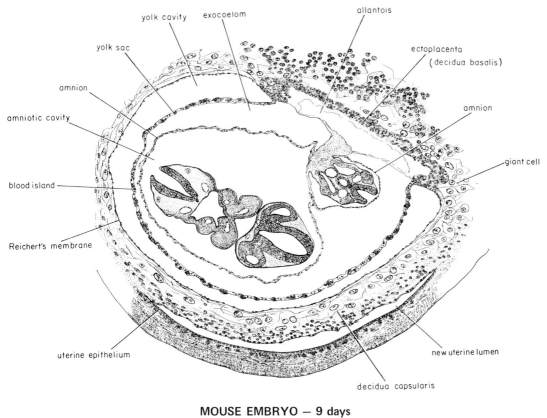

MOUSE EMBRYO — 9 days
(showing membranes)

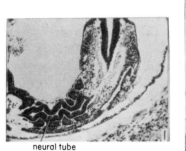

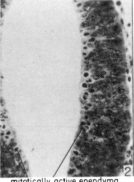

neural tube
(tortuous due to body flexures and torsions)

mitotically active ependyma

NEURAL TUBE OF MOUSE EMBRYO — 9 days

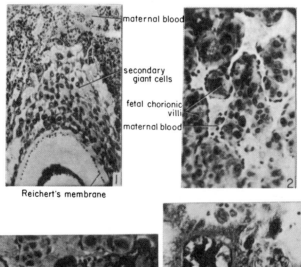

maternal blood

secondary
giant cells

fetal chorionic
villi

maternal blood

Reichert's membrane

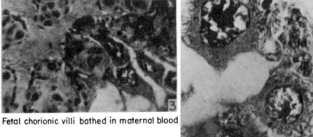

Fetal chorionic villi bathed in maternal blood

Giant trophoblast cells

IMPLANTATION SITE — MOUSE EMBRYO 9 days

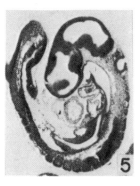

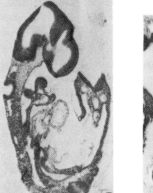

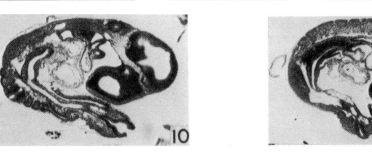

Photographs of Mid- and Para-sagittal sections of 9 day mouse embryo
showing degree of cephalic flexures and the proximity of the head and
tail. Note appearance of some of the major organ systems.

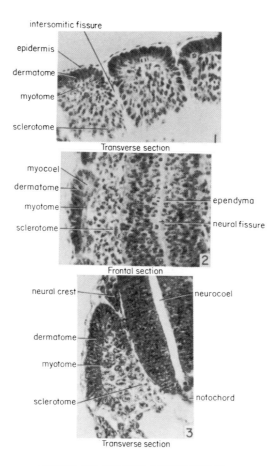

**STATE OF SOMITE FORMATION
IN 9 DAY MOUSE EMBRYO**

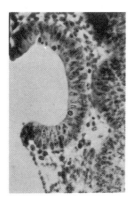

**OTIC VESICLE OF
MOUSE EMBRYO — 9 days
(showing active mitosis)**

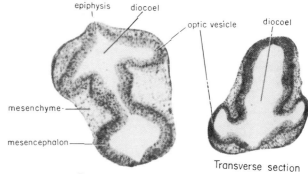

OPTIC VESICLES OF MOUSE EMBRYO — 9 days

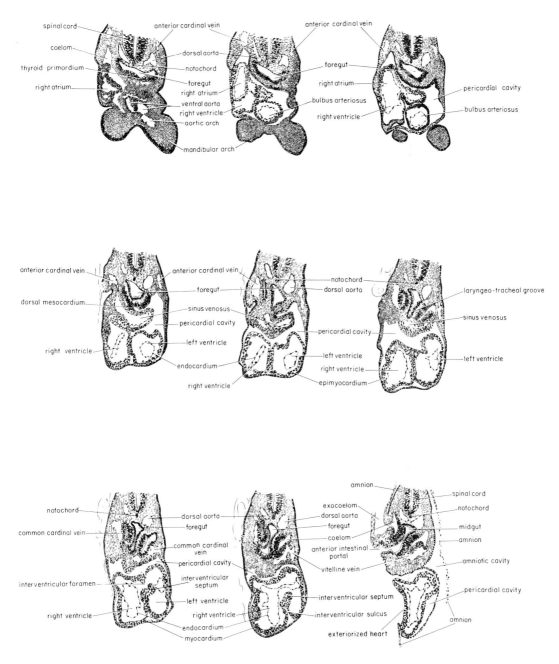

DEVELOPMENT OF THE MOUSE HEART AT 9 GESTATION DAYS
Serial X-Sections

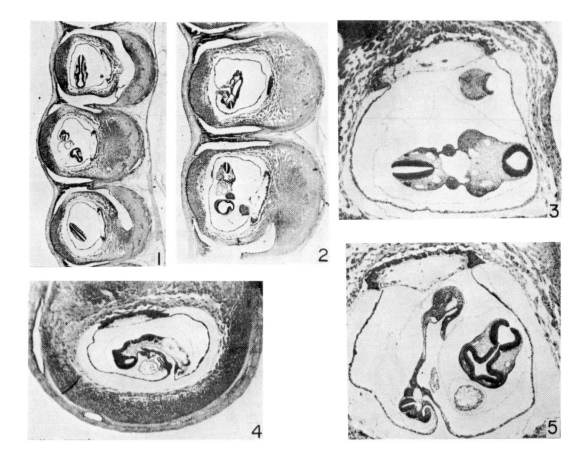

MOUSE EMBRYO — 9 days
(within the uterus at various magnifications)

Figs. 1, 2 — Longitudinal sections of gravid uterus of 9 day mouse gestation showing embryos in various transections and the occluded uterine cavity. Fig. 2 is at higher
magnification than Fig. 1.

Fig. 4   — Still higher magnification of section of 9 day embryo in utero. Note anterior
intestinal portal leading into foregut, and prominent heart beneath the head.

Figs. 3, 5 — Enlarged views of embryonic sections while in the uterus, showing relation to
chorionic placenta. Note in Fig. 5 three separate sections through the central
nervous system, indicating degree of flexion and torsion. Note also anterior intestinal portal.

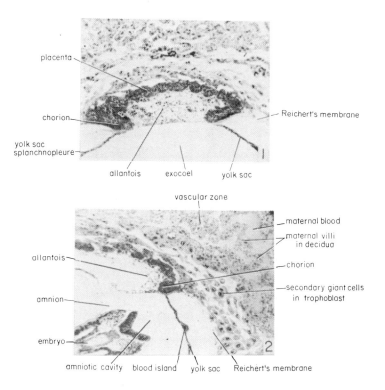

placenta

chorion

yolk sac
splanchnopleure

Reichert's membrane

allantois   exocoel   yolk sac

vascular zone

maternal blood

maternal villi
in decidua

allantois

chorion

secondary giant cells
in trophoblast

amnion

embryo

amniotic cavity   blood island   yolk sac   Reichert's membrane

DEVELOPING PLACENTA OF MOUSE — 9 days

D. THE 9½ DAY EMBRYO: The mouse embryo is now embarked on developmental changes that are characteristic of all vertebrate embryos but
with rather minor variations.   However, since its development is telescoped
into such a brief period, the changes occur at accelerated rates compared to
those in many mammals.

Embryos at this stage may be dissected away from their membranes quite
easily and examined whole, with transmitted light.   If they are fixed and cleared
in oil of wintergreen, their inner structures show up even better.   Sections in
various planes are necessary to establish the extent of organ differentiation,
however.

The reversal of the original lordosis of the back now brings the telencephalon and tail into close approximation, with the dorsal side arched outward.
The shape might be called a kyphosis curve, a C with the ends somewhat overlapping.   In frontal view the tail overlaps and is generally situated to the right
of the head.   The head is far in advance of the rest of the body in development,
and 21 to 25 somites are apparent through the surface skin, gradually tapering
off toward the tail, with the more anterior ones always the further developed.
The posterior neuropore is closing.

The four major divisions of the brain are also apparent from the surface
since the brain vesicles are beginning to separate.   Most prominent, of course,

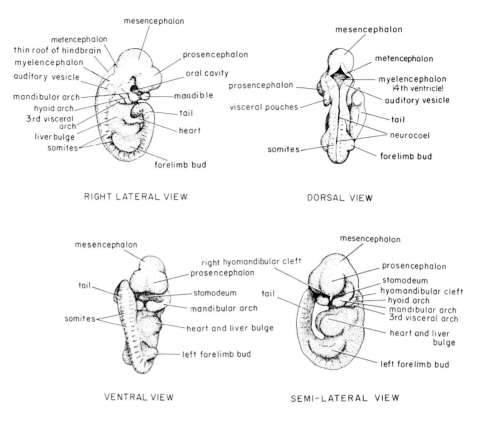

RIGHT LATERAL VIEW

DORSAL VIEW

VENTRAL VIEW

SEMI-LATERAL VIEW

MOUSE EMBRYO — 9.5 g.d.

are the two vesicles of the prosencephalon. The single mesencephalon is at the peak of the head, followed by the short metencephalon and finally the elongated myelencephalon. The last region is identified by the otic vesicles, which are now all but covered over. Crowded between the head and the tail are bulges of at least three pairs of visceral arches, the bulbous and exteriorized heart, and the liver. The forelimb buds show but, owing to cephalization, may appear to be placed almost at mid-body level.

In sections the brain cavities exhibit characteristic wall structures, thick in the dorsal walls of mesencephalon and the metencephalon and thin in the myelencephalon. Ventrally the infundibulum extends toward Rathke's pocket, and nearby are the paired optic vesicles as lateral projections. The spinal cord is well differentiated.

The primitive streak has disappeared by being incorporated into the embryo. The more anterior somites have dermatome, myotome, and some sclerotome but the myocoel is not visible unless in the most anterior ones. Much mesenchyme is scattered about between organs and blood vessels. The lateral thyroid diverticula appear, and just posteriorly the primordia of the pronephric ducts and tubules. The primordial germ cells are migrating via the mesentery toward their final residence in the gonads.

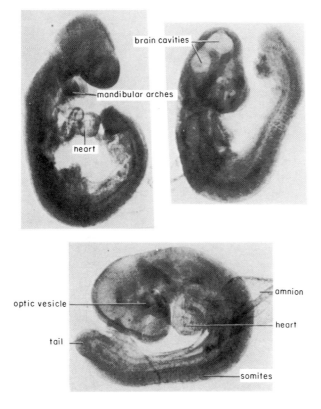

9.5 DAY MOUSE EMBRYO (2.25 mm)
(Photographs — Whole Embryos)

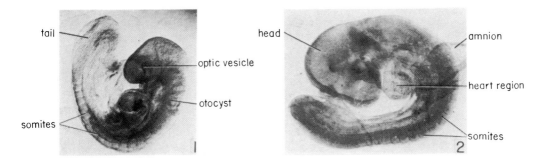

WHOLE MOUSE EMBRYO — 9.5 days
(showing approximation of head and tail)
(Photographs — Whole Embryos)

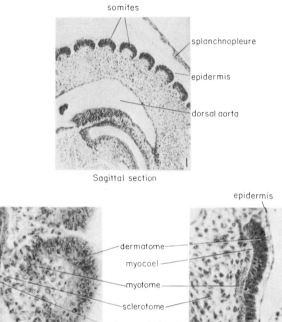

Sagittal section

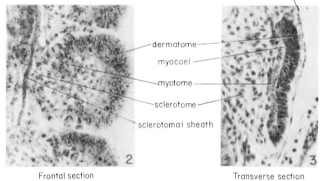

Frontal section   Transverse section

SOMITES OF MOUSE EMBRYO at 9.5 days

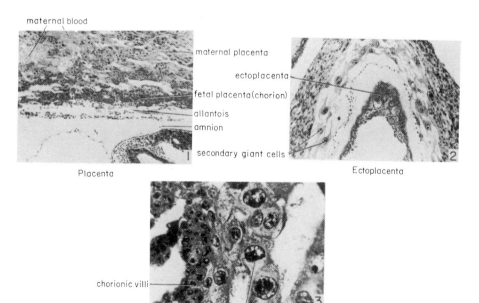

Placenta   Ectoplacenta

Giant cells

PLACENTAL RELATIONS IN MOUSE EMBRYO at 9.5 days
(Photographs)

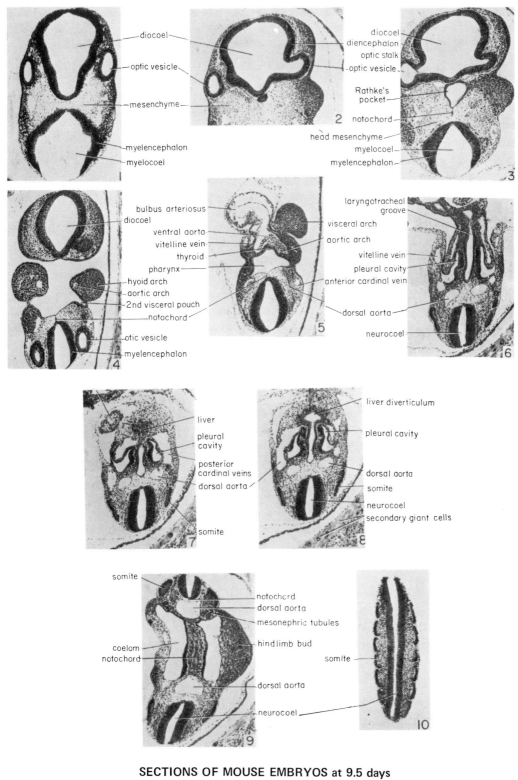

SECTIONS OF MOUSE EMBRYOS at 9.5 days
(Photographs)

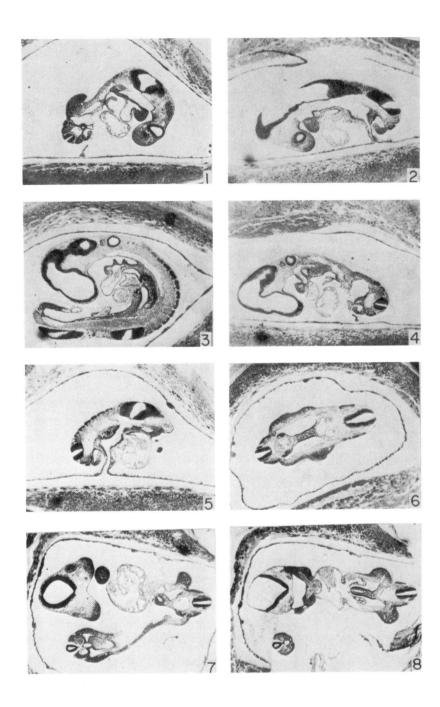

SECTIONS OF MOUSE EMBRYOS IN UTERUS at 9.5 days
(Photographs)

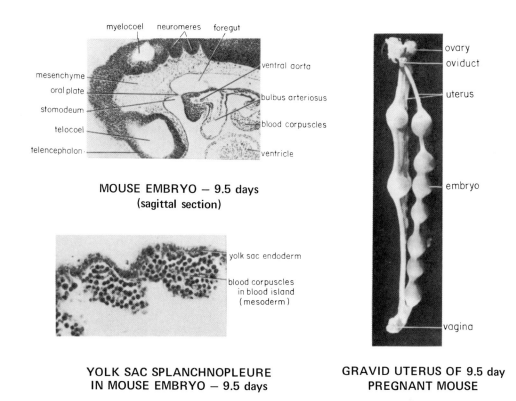

MOUSE EMBRYO — 9.5 days
(sagittal section)

YOLK SAC SPLANCHNOPLEURE
IN MOUSE EMBRYO — 9.5 days

GRAVID UTERUS OF 9.5 day
PREGNANT MOUSE

The gut, by this time, is rather a simple tube, wider at both ends than in the middle. The primordia of the lungs, liver, and dorsal pancreas are small but discernible. Lateral to the anterior region of the gut is a thick epithelial-like plate of cells, temporarily well-defined, called the anterior splanchnic mesodermal plate (ASMP). The cytoplasm of the constituent cells is rather basophilic while the nuclei resemble those of the loose mesenchyme. It has been suggested that this transient, thick epithelium may have some growth and morphogenetic function, particularly in the development of the stomach, greater omentum, and spleen. Shortly after the embryo has turned on its axis the anterior gut and coelom close as far back as the primordium of the liver. Posterior to the ASMP the splanchnic mesoderm, which covers the wide anterior portion of the intestine, is several layers in thickness, composed of loosely arranged mesenchymal cells. More posteriorly the mesoderm is only one cell thick, formed in a plate-like arrangement. From the 16th intersegmental artery posteriorly the splanchnic mesoderm lateral to the gut, and also covering the aorta and the umbilical arteries, is organized into a compact epithelial plate which is thin but resembles the ASMP. The posterior boundary of the coelom is formed partly by the roots of the umbilical artery as they pass around the gut. The coelom may be seen to extend caudally and ventrally a short distance lateral to the umbilical roots, thus forming a pocket on each side. The vitelline vein enters the body from the left, a branch passing to the right on the ventral side of the gut, but both branches go directly to the heart through the newly

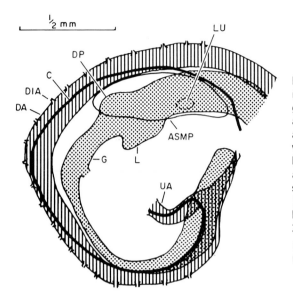

Parasagittal reconstruction of an early 9½ day mouse embryo showing the dorsal aorta and gut projected onto a parasagittal plane. The aorta is a single median structure except at the anterior and posterior ends, where it is double, with the gut lying in part between the two branches. ASMP, extent of anterior, dorsal, and posterior borders of the coelom; DA, dorsal aorta, DIA, dorsal intersegmental artery; DP, dorsal pancreas; G, gut; L, liver; LU, lung bud; UA, umbilical artery. Embryo measured 2.4 mm.

(Courtesy Margaret C. Green, 1967 Developmental Biology 15:62-89, Academic Press.)

forming liver on either side of the gut. Shortly a new anastomosis forms between the left and right branches around the dorsal side of the gut and the central connection is lost. Thus, the vitelline vein passes dorsal to the gut from left to right and then into the liver. The anastomosis occurs just posterior to the dorsal pancreas, and anterior to the posterior end of the ASMP.

The embryonic-maternal relations should be examined at this time in an attempt to determine the area where the two sources of tissue interdigitate. The yolk sac marks the inner boundary of Reichert's membrane. Outside Reichert's membrane some scattered giant cells lie between the embryo and the decidua.

E. THE 10 DAY EMBRYO:

The 10 day mouse embryo is structurally comparable to the 96 hour chick embryo. It has 26 to 28 somites, measures about 3.8 mm from crown to rump in total length. If straightened out it might well be twice the crown-rump length, and is enlarging rapidly in almost all directions, changing from a slim and elongate shape to an almost bulbous one. This is due largely to expansion of the brain vesicles. Most obvious at this time is the further development and ventral curvature of the head region with the mesencephalon forming its main (and most anterior) bulge. The midbody region is beginning to round out and incorporate its earlier bulges. The tail is still curled to the right side.

In cleared whole specimens many internal structures can be discerned, such as the visceral arches, the heart, the liver, and the rudimentary mesonephros. The otic vesicles are now closed to the outside. The olfactory discs are invaginating to pits, an internasal cleft is forming between them, and triangular cells of Cajal of the accessory olfactory bulb appear. The paired telencephalic vesicles continue to expand, but neither is as prominent as the median mesencephalon. The isthmus between the metencephalon and myelencephalon is deepened, and the roof of the latter is very thin.

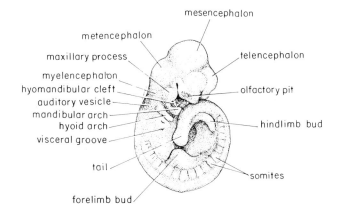

RIGHT LATERAL VIEW

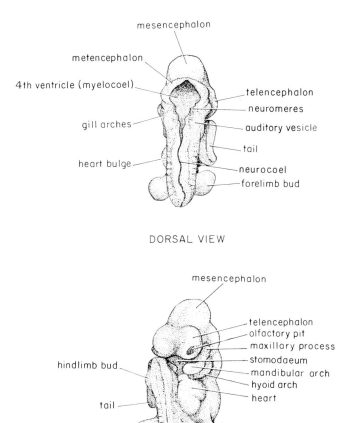

DORSAL VIEW

VENTROLATERAL VIEW

MOUSE EMBRYO — 10 g.d.

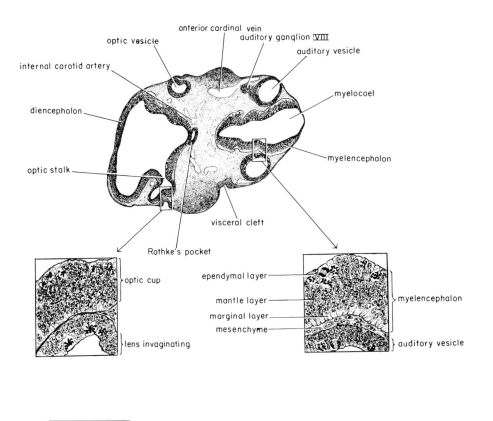

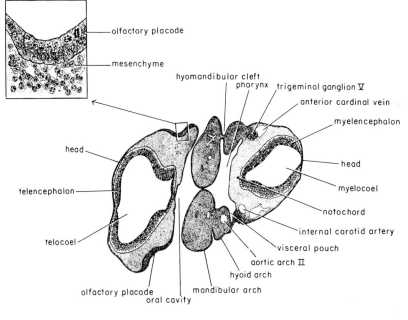

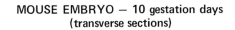

MOUSE EMBRYO — 10 gestation days
(transverse sections)

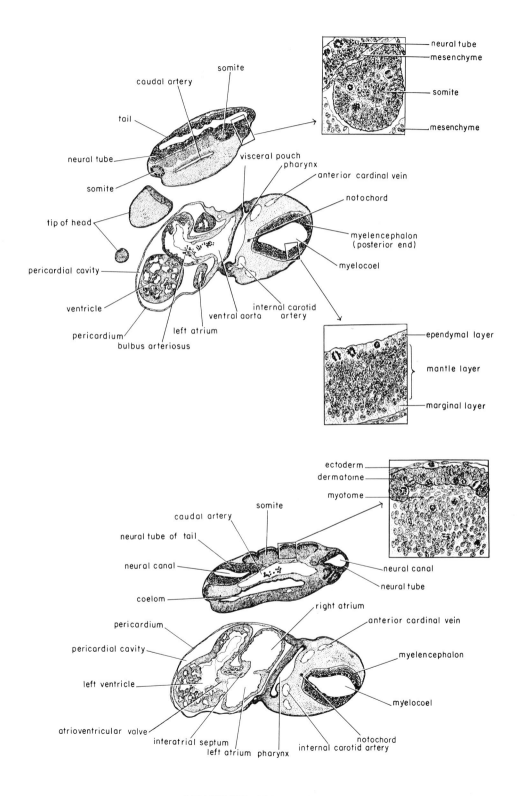

MOUSE EMBRYO — 10 g.d.

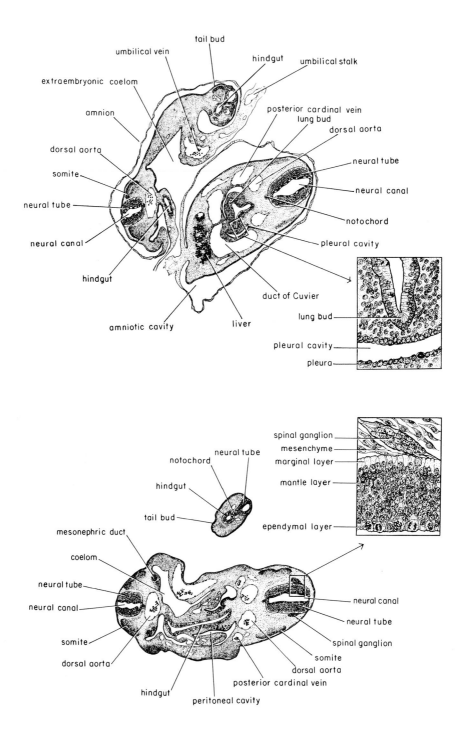

MOUSE EMBRYO — 10 g.d.

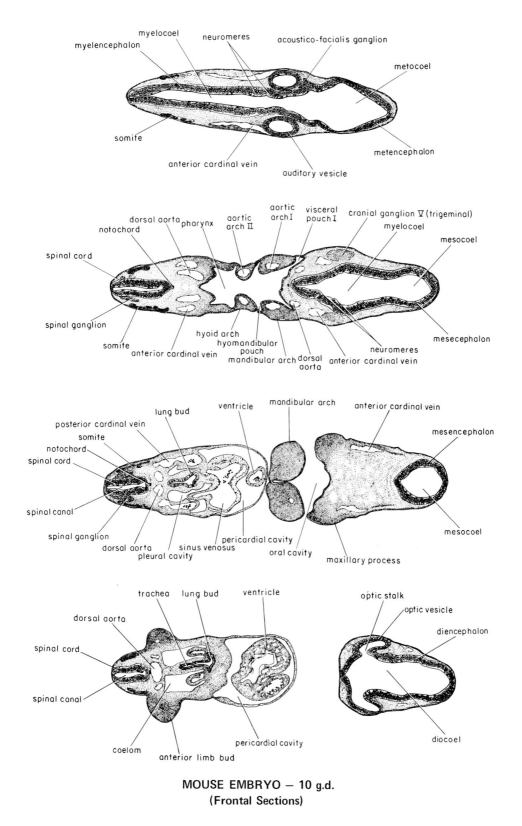

MOUSE EMBRYO — 10 g.d.
(Frontal Sections)

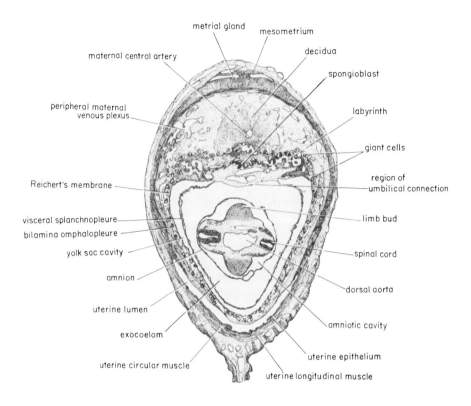

PLACENTAL RELATIONS OF THE 10 DAY MOUSE EMBRYO

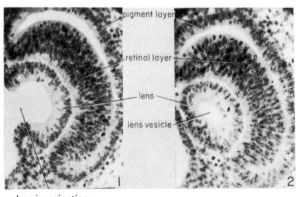

EYE DEVELOPMENT IN MOUSE EMBRYO at 10.0 days

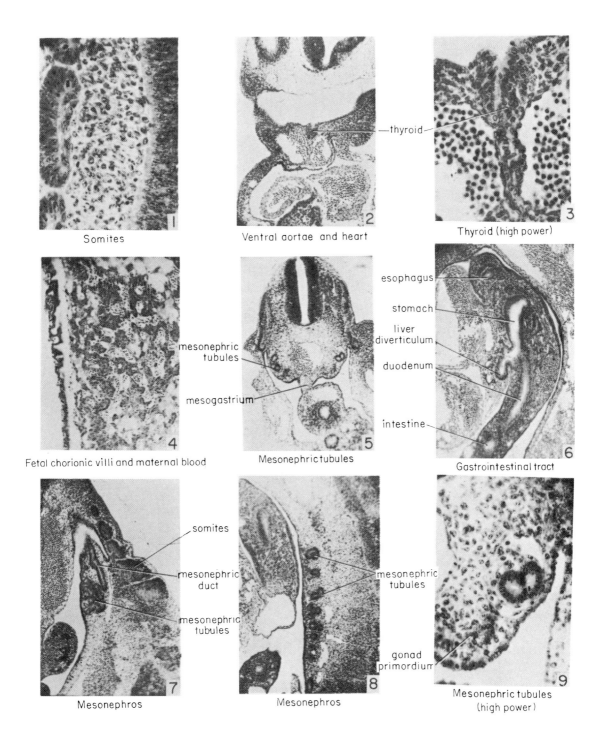

1 Somites

2 Ventral aortae and heart

3 Thyroid (high power)
—thyroid—

4 Fetal chorionic villi and maternal blood

5 Mesonephric tubules
mesonephric tubules
mesogastrium

6 Gastrointestinal tract
esophagus
stomach
liver diverticulum
duodenum
intestine

7 Mesonephros
somites
mesonephric duct
mesonephric tubules

8 Mesonephros
mesonephric tubules

9 Mesonephric tubules (high power)
gonad primordium

MOUSE EMBRYO at 10.0 days
(Photographs)

In sections visceral pouch III can be identified indicating that the visceral arches and pouches also develop in sequence from anterior toward posterior. The first three pairs of aortic arches are visible. The diverticulum has expanded and begins to bifurcate into primitive lung buds, later to form bronchi, joining the foregut at the laryngotracheal groove. The liver first appears. Although the oral (stomodeal) membrane has broken through at 9 days gestation the cloacal (proctodeal) membrane has just formed.

The heart is most important because it removes metabolic wastes from the newly formed and forming organs and brings requisite nutrition to them. It is therefore differentiated before the other organs and functions regularly with an ever-increasing volume of corpuscles and fluid. The corpuscles come from hematopoietic centers in the yolk sac splanchnopleure. Serial sections through the heart region at 10 days show not only the four chambers but myocardial development so well advanced that the atria can be easily distinguished from the ventricles, which appear to be filled with trabeculae. The endocardial cushion and interventricular septum are being completed, and a thick pericardial membrane encloses all. In more posterior sections the proximity and actual vascular relations of the liver and heart can be seen. At about mid-body level the pronephric ducts and tubules can be located. The paired pronephroi project as bulges into the coelom, connecting with it via the ciliated nephrostomes.

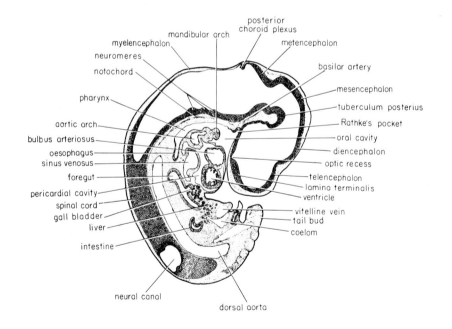

MOUSE EMBRYO — 10 days
(mid-sagittal)

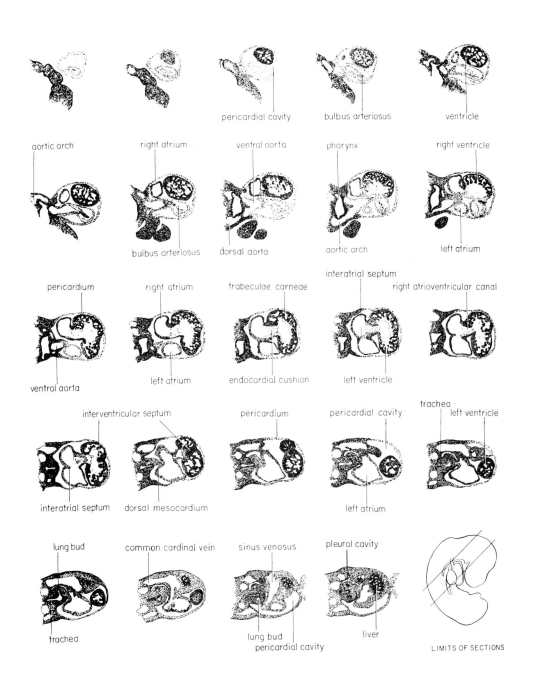

**DEVELOPMENT OF THE MOUSE HEART
AT 10 GESTATION DAYS**
(Serial Sections)

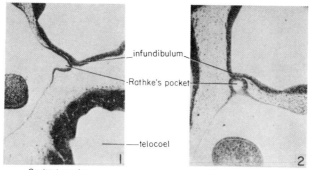

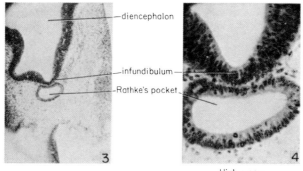

**PITUITARY GLAND DEVELOPMENT
IN MOUSE EMBRYO at 10.0 days**

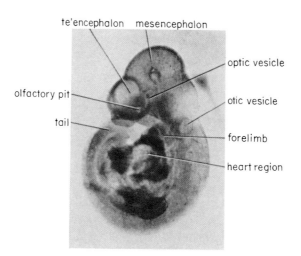

MOUSE EMBRYO at 10.0 days

F. THE 10½ DAY EMBRYO: At $10\frac{1}{2}$ days the embryo has 29 to 36 somites and measures 5.2 mm from crown to rump. Again, its total straightened length might be twice the crown-rump length. The forelimb buds are growing rapidly, the hindlimb buds make their appearance, and four visceral arches can be distinguished.

Sections show that the outer surface of each optic vesicle has invaginated to form an optic cup which is being pinched off from the cavity of the diencephalon to form an optic stalk. In each cup appears a lens placode, a thickening of the ectoderm. The lens placode may even be pinched off and contain a vesicle. The inner (retinal) layer of the double-layered optic cup becomes thick with neuroblasts, while the outer (pigmented) layer (closer to the brain) becomes thinner. Numerous mitoses are seen in each layer. The endolymphatic duct, a remnant of the opening of the otic vesicle leads ventro-mesially toward the vesicle, closely associated with a cluster of neuroblasts. These give rise to the acoustic nerve (cranial nerve VIII). Slightly anteriorly is the facial nerve, (cranial nerve VII). The olfactory pits have deepened.

The spinal ganglia are metameric and resemble superficially the cranial ganglia, which are actually quite different.

The dermatome of the anterior somites is very thick, and the myotome (lining the myocoel) is thickening. The somites appear to have moved away slightly from the midline leaving the spongy sclerotome behind. Overlying the somites is the integumentary ectoderm, which is very thin at this time.

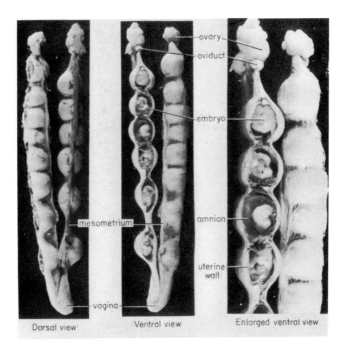

MOUSE EMBRYOS IN UTERUS at 10.5 days

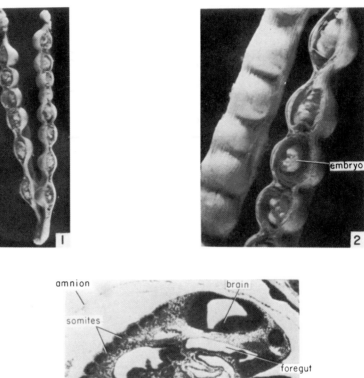

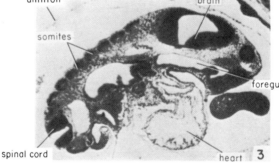

MOUSE EMBRYO AT 10.5 DAYS DEVELOPMENT

Fig. 1 — Paired uteri of gravid mouse at 10.5 days gestation, with embryos exposed in situ. Note ovaries above and vagina below, where bicornuate uteri converge.

Fig. 2 — Enlarged view of uterus showing (right) embryos in situ. Note various positions of embryo in muscular uterus.

Fig. 3 — Section of 10.5 day embryo in situ, showing much coiled position so that both brain and caudal neurocoel are shown in the same field.

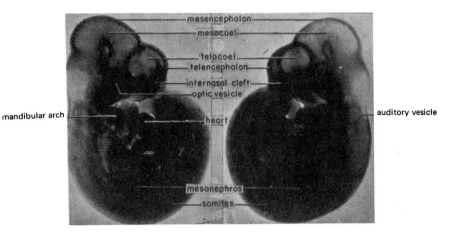

10.5 DAY MOUSE EMBRYO (4.75 mm)

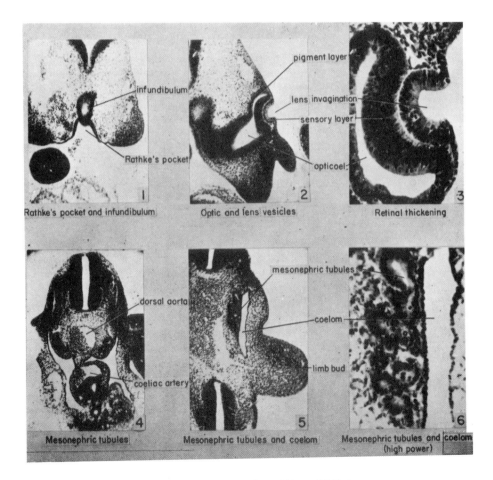

SECTIONS OF MOUSE EMBRYO at 10.5 days
(Photographs)

Sagittal sections show a wide open mouth, with no remains of the stomodeum and a trachea separate from and ventral to the foregut (esophagus).

By this time the gut has elongated and the region between the liver, pancreas, and lungs has grown enormously. The stomach is now clearly distinguished by a narrowing of its lumen both anterior and posterior to it. The
dorsal mesentery of the stomach has enlarged toward the right side and forms a
fold which will become the greater omentum. The space between this fold and
the forming stomach becomes the omental bursa. The stomach now lies slightly
to the left of the midline. The endodermal portion of the dorsal pancreas has
enlarged and grown into the fold of the dorsal mesentery, extending caudally
from the stomach on the left side. The endodermal portion of the ventral pancreas has begun to bud off from the gut just posterior to the liver, and the biliary ducts are enlarging as they grow into the liver. The anterior part of the
ASMP has disappeared, but may still be seen in the vicinity of the stomach and
pancreas. The intestine is somewhat elongated, and a loop is formed to the left.

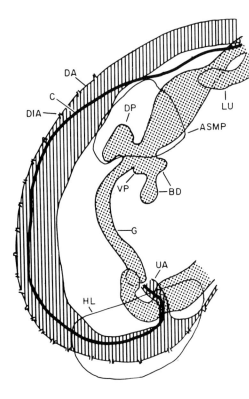

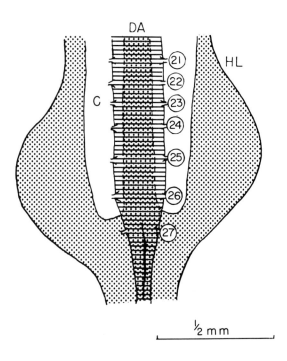

Parasagittal section reconstruction of 10½ day
mouse embryo. C, border of coelom; DA,
dorsal aorta; DIA, dorsal intersegmentary
artery; DP, dorsal pancreas; G, gut; HL, hind
limb bud; BD, biliary ducts; LU, lung; UA,
umbilical artery; VP, ventral pancreas. Embryo measures 3.6 mm.

(Courtesy Dr. Margaret C. Green, 1967, Developmental Biology 15:62-89, Academic Press.)

Frontal section of 10½ day mouse embryo,
reconstructed, to show structures related to
the hind limb region projected onto a dorsal
plane. C, coelom; DA, dorsal aorta; HL, hind
limb bud. Numbers are intersegmental levels.

(Courtesy Dr. Margaret C. Green, 1967, Developmental Biology 15:62-89, Academic Press.)

The splanchnic mesoderm has increased in thickness, but appears to be loose mesenchyme covered by a single layer of epithelium without the plate-like structure seen earlier. It is probable that this splanchnic mesoderm interacts with the endoderm of the gut to promote its growth.

Already the liver seems to be establishing itself as a hematopoietic center, supplementing the vascular centers in the yolk sac splanchnopleure. Primary intestinal loops appear.

The heart has the interventricular septum, the septum primum, the atrioventricular canals and the right venous valve primordium. Paired aortic arches III, IV, and VI are formed. Lateral to the single large dorsal aorta, which is usually engorged with corpuscles, are the pronephroi. Although they never function they give rise to some genital structures. Frontal sections show the paired pronephric ducts, connected at their anterior ends with the pronephric tubules and at their posterior ends with the cloaca. The mesonephroi continue to develop.

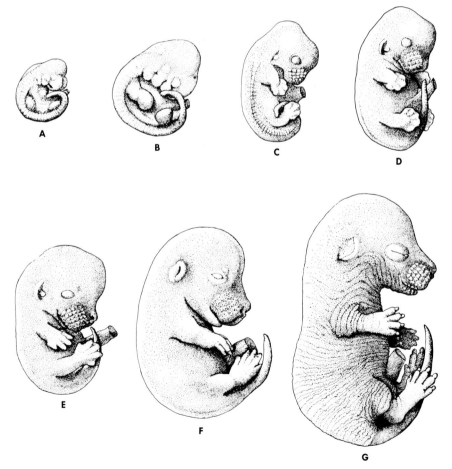

DIAGRAMS SHOWING DEVELOPMENT OF MOUSE EMBRYO
from 10½ to 16½ days

(From R. Rugh, "Vertebrate Embryology", Harcourt, Brace & World, Inc., New York, 1964.)

G. THE 11 DAY EMBRYO: The 11 day mouse embryo is comparable to the 30 day human embryo.   It has 37 to 42 well defined somites and measures 6.2 mm from crown to rump.   The body in general still shows kyphosis and cephalization - the head is about one-third the total body length - but is filling out.   There is an umbilical hernia, which later recedes.   The tail is still growing and much coiled, so that the olfactory pits, (now nasal chambers) are close to its base.   The forelimb buds appear to be curved and have knoblike extremities.   The hindlimbs are still buds.   The eye parts are usually visible from the surface.

Brain differentiation continues, with the walls of the entire central nervous system actively proliferating neuroblasts which begin to occlude some of the neural cavities.   The floor of the diencephalon may project as a sac.   The epiphysis first appears as a median dorsal evagination from the diencephalon.

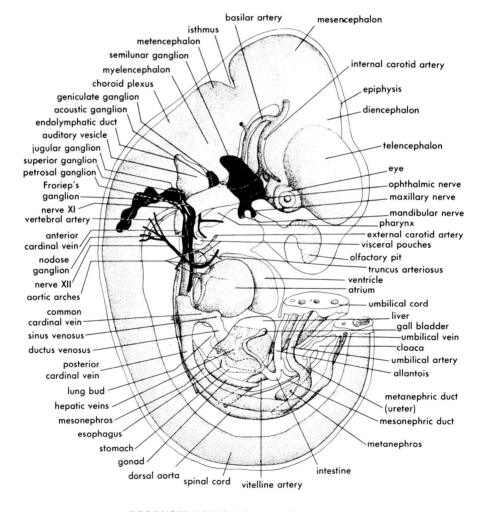

**RECONSTRUCTION OF MOUSE EMBRYO
AT 11 DAYS GESTATION**

(From R. Rugh, "Vertebrate Embryology", Harcourt, Brace & World, Inc., New York, 1964.)

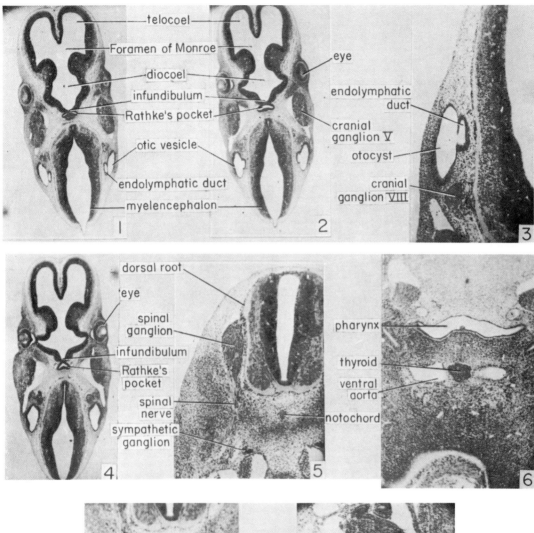

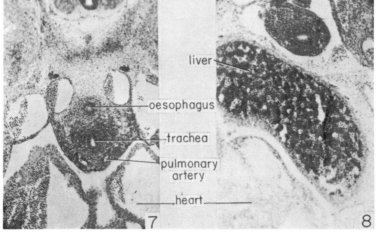

MOUSE EMBRYO SECTIONS at 11 days
(Photographs)

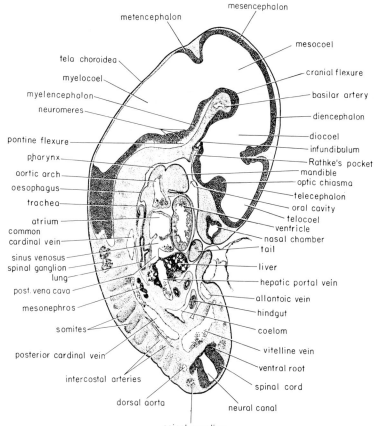

MOUSE EMBRYO — 11 days
(mid-sagittal)

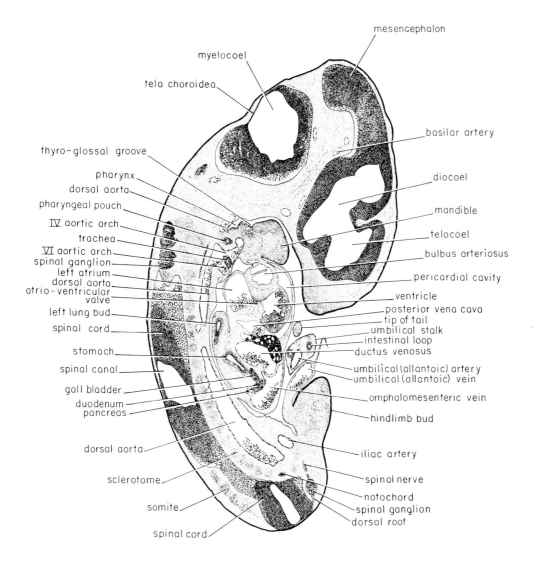

MOUSE EMBRYO — 11 days
(para-sagittal)

Rathke's pocket and the infundibulum are contiguous, but there is no cellular differentiation in them.   In the roof of the myelencephalon is the thin and vascular posterior choroid plexus.   The isthmus between the metencephalon and myelencephalon has further deepened.   Owing to the extreme flexure of the brain around the tip of the notochord, a section can be cut that shows it apparently separating the diencephalon from the metencephalon.   Lateral to the hindbrain cranial ganglia V to IX suddenly develop.   The largest aggregation of neuroblasts is lateral to the pharynx, where the trigeminal (ganglion V) forms.   Its branches issue to the ophthalmic, maxillary, and mandibular regions.   A separate group of ganglia (X to XII) controls the visceral organs and is less compact. Tucked between these two areas of neuroblastic activity on each side is the enlarging otic vesicle, which has lost all connections with the outside.   The epithelial lining of the semicircular canals first appears. The original invagination of ectoderm from the surface, now the lining of the endolymphatic duct, is elongated and joins a vestibular portion, which in turn joins a cochlear portion. Neither is at all differentiated.   These may be identified as the ultimate saccule and utricle.   Mitral cells arise in the olfactory bulbs, along with the triangular cells of Cajal in the accessory olfactory bulbs.   Some superficial tufted (sensory) cells are also apparent.   Olfactory nerves reach the telencephalon.   The vomeronasal organ is visible.   The spinal cord begins to display typical organization, with ventral horns and ependyma.   There is as yet no sharp distinction between white matter and grey matter.   The spinal ganglia are well formed, with both dorsal and ventral roots.   Near the heart they give off sympathetic branches.   At about heart level, a very small single cluster of cells beneath the spinal cord is the sole remnant of the notochord.

The thyroid diverticulum may still retain a very small tubular connection with the pharynx, which is lost during this day.   Around the diverticulum are clusters of gland cells.   These quickly aggregate into lobules and become highly vascular.   The parathyroid, and thymus glands and the ultimobranchial bodies begin to form.   The dorsal and ventral pancreatic rudiments are fusing.

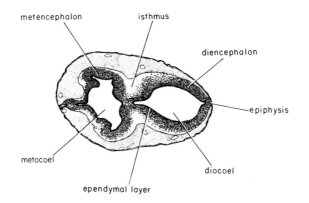

MOUSE EMBRYO — 11 gestation days
(transverse sections)

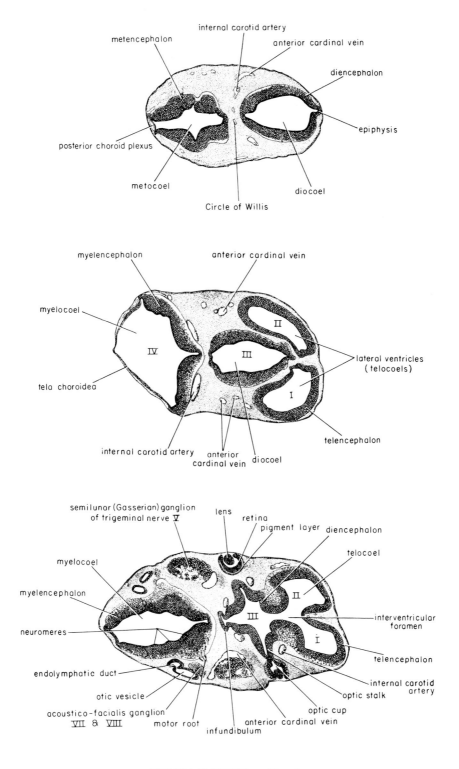

MOUSE EMBRYO — 11 g.d.
(transverse sections)

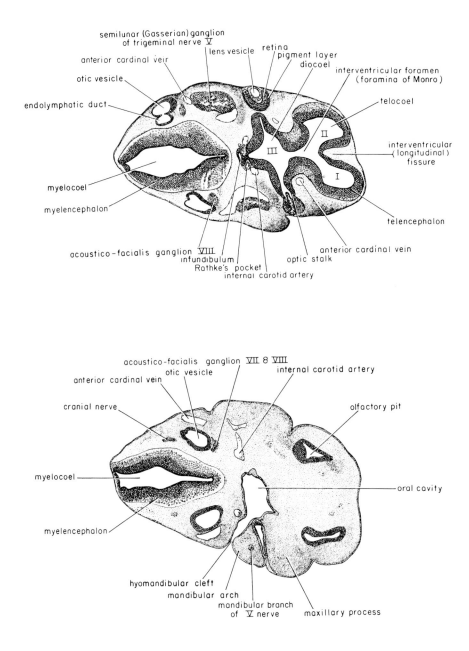

MOUSE EMBRYO — 11 g.d.
(transverse sections)

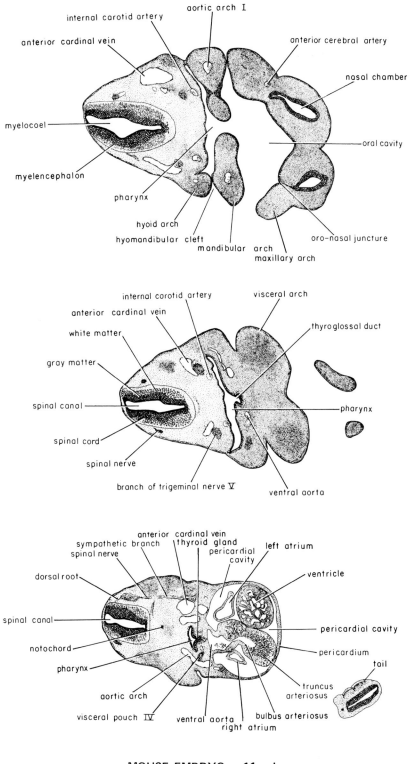

MOUSE EMBRYO — 11 g.d.
(transverse sections)

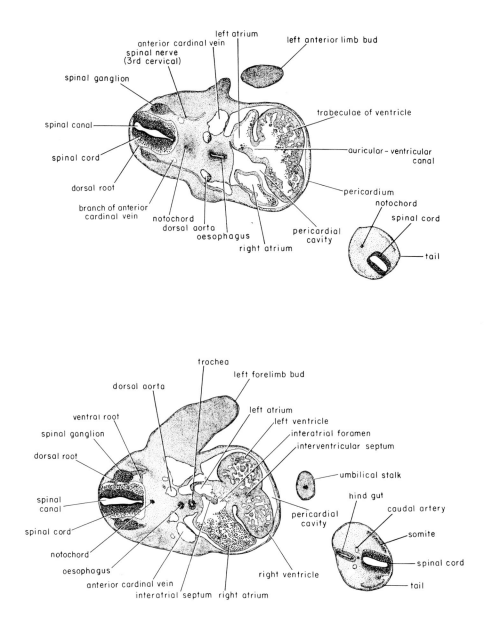

MOUSE EMBRYO — 11 g.d.
(transverse sections)

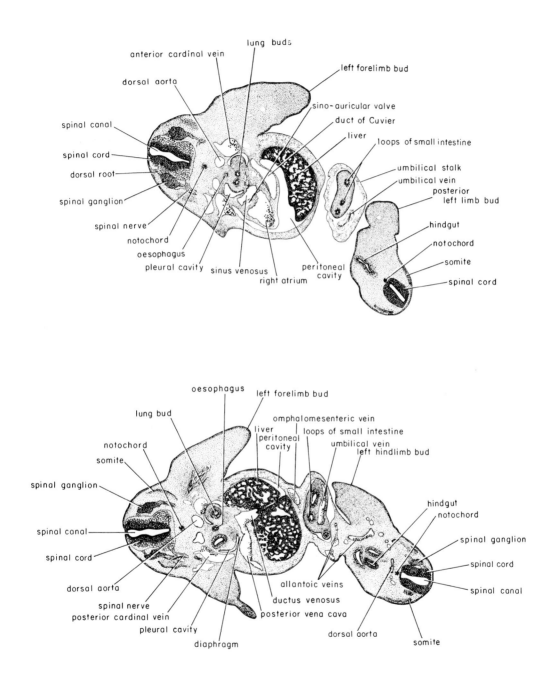

MOUSE EMBRYO — 11 g.d.
(transverse sections)

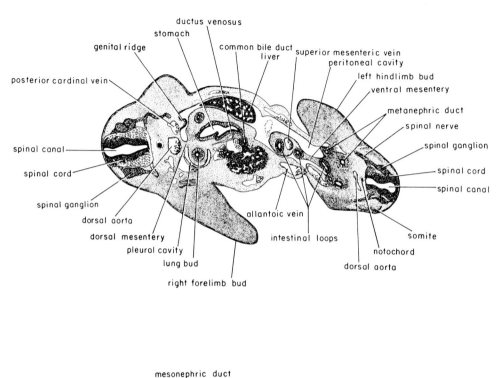

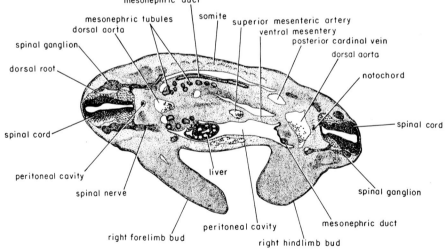

MOUSE EMBRYO — 11 g.d.
(transverse sections)

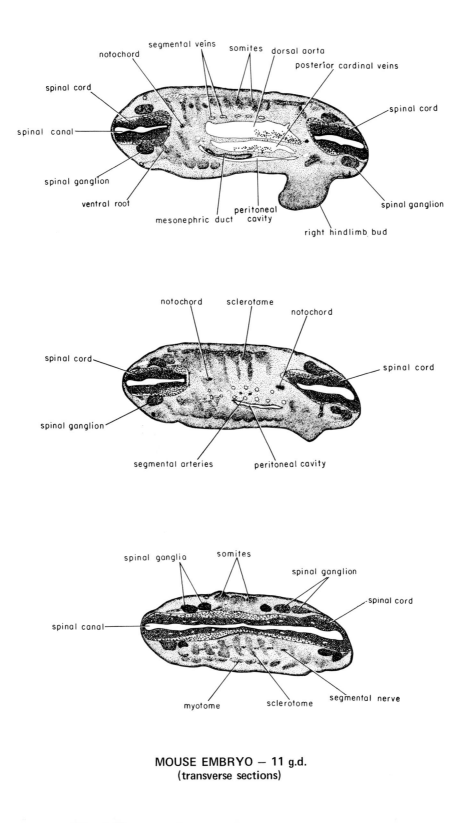

MOUSE EMBRYO — 11 g.d.
(transverse sections)

The thick columnar layer of cells comprising the anterior splanchnic mesodermal plate (ASMP) now may be seen over the pancreas and the posterior part of the stomach. It is in the process of disappearing except from the dorsal part of the mesenteric fold and the posterior part of the omental bursa on the side adjacent to the stomach.

The heart continues toward completion, with the interatrial foramen much constricted and the endocardial cushion fused. The ventricular walls are much thickened, whereas the atrial walls appear to be thinner and more expanded. The dorsal mesocardium, which was formed with the heart and supports it for some time, is still present at its extremities, but the ventral mesocardium has ruptured to give a single pericardial cavity. The diaphragm is a new development, separating the pericardial cavity from the peritoneal cavity. The pleural cavity (now cut off from the coelom) can be seen at about the level of the sinus venosus and diaphragm.

Internal carotid arteries and anterior cardinal veins are visible anteriorly. Since they are all thin-walled at this time, their positions rather than their structure distinguish them. The circle of Willis, which connects the internal carotid arteries with the basilar arteries around the forming pituitary gland, (Rathke's pocket and infundibulum combined) is also visible. More posteriorly are the paired lung buds, the right one of which is developing bronchial areas, the umbilicus, allantoic veins, and metanephric ducts. Numerous mesonephric tubules lie laterally near the peritoneal cavity, and posteriorly the mesonephric ducts lead to the urogenital sinus (cloaca). The subcardinal veins are forming ventrally to the mesonephroi. The most posterior sections, which are likely to be frontal rather than transverse because of the twisting of the body, clearly demonstrate the metameric arrangement of the somites and the spinal ganglia. Since many of the organ systems are longitudinally distributed, frontal sections are extremely instructive. They show the neural tube with its typical cellular differentiation; the relation of pharynx to trachea to lung buds; the descending aorta with its branching segmental arteries; and the paired excretory (mesonephric and metanephric) ducts.

It is quite obvious, as one examines the living or cleared whole specimen at 11 days or sections of it (in any of the three planes), that the embryo is now an intricate mosaic of interrelated developing organ systems, all integrated for the proper functioning of the organism as a whole. Time and place relationships must be adhered to, or else there will be defective form and, in consequence, defective function. The vast majority of embryos will conform to the necessary pattern and arrive at a state of independent existence at about the same time.

From this stage on, general descriptions will be provided at daily intervals through 16 days, and thereafter certain organ systems will be independently described to show how they originate and develop. The purpose of this organization is to impress upon the student that the embryo is not made up of separate and independent parts that can be fitted together at a certain time but that its development is continuous and involves all its organ systems, some of which appear early and others late. If the student is aware of this situation, we can finally study specific organ systems without the risk of separating, in his mind, the developmental histories of parts that must be related to the whole.

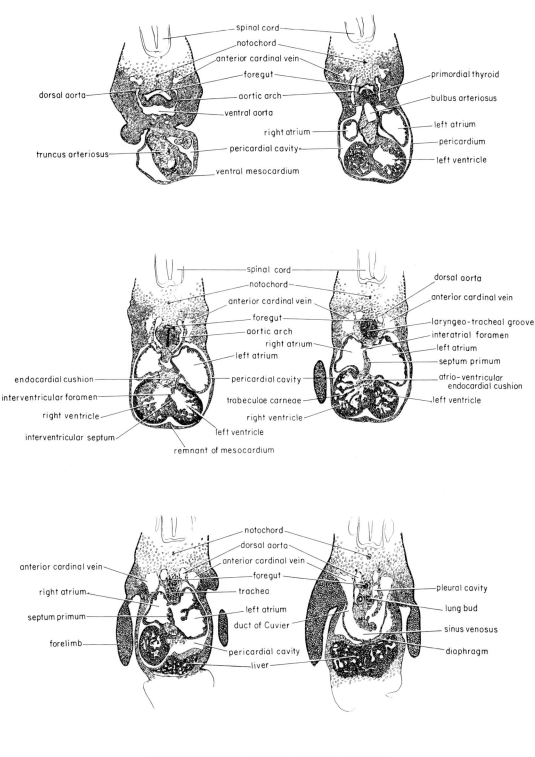

**DEVELOPMENT OF THE MOUSE HEART
AT 11 GESTATION DAYS**

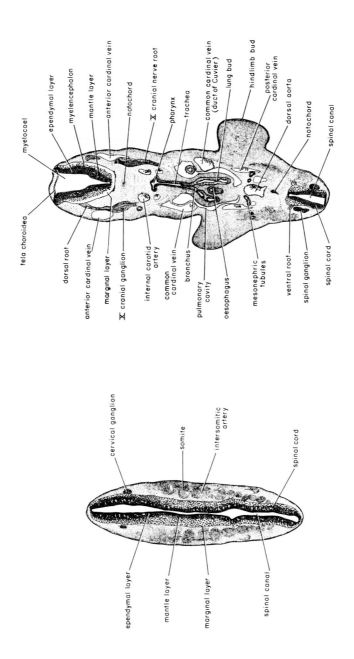

MOUSE EMBRYO – 11 gestation days
(frontal sections)

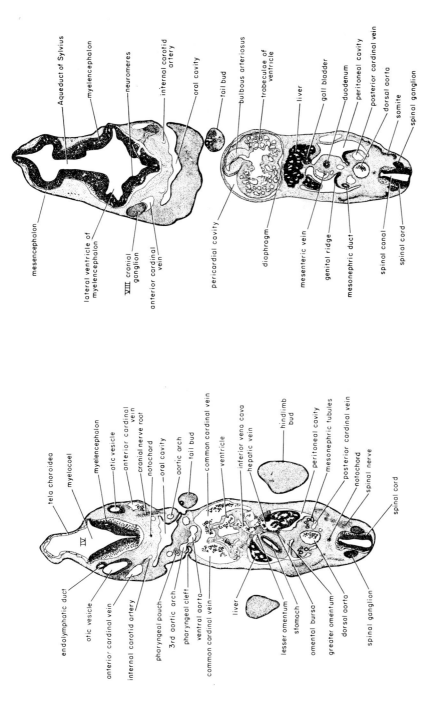

MOUSE EMBRYO — 11 gestation days
(frontal sections)

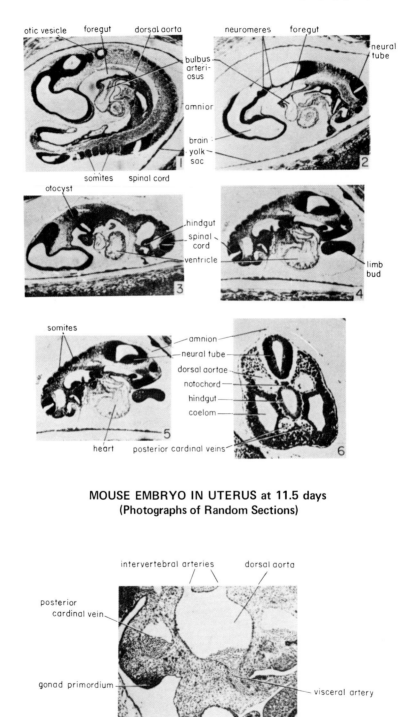

MOUSE EMBRYO IN UTERUS at 11.5 days
(Photographs of Random Sections)

SECTION THROUGH 11.5 DAY MOUSE EMBRYO

Showing dorsal aorta with two dorsally-directed intervertebral arteries and a single ventrally-directed visceral
artery. Note also posterior cardinal vein in mesonephros.

H. THE 12 DAY EMBRYO: The 12 day mouse embryo is comparable in development to the 36 day human embryo. It has from 43 to 48 somites and measures 7.2 mm from crown to rump. The body is still coiled upon itself, and the cranial and cervical flexures are advanced, but the back is beginning to straighten. All four limb buds are prominent, although the anterior pair is slightly better developed than the posterior pair. The extremities are flared out into plates with indications of digital divisions. This is a critical period for digit formation, from mesenchyme to precartilage.

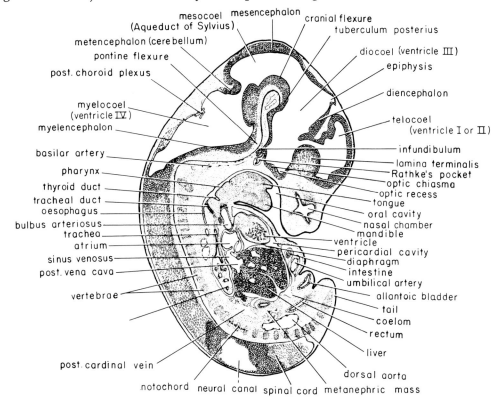

MOUSE EMBRYO — 12 days
(mid-sagittal)

The external ear and eye regions can be located. The vestibular and cochlear portions of the otic vesicles are much enlarged. In the narrow optic stalk behind each optic cup an artery may be seen. The lenses are now below the surface, and vitreous humor is forming behind them. A fibrous portion is differentiating posteriorly to the vesicle in each. The retinal layer of the optic cups is very thick, stretching the pigmented layer until it is much thinner than before. Short axon cells of Cajal arise in the olfactory bulbs.

Mid-sagittal sections show the brain convolutions to be extensive, involving thickenings and thinnings at the various levels. Eventually most of the vesicles will be obliterated with tissue. The roof of the mesencephalon is especially thick, as are the roofs of the telencephalic vesicles. The infundibulum seems to be encircling and incorporating Rathke's pocket just posterior to a prominent

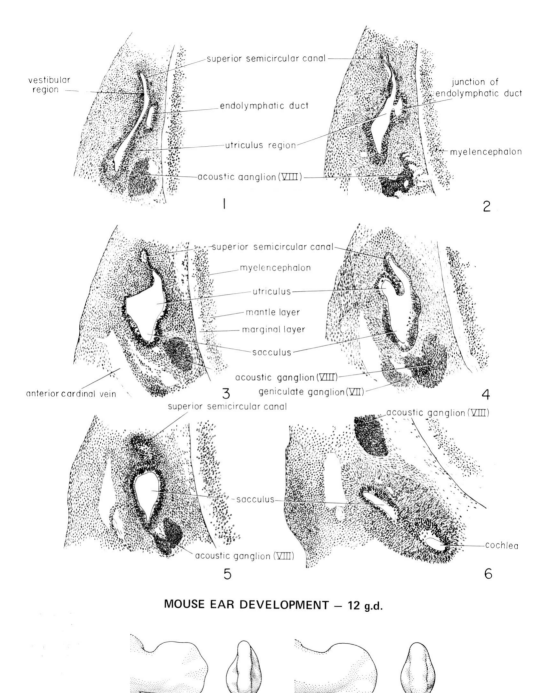

MOUSE EAR DEVELOPMENT — 12 g.d.

FORE LIMB   *HIND LIMB*

**RIGHT FORE AND HIND LIMB OF 12 DAY
NORMAL MOUSE EMBRYO**

(Grüneberg 1961, 1963)

thickening in the floor of the telocoel, the lamina terminalis. Many neurons are present in the dorsal and ventral, cochlear, basal, optic, lateral, cuneate, and lateral reticular nuclei, indicating that neurodifferentiation has begun.

Mid-sagittal sections also show an open mouth, with a transverse groove in the mid-ventral floor marking the posterior limit of the tongue. Just posterior to this is the pharynx, in the mid-ventral floor of which are the remnant of the thyroid duct and then the tracheal duct, leading posteriorly to the trachea, the bronchi, and the lung buds. Parathyroid III and the thymus have separated from visceral pouch III. The liver, is a very large and highly vascular organ, surrounding a large posterior vena cava which joins the sinus venosus. More posteriorly are the fused dorsal and ventral pancreatic rudiments, and the slightly coiled intestine, leading to the origin (at hindgut level) of the allantois and the umbilical vessels that leave the body via the umbilical cord.

Slightly off-center sagittal (parasagittal) sections show the metameric and mesenchymal vertebrae, derived from sclerotome. These extend almost to the tip of the tail. Parasagittal sections also show the degeneration of the pronephric tubules (but not the pronephric ducts of the female), the gradual regression of the mesonephric tubules (but not the mesonephric ducts of the male), and the more posterior paired metanephric cell masses and short (urogenital sinuses) ducts (ureters) leading to the cloaca.

Aortic arches III and IV continue to enlarge, V is gone, and right VI is reduced. The ventricles of the heart appear to be almost occluded by the expanding trabeculae. The interventricular foramen persists as the interventricular septum grows cephalad. Sections of the liver may appear to be within the pericardial cavity, but more anterior or posterior sections show that the diaphragm separates the pleural and peritoneal cavities.

Transverse sections through the anterior end of the embryo show the nasal septum separating the paired nasal chambers; the thickened Jacobson's organ in each chamber; the posterior nares or nasal choanae, which connect the nasal chambers with the pharynx; and the laterally directed nasolacrimal ducts on either side of the obviously muscular tongue.

The surface of each thymus is covered by a basement membrane, which separates the gland from the surrounding mesenchyme. Some thymus cells are ciliated and appear to be derived from pharyngeal epithelium. They contain thin-walled vesicles.

At the level of the liver and intestinal loops are the mesonephric ducts and adjacent to them and bulging into the peritoneal cavity are the genital ridges. The vitelline vein may be seen at this time near the most posterior level of the liver, as a broad venous channel leading to the heart. Sex of the embryo can be determined from the initial arrangement of these cells. The ridges may be recognizable as testes with their epithelial cords, or as ovaries with their nests of ova. The proximity of the ridges and the ducts suggests their ultimate relationship in the male. The subcardinals anastomose to form the renal portal veins. Still further posteriorly the metanephric ducts, associated with the nearby but as yet undeveloped metanephroi, enter the cloaca.

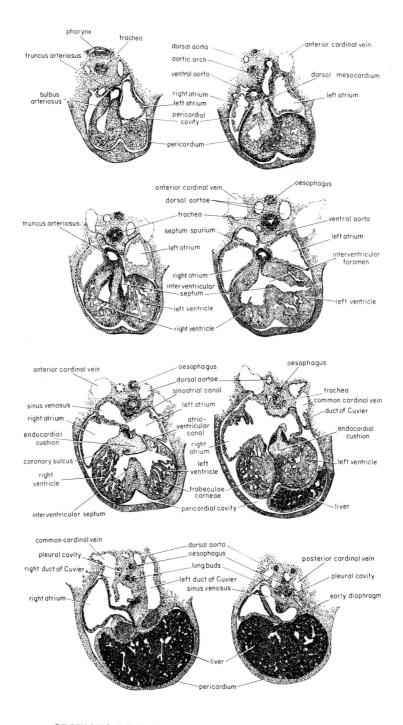

**SECTIONS OF THE 12 DAY MOUSE EMBRYO HEART
IN SEQUENCE**

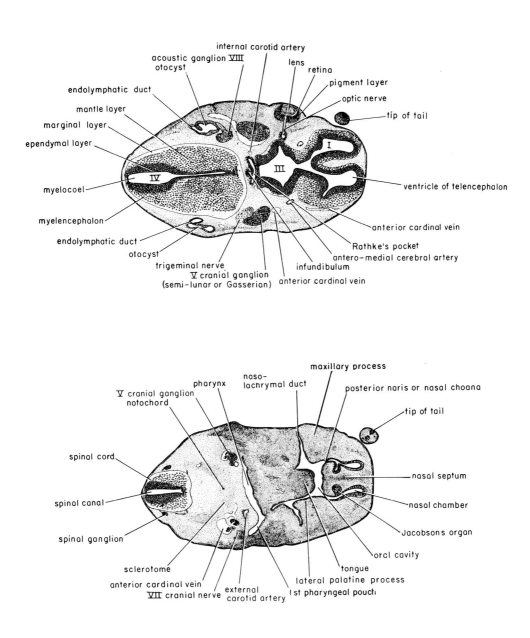

MOUSE EMBRYO — 12 gestation days
(transverse sections)

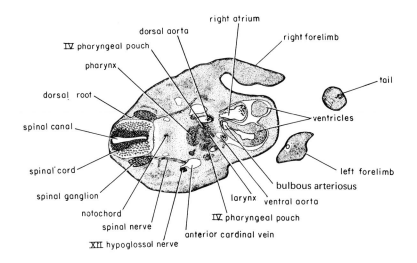

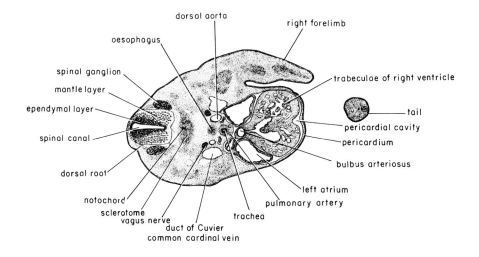

MOUSE EMBRYO — 12 g.d.
(transverse sections)

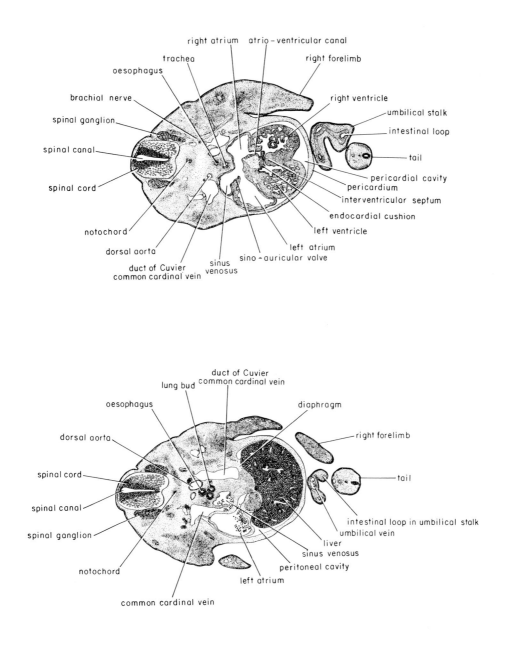

MOUSE EMBRYO — 12 g.d
(transverse sections)

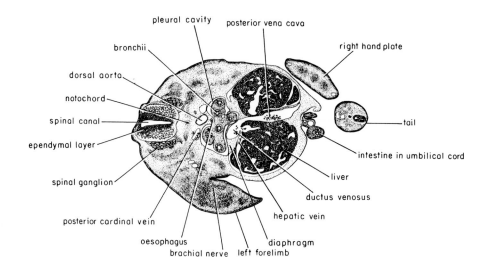

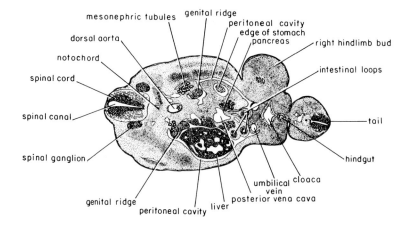

MOUSE EMBRYO — 12 g.d.
(transverse sections)

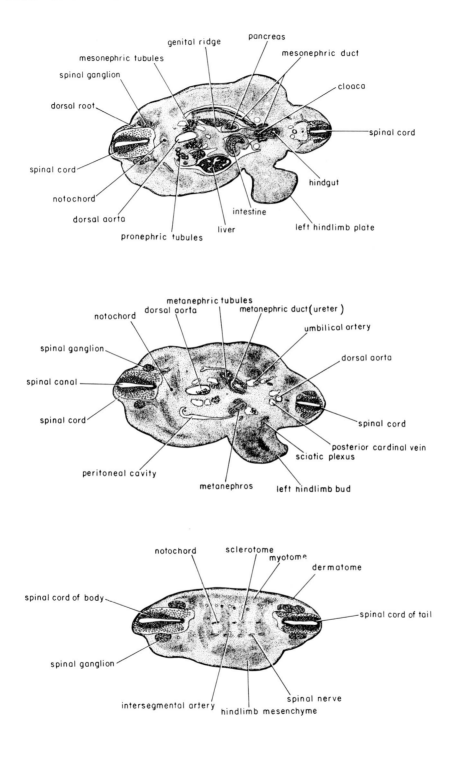

MOUSE EMBRYO — 12 g.d.
(transverse sections)

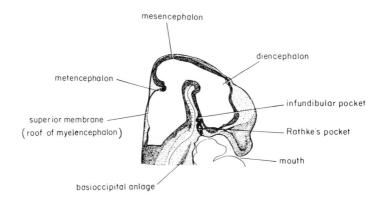

MOUSE HEAD at 12.5 g.d.
(sagittal section)

(Redrawn from Forsthoefel 1963)

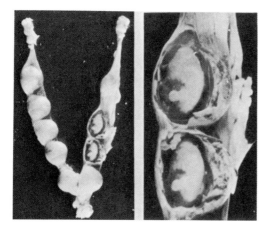

MOUSE EMBRYOS IN UTERUS
at 12.5 days

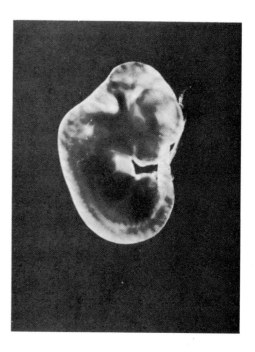

MOUSE EMBRYO at 12.5 days

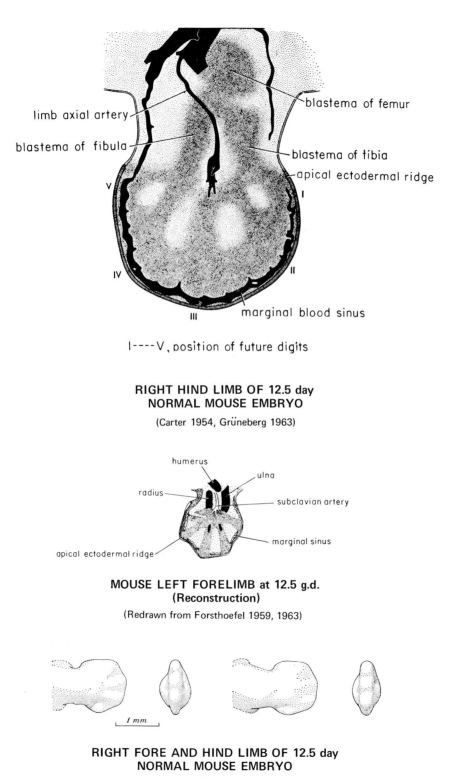

limb axial artery

blastema of fibula

blastema of femur

blastema of tibia

apical ectodermal ridge

marginal blood sinus

I----V, position of future digits

**RIGHT HIND LIMB OF 12.5 day
NORMAL MOUSE EMBRYO**

(Carter 1954, Grüneberg 1963)

humerus

ulna

radius

subclavian artery

apical ectodermal ridge

marginal sinus

**MOUSE LEFT FORELIMB at 12.5 g.d.
(Reconstruction)**

(Redrawn from Forsthoefel 1959, 1963)

1 mm

**RIGHT FORE AND HIND LIMB OF 12.5 day
NORMAL MOUSE EMBRYO**

(From Grüneberg 1962)

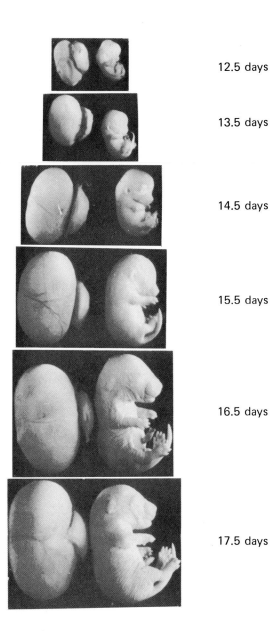

12.5 days

13.5 days

14.5 days

15.5 days

16.5 days

17.5 days

**MOUSE FETUS WITHIN ITS CHORIONIC VESICLE AND
LITTER MATE AFTER REMOVAL FROM ITS VESICLE
TO SHOW RELATIVE SIZES AND PLACENTAE.***

*Note: Capillaries on surface of vesicle and discoidal type of
attached placenta. The contained embryos are obviously
crowded with the approximation of head and tail.

I. THE 13 DAY EMBRYO: The 13 day mouse embryo is comparable to the 38 day human embryo. It has 52 to 60 somites and measures 9.4 mm from crown to rump. The head still comprises more than one-third of the total body volume. The back is straightening. The tip rather than the base of the tail now touches the side of the face, so that the tail seems to be shortening. Actually, since it has acquired all but the 5 most caudal somites, it has almost reached its maximum length. The somites still show through the integument, but the skin layers are differentiating. The hair follicles of the face (vibrissae) appear in rows. The external auditory meatus is circular and easily visible on each side of the head with a covering by the early pinna. The eyes appear on the surface as ovals but their inner structures cannot be seen. The limb buds have grown further, with differentiation of their extremities into early (mesenchymal) digits, although these are not distinct. There is chondrification of the ribs and the humerus. The umbilical hernia is receding into the body, and the mammary glands are cup-shaped.

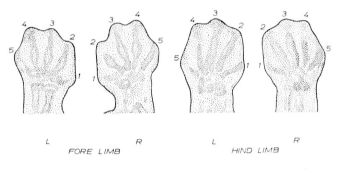

FEET OF NORMAL 13 DAY MOUSE EMBRYO

(Grüneberg 1961, 1963)

Mid-sagittal sections show further thickening of the brain walls, particularly in the laminia terminalis, dorsal thalamus, tuberculum posterius, roof of the mesencephalon, and floor of the myelencephalon. The first two ventricles (the telocoels) are expanding and will cover part of the diencephalon and third ventricle. The anterior choroid plexus has expanded laterally into the first two ventricles. The thin roof of the myelencephalon (the fourth ventricle) is much folded into the myelocoel, and becoming the vascular posterior choroid plexus. The spinal cord extends almost to the tip of the tail, and paired spinal ganglia extend to the level of somite 40. The eyes have lost their lens vesicles by the close apposition of lens epithelium to lens fibers, and behind the lenses vitreous humor is accumulating; it may be loosely cellular. Surrounding each eye is the early formation (mesenchyme) of the eye muscles (there is no evidence of cartilage or bone yet). Both above and below the eye, the folds will be the lids. The retina has further thickened (to about 10 times the thickness of the pigmented layer) but does not yet exhibit any indications of the various cell layers to be developed. There are nerve fibers in the optic stalks, and oculomotor nuclei are forming. Each ear, surrounded by mesenchyme and early cartilage, comprises a long endolymphatic duct connected laterally with the ampulla and mesially with the utricle, saccule, and elongated and coiled cochlea.

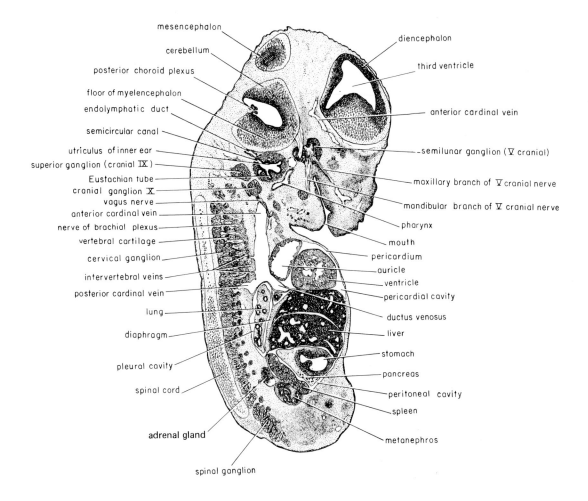

mesencephalon
cerebellum
posterior choroid plexus
floor of myelencephalon
endolymphatic duct
semicircular canal
utriculus of inner ear
superior ganglion (cranial IX)
Eustachian tube
cranial ganglion X
vagus nerve
anterior cardinal vein
nerve of brachial plexus
vertebral cartilage
cervical ganglion
intervertebral veins
posterior cardinal vein
lung
diaphragm
pleural cavity
spinal cord
adrenal gland
spinal ganglion

diencephalon
third ventricle
anterior cardinal vein
semilunar ganglion (V cranial)
maxillary branch of V cranial nerve
mandibular branch of V cranial nerve
pharynx
mouth
pericardium
auricle
ventricle
pericardial cavity
ductus venosus
liver
stomach
pancreas
peritoneal cavity
spleen
metanephros

MOUSE EMBRYO — 13 days
(para-sagittal)

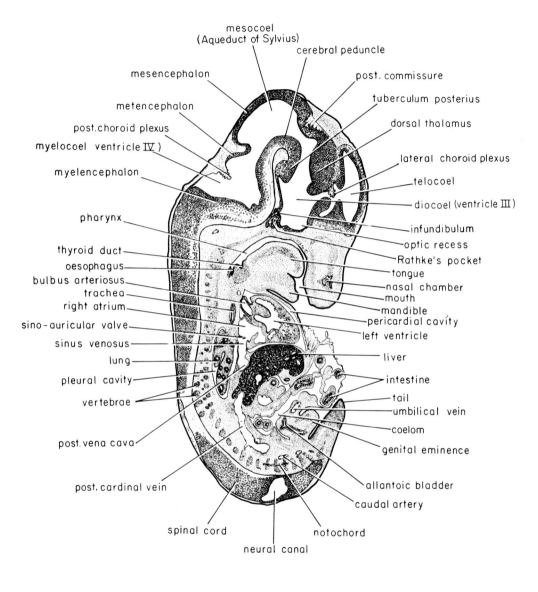

mesocoel
(Aqueduct of Sylvius)
cerebral peduncle
mesencephalon
post. commissure
metencephalon
tuberculum posterius
post. choroid plexus
dorsal thalamus
myelocoel ventricle IV)
lateral choroid plexus
myelencephalon
telocoel
diocoel (ventricle III)
pharynx
infundibulum
optic recess
thyroid duct
Rathke's pocket
oesophagus
tongue
bulbus arteriosus
nasal chamber
trachea
mouth
right atrium
mandible
sino-auricular valve
pericardial cavity
sinus venosus
left ventricle
lung
liver
pleural cavity
intestine
vertebrae
tail
umbilical vein
coelom
post. vena cava
genital eminence
post. cardinal vein
allantoic bladder
caudal artery
spinal cord
notochord
neural canal

MOUSE EMBRYO — 13 days
(mid-sagittal)

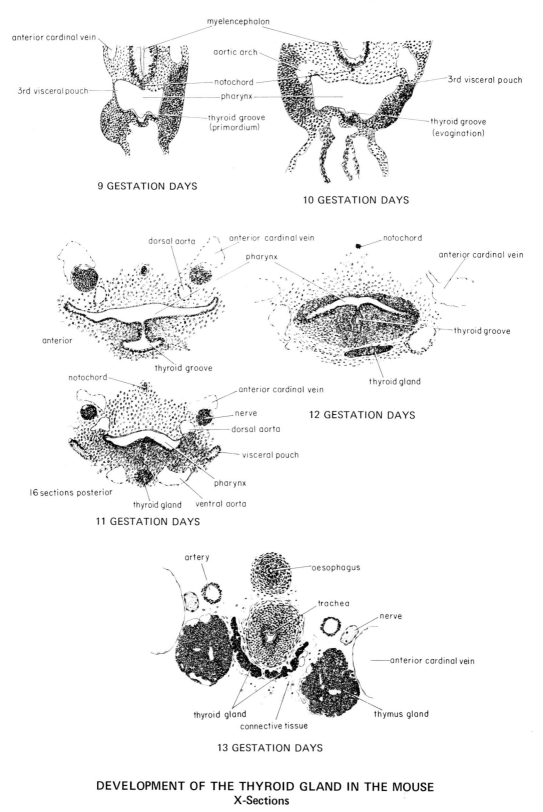

9 GESTATION DAYS

10 GESTATION DAYS

12 GESTATION DAYS

11 GESTATION DAYS

13 GESTATION DAYS

DEVELOPMENT OF THE THYROID GLAND IN THE MOUSE
X-Sections

The thyroid is a definitely lobular horseshoe shaped structure ventral to the trachea. Nearby are the paired thymus glands, adjacent to the anterior cardinal veins. Some thymus cells are differentiated into stromal elements with pseudopodia, which seem to encircle or engulf other cells. Some of these may be advance lymphopoietic cells. Between the thymus glands is loose connective tissue. Both the lungs and the liver have enlarged considerably, and the lungs show lobular development. The cells of the spleen may now be seen accumulating under the epithelium of the posterior part of the mesenteric fold. The anterior splanchnic mesodermal plate persists longest in the region of spleen development. The gut now appears more as it will in the adult, and the vitelline vein, passing dorsally from the umbilicus, loops around the duodenum on its path to the liver. The posterior vena cava arises from the right posterior cardinal vein but at this time the paired posterior cardinals are large, lying on either side of the dorsal aorta and deriving their venous blood from the more posterior regions of the developing embryo. Between 12 and 13 days a connection had been formed between the right posterior cardinal vein and channels in the liver which lead to the heart. Then the anterior portion of the right posterior cardinal vein (anterior to the junction with the left branch) decreased in size, as did the posterior portion of the left posterior cardinal so that by 14 days the venous drainage from the trunk caudal to the liver is largely through the remaining portions of both the posterior cardinals which have joined to form a single vessel, the posterior vena cava. The hepatic portion of the posterior vena cava is formed largely from the vitelline vein. The muscular primordia and the thickening of the submucosa of the esophagus are apparent, as are vacuoles in the stomach epithelium and beginning of intestinal villi. A genital eminence may be found just beneath the tail, closely associated with the origin of the allantois.

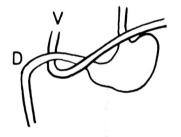

Diagrammatic representation of the relation of the vitelline vein to the duodenum in the 13½ day mouse embryo, ventral view. D, duodenum; V, vitelline vein.

(Courtesy Dr. Margaret C. Green, 1967, Developmental Biology 15:62-89, Academic Press.)

Dorsally many vertebral cartilages are seen, each separate from its neighbor and derived from the scleratome of two adjacent somites. Mesenchymal concentrations, particularly in the head, are precartilage masses. The aortic, pulmonary, and interventricular septa and the atrioventricular valves are complete, and the right dorsal aorta between aortic arches III and IV has disappeared. The segmental arteries and veins are well formed. Enucleated cells are only about 1% of the red blood cells. The metanephroi are prominent, but no vestiges of the pronephroi or mesonephroi remain except the ducts, which are present according to sex. A suprarenal (adrenal) gland is located just anterior to each metanephros. Cellular differentiation has progressed to the point that sex can be predicted with certainty. Serial sagittal sections are highly instructive particularly for following the various blood vessels, which are now numerous.

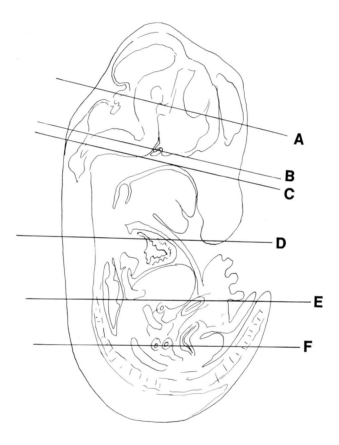

A

B
C

D

E

F

Guide lines represent positions of the following sections.

(Series from Rugh: "Vertebrate Embryology,"
Harcourt, Brace & World, 1964.)

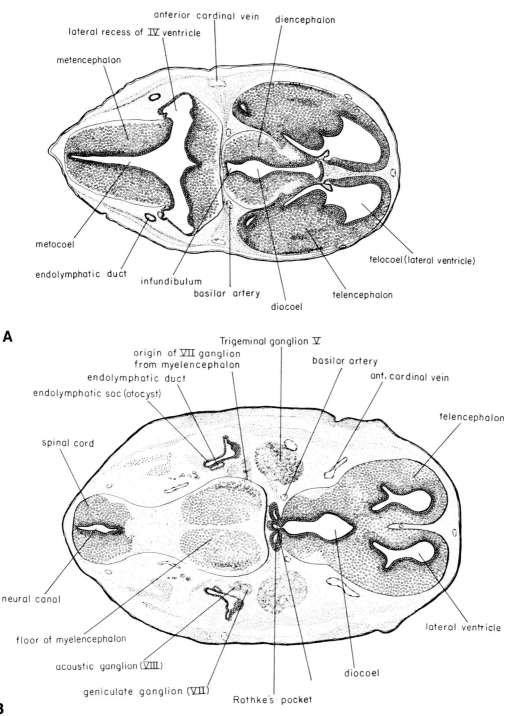

A

anterior cardinal vein

diencephalon

lateral recess of IV ventricle

metencephalon

metocoel

endolymphatic duct

infundibulum

basilar artery

diocoel

telencephalon

telocoel (lateral ventricle)

B

Trigeminal ganglion V

origin of VII ganglion
from myelencephalon

basilar artery

endolymphatic duct

ant. cardinal vein

endolymphatic sac (otocyst)

telencephalon

spinal cord

neural canal

floor of myelencephalon

acoustic ganglion (VIII)

geniculate ganglion (VII)

Rathke's pocket

diocoel

lateral ventricle

MOUSE EMBRYO — 13.5 g.d.
(transverse sections)

(From R. Rugh, "Vertebrate Embryology", Harcourt, Brace & World, Inc., New York, 1964.)

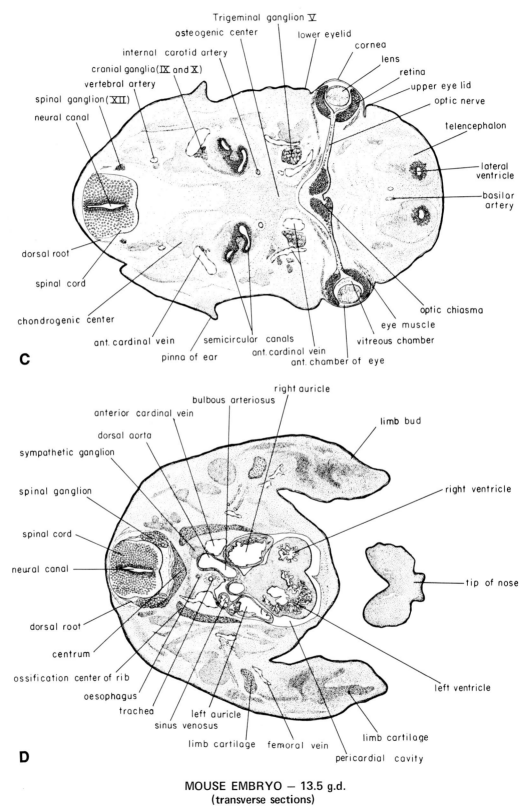

MOUSE EMBRYO — 13.5 g.d.
(transverse sections)

(From R. Rugh, "Vertebrate Embryology", Harcourt, Brace & World, Inc., New York, 1964.)

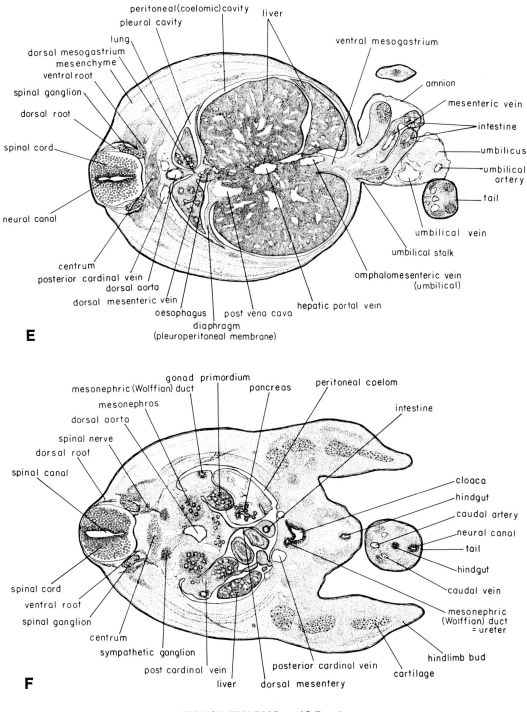

E

peritoneal(coelomic)cavity
pleural cavity
lung
dorsal mesogastrium
mesenchyme
ventral root
spinal ganglion
dorsal root
spinal cord
neural canal
centrum
posterior cardinal vein
dorsal aorta
dorsal mesenteric vein
oesophagus
post vena cava
diaphragm
(pleuroperitoneal membrane)
hepatic portal vein
omphalomesenteric vein
(umbilical)
umbilical stalk
umbilical vein
tail
umbilical artery
umbilicus
intestine
mesenteric vein
amnion
ventral mesogastrium
liver

F

gonad primordium
mesonephric (Wolffian) duct
mesonephros
dorsal aorta
spinal nerve
dorsal root
spinal canal
spinal cord
ventral root
spinal ganglion
centrum
sympathetic ganglion
post cardinal vein
liver
dorsal mesentery
posterior cardinal vein
cartilage
hindlimb bud
mesonephric
(Wolffian) duct
= ureter
hindgut
caudal vein
tail
neural canal
caudal artery
hindgut
cloaca
intestine
peritoneal coelom
pancreas

MOUSE EMBRYO — 13.5 g.d.
(transverse sections)

(From R. Rugh, "Vertebrate Embryology", Harcourt, Brace & World, Inc., New York, 1964.)

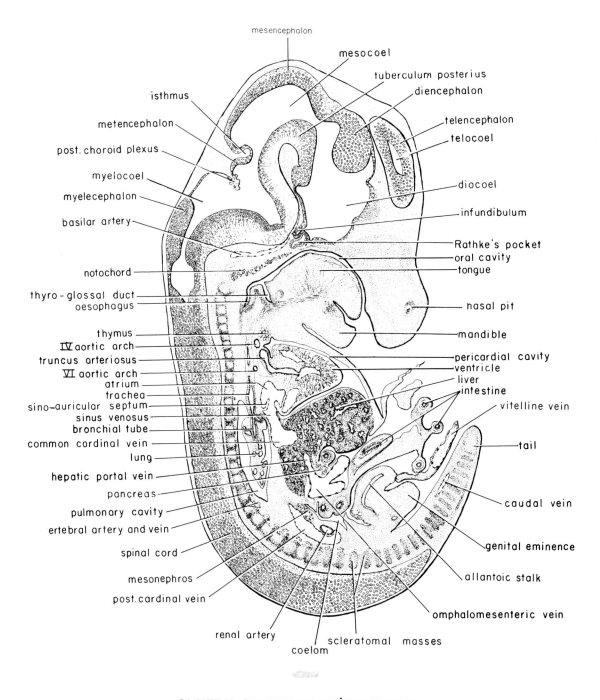

SAGITTAL SECTION OF MOUSE EMBRYO
(13.5 gestation days)

(From Rugh: "Vertebrate Embryology,"
Harcourt, Brace & World, 1964.)

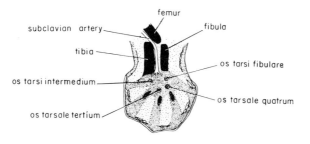

femur

subclavian artery

fibula

tibia

os tarsi fibulare

os tarsi intermedium

os tarsale quatrum

os tarsale tertium

MOUSE LEFT HINDLIMB at 13.5 g.d.
(Reconstruction)
(Redrawn from Forsthoefel, 1959, 1963)

Dorsal View

Ventral View

Same Embryos Exposed

Enlarged View in Amnion

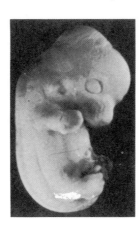

Embryo with Amnion Removed

MOUSE EMBRYOS IN UTERUS at 13.5 days

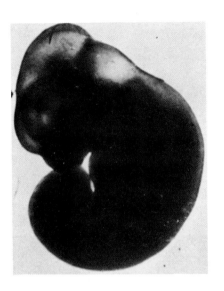

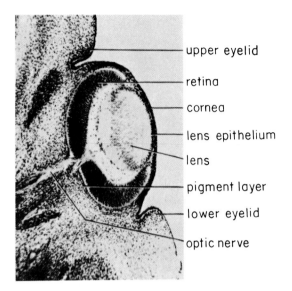

WHOLE MOUSE EMBRYO
at 13.5 days

MOUSE EMBRYONIC EYE
at 13.5 days

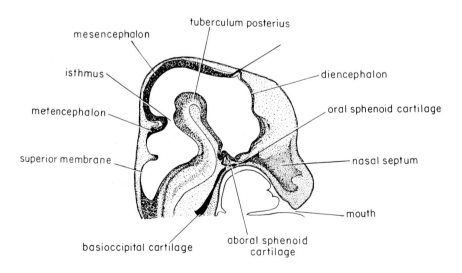

MOUSE HEAD at 13.5 g.d. (Sagittal section)

Showing early chondrification.

(Redrawn from Forsthoefel 1963)

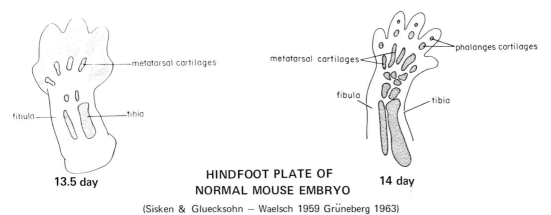

metatarsal cartilages

phalanges cartilages

metatarsal cartilages

fibula

tibia

fibula

tibia

fibula

tibia

13.5 day

**HINDFOOT PLATE OF
NORMAL MOUSE EMBRYO**

14 day

(Sisken & Gluecksohn — Waelsch 1959 Grüneberg 1963)

J. THE 14 DAY EMBRYO: The 14 day mouse embryo compares well with the 6 weeks old human embryo. It has 61 to 62 somite pairs and measures $10\frac{1}{2}$ mm from crown to rump. By this time major congenital anomalies can no longer be produced by trauma or other insult to the fetus because most of the major organs have been laid down, and future development is largely a matter of refinement and cellular differentiation. The only systems that remain particularly vulnerable to trauma is the skeletal system, in which cartilage and then ossification centers begin to form at about 12 days, and the nervous system which is made up largely of differentiating neuroblasts. Radiation exposure of the 14 day embryo generally results in structural changes in the skeleton (stunting) and may affect the functioning of the nervous system. The first seven ribs are chondrified. The first cartilage is seen in the otic capsules. The region of the developing frontal bones and the zygomatic arch show ossification centers. We are therefore now describing a topographically complete embryo.

The main changes in contour occur in the head. The face assumes a pig or rodent-like shape, with a snout and forehead bulges. The back continues to straighten. The appendages look like paddles, with fanlike extremities having partially separated digits, each external auditory meatus is open, and partially covered by a single-scalloped pinna, and each eyeball is larger and more distinct than at 13 days, but still undistinguished from the surface and not yet covered by the lids. The external nares are open. The skin layers begin to appear with occasional concentration of cells that later give rise to hair follicles, like those already on the face. The facial hair shafts, roots and bulbs, and papillae are all recognizable. The cloacal membrane has ruptured to form the lining of the anal chamber.

The brain can be dissected out, and the major parts identified. There are paired telencephalic lobes (cerebral hemispheres) between which is the epiphysis (pineal). Posterior to them is the single, bulbous mesencephalon. The metencephalon (cerebellum) is laterally expanded and thick-walled but extends posteriorly only to the thin-walled posterior choroid plexus of the myelencephalon. The myelencephalon tapers into the spinal cord. Lateral or ventral views show the olfactory lobes, optic chiasma, and pituitary. The brain is about 6 mm in length. The skin of the face, in contrast with that over the body, does show vibrissae formation which begins at 13 days, and is actively developing at 14 days.

The pineal gland exhibits cellular differentiation. It grows into the head mesenchyme from the diencephalon, closely associated with a cerebral vessel. The pituitary, especially the anterior lobe (from the infundibulum) is glandular-looking, although it has no follicles or secretion yet. It somewhat surrounds the intermediate lobe. The sphenoid cartilage and sella turcica into which the pituitary grows are visible. A lens and its covering epithelium are now suspended in each optic cup and so the iris is formed, as well as an anterior chamber between lens and cornea. The fibrous structure of the lens continues to differentiate. The posterior wall of the optic cup is the much thicker retina, while nerve fibers may be seen passing in from the optic nerve to line the optic cup. The optic fibers connect with the brain. The ears have grown to six times the length of the otic vesicles at 9 days. Semicircular canals, ampulla, utricle, saccule, and a remnant of the endolymphatic sac are present in each one. The ear areas are invested with loose mesenchyme, which will develop into encapsulating otic cartilage.

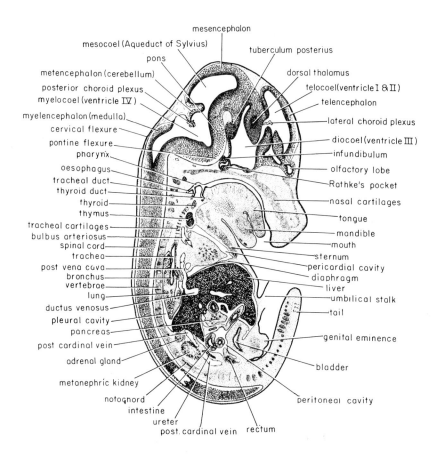

MOUSE EMBRYO — 14 days
(mid-sagittal)

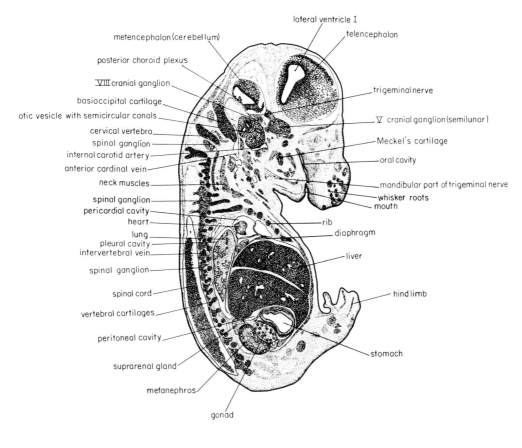

MOUSE EMBRYO — 14 days
(para-sagittal)

Transverse striated muscle fibers are apparent in the tongue, and the beginnings of the incisors with dental papillae, and outer and inner enamel, in the lateral oral epithelium closely associated with Meckel's cartilage and early ossification. The first molar bud is visible on each side. Salivary glands are supplied with secretory ducts.

The liver continues to grow and become more vascular, and the pancreas shows histological differentiation and vascularization. Posterior to the liver is the large and thick-walled stomach. Intestinal loops are plentiful, with their first villi.

The heart is essentially complete. The foramen ovale persists, and the aortic pulmonary semilunar valves appear. Aortic arch III branches into the lingual and carotid arteries. The left aortic arch IV (the systemic arch) receives the ductus arteriosus from left aortic arch VI, gives off the left subclavian and vertebral arteries, and continues as the dorsal aorta. The diaphragm is completed by the closing of the pleuroperitoneal canals. Enucleated cells may constitute 25% of the red blood cells.   The mesonephroi are much reduced except for the parts (ducts) giving rise to the epididymi in the male, and the ureters open into the cloaca (urogenital sinus).   The oviducts (pronephric ducts) cross over the ureters toward the pelvis, and the ovaries and testes become quite vascular.   The metanephroi are paired, compact, masses of tubules and ducts, close to the gonads and suprarenal glands.

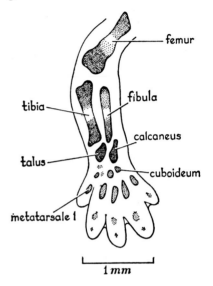

**CARTILAGINOUS SKELETON
OF LEFT HIND LIMB OF 14.5 day
NORMAL MOUSE EMBRYO**

(Searle 1963, Grüneberg 1963)

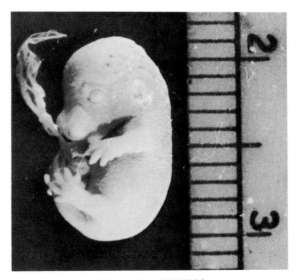

ENTIRE LITTER   SIZE IN MILLIMETERS

MOUSE EMBRYOS at 14.5 days

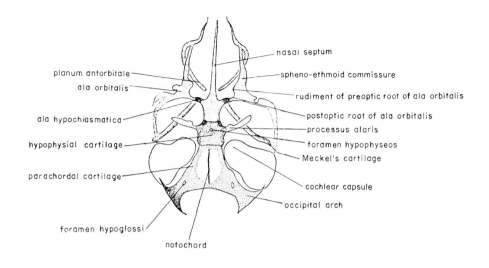

**DORSAL VIEW OF CHONDROCRANIUM OF
NORMAL 14.5 day MOUSE EMBRYO**

(Grüneberg 1953, 1963)

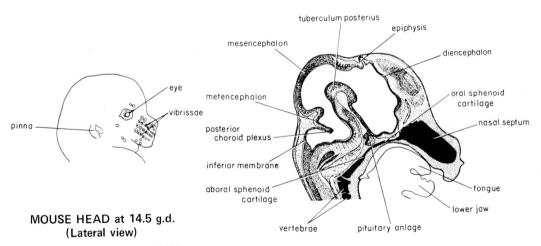

**MOUSE HEAD at 14.5 g.d.
(Lateral view)**

(Redrawn from Forsthoefel 1963)

**MOUSE HEAD at 14.5 g.d.
(Sagittal section)**

(Redrawn from Forsthoefel 1963)

K. THE 15 DAY EMBRYO: The 15 day mouse embryo corresponds in development to a 55 day human embryo. It has its full quota of 65 pairs of somites and measures about 12 mm, from crown to rump.

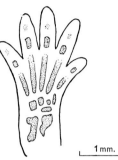

LEFT HINDFOOT OF 15 day NORMAL MOUSE EMBRYO

Methylene-blue preparations.

1 mm.   (From Grüneberg 1956)

Superficially the head is gradually taking on the rodent shape. On each side the external auditory meatus is a clearcut pit with the pinna all but overlapping it from behind. The eyes are partially visible through the skin. The appendages are much longer and better developed than at 14 days, with the digits almost completely separated, having small extensions at their tips. The vibrissae on the face are easily recognized. Hair follicles are extensively distributed elsewhere over the body. The skin seems to be growing faster than the body, as it is greatly wrinkled. Since some of the intestinal loops still project into the umbilical cords the umbilical hernia remains. Owing to further straightening of the back, the tail seems shorter than before, curling between the hind legs toward the face but not touching it.

The brain appears as it did at 14 days except that each part is further enlarged and some are thickened. The cerebrum seems to be growing fastest. It is obvious that neuroblasts are proliferating rapidly to fill the cerebral vesicles and thus to produce a substantive brain. Trauma at this time can lead to cessation of this proliferation and hydrocephalus. The cerebellum is fused at the mid-line. The eyelids are advancing over the eyes but more significant is the development of inner and outer nuclear layers in the retina. The ears have become complicated ramifications of canals. The olfactory organs have not changed radically but have expanded, with the lining epithelium thickening. They are embedded in the nasal and ethmoid cartilages. The anterior, posterior, and intermediate lobes of the pituitary are present but are cytologically undifferentiated. An infundibular recess remains as does the lumen of Rathke's pocket between the anterior (pars distalis) and intermediate lobes. The pineal gland shows signs of glandular function. A stalk still connects it with the diencephalon.

Mid-body transverse sections show that the spinal cord is much like that of the adult. Marginal, mantle, and ependymal layers are easily distinguished. Thickest now is the mantle layer. The ependymal layer is crowded around the neural canal, full of mitoses. The spinal root ganglia are now well developed and the dorsal and ventral roots are properly associated with the cord. Below the cord is the remnant of the notochord, which is being surrounded and incorporated by the sclerotome of the centrum. The notochord generates from anterior to posterior, so that it is visible later in posterior sections rather than in anterior ones. The ribs are so well ossified that they compose a substantial rib cage, which gives the body a new rigidity.

The sternal cartilage is new, and cartilage is forming in the walls of the trachea. The thyroid is well advanced, and the thymus glands are enlarging rapidly, even faster than the thyroid. The liver appears to fill about half the body cavity. Among its cells is the gall bladder, and close to it the highly glandular pancreas. Many lymph gland cells appear to be definite lymphocyte precursors, and are small thymic lymphocytes. Stellate reticular cells, columnar cells in acinar formation and containing glycogen, and hypertrophied cells may be found. For the first time parenchymal blood vessels are seen. Nucleated cells constitute only about 5% of the red blood cells. Posteriorly, at the base of the tail, the urethra enters the bladder.

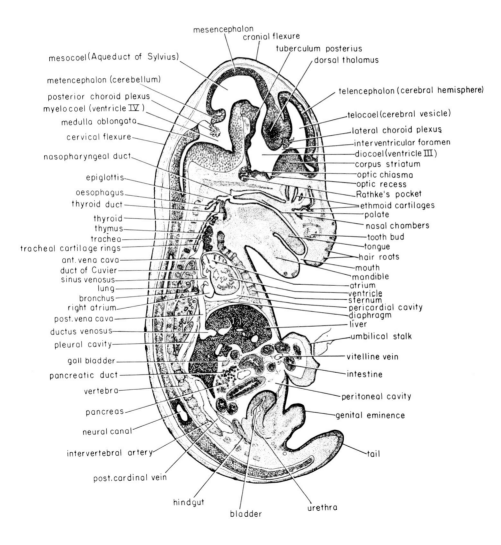

MOUSE EMBRYO — 15 days
(mid-sagittal)

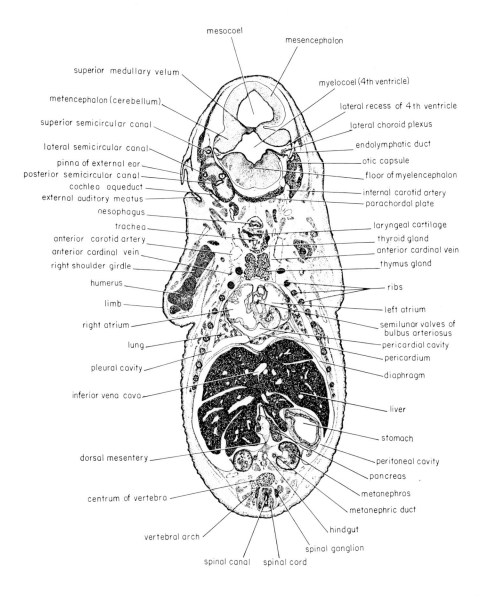

MOUSE EMBRYO — 15 days
(frontal section)

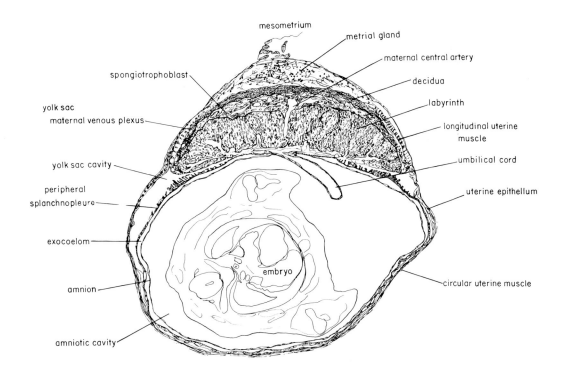

MOUSE PLACENTA at 15 days

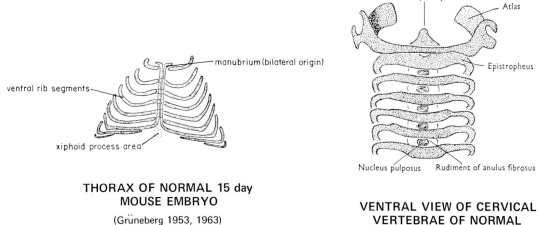

THORAX OF NORMAL 15 day
MOUSE EMBRYO

(Grüneberg 1953, 1963)

VENTRAL VIEW OF CERVICAL
VERTEBRAE OF NORMAL
15 day MOUSE EMBRYO

(Grüneberg 1953, 1963)

L. THE 16 DAY EMBRYO: The 16 day mouse embryo is comparable to the 10.4-week human embryo. It measures about 15 mm from crown to rump, is recognizable as a mouse, and moves actively within the uterus. It is sufficiently straightened out that its fore- and hindlimbs no longer touch, and its skin is extensively wrinkled, with hair follicles. The eyelids have fused over the eyes and the pinnas cover the auditory meatuses.

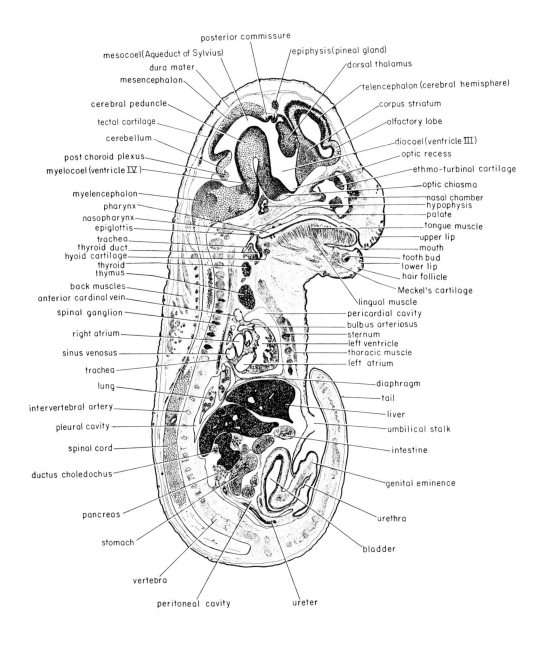

MOUSE EMBRYO — 16 days
(mid-sagittal)

Sagittal sections show that the pineal is almost entirely separated from the diencephalon, the cerebral peduncle is growing upward to fill the mesocoel, the corpora striata are extremely thick, and the corpus callosum is formed. The brain, except for the cerebellum, is completing its development. The nasal chambers are extensive.

The most apparent changes relate to the striated muscles and the skeleton, which are present throughout the embryo. Muscles of the tongue and back are most prominent, while the cranial, sternal, and vertebral regions exhibit many ossification centers. The laryngeal and tracheal cartilages are developing.

The thyroid is a bilobed, highly vascular organ with the parathyroids attached to it and the ultimobranchial bodies disappearing within its substance. The thymus is several times the size of the thyroid and is found just cephalad to the heart. The lungs have alveoli. Parasagittal sections show the metanephroi with well-formed glomeruli. The bladder is a large muscular chamber, closely associated with the ureters. Cartilage is changing to bone in the appendages, so that they look like fore- and hindlimbs, with integument folded at their bases like accordians awaiting their further growth. The hyaloid and retinal arteries enter the eyes with the optic nerves. Approximately 1% of the red blood cells remain nucleated, in the immature state.

The centrum is ossified. Except for further ossification, development of full musculature, and further growth of the brain and hematopoietic centers, the mouse is now essentially complete. From this point on, certain organ systems will be described as they develop, and independently of the adnexa or the embryo as a whole. Description will be given first of the changes observable externally and then of the histological changes.

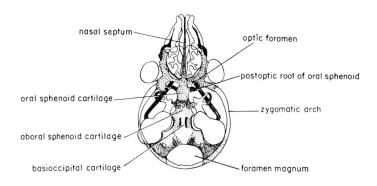

MOUSE CHONDROCRANIUM at 16.5 g.d.
(Ventral view)

(Redrawn from Forsthoefel 1963)

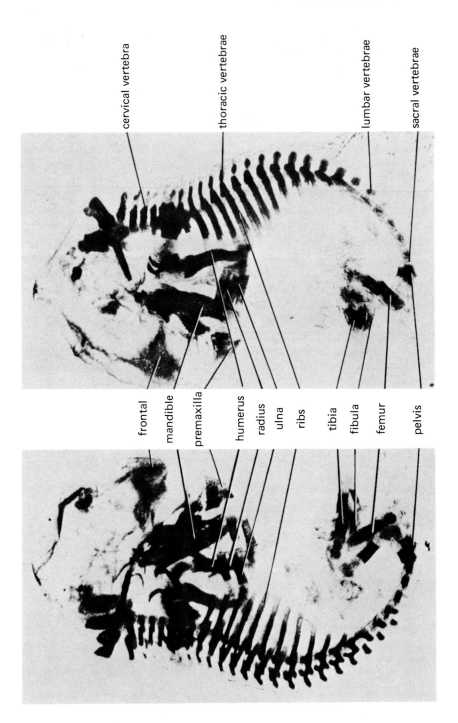

SKELETON OF MOUSE EMBRYO — 16.5 days

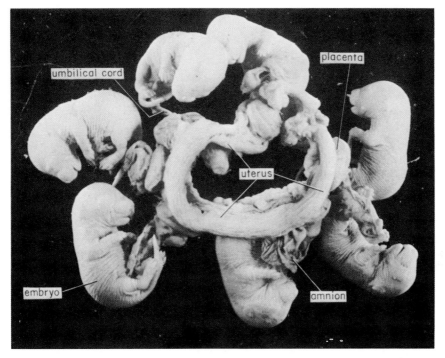

DORSAL VIEW

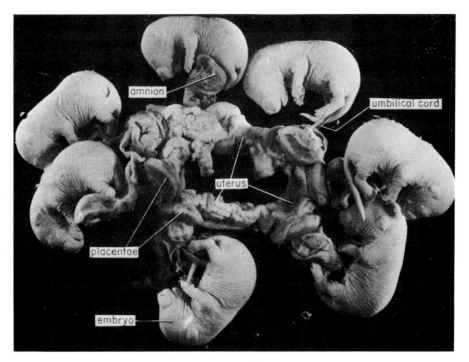

VENTRAL VIEW

MOUSE LITTER AT 16.5 gestation days

ORGANOGENY

EXTERNAL CHANGES

BODY CONTOUR: The accompanying photograph shows a series of mouse embryos from 10 to 18 days gestation. Such a series demonstrate the chronological changes in general topography, emphasizing (1) change in body posture from a much-coiled C shape at 10 or 11 days to the late fetus of 16 days, when the back is much straightened; (2) change from a lobular head with pronounced visceral arches to an elongated head with snout and vibrissae and no superficial evidence of the arches; (3) the early appearance of the otic and the optic vesicles at 9 days, both to be partially covered by 15 days; (4) growth of the appendages, from appearance of the forelimbs at 9 days and that of the hindlimbs at 10 days to digital development at 16 days; (5) excessive wrinkling of the fast growing skin at 16 days in anticipation of the filling out of the next 3 days; (6) incorporation of the bulging visceral and umbilical areas of the early

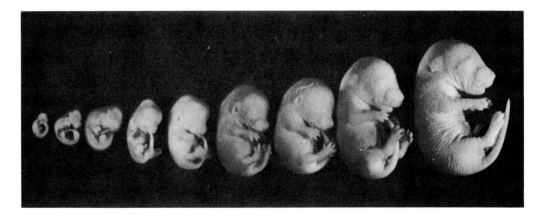

MOUSE EMBRYOS FROM 10 to 18 GESTATION DAYS

stages, into the slender abdomen of the newborn; (7) early growth of the tail, so long with its somites that it is coiled to the right of the head at 10 days, and the changes leading to a relatively short motile tail lacking any vestige of the somites and no longer touching the face. One cannot see, in these illustrations, the changes in skin texture and color. At times the skin is quite translucent and pink. The texture and color of the skin are often indications of the newborn's general health. Between 10 and 18 days gestation, the embryo increases in crown-rump length from an average of 4.08 to 20.31 mm and in weight from about 0.0061 to 1.087 grams.

A transient ectodermal ridge (VER) about 300 μ long arises on the ventral aspect of the tail tip, where the ectoderm is thickened and columnar rather than squamous. It gives a positive alkaline phosphatase reaction suggesting cell hypertrophy. The VER originates at about 9 days gestation from the cloacal membrane, where the ectoderm and endoderm meet. It reaches its maximum development at 10 days, is less obvious at 11 days, and disappears by 12 days. It is believed by some to be a stimulating organ for the production of paraxial mesoderm.

APPENDAGES: The most critical period for appendage development seems to be at 12 days, when there are many precartilage concentrations of mesenchyme. This corresponds to about 36 days for the human embryo, a time when the drug thalidomide administered to a pregnant woman, has its most devastating effect on human embryonic limbs. In the mouse the sclerotome makes no contribution to the skeleton of either girdle or limbs.

At 9 days no external evidence of forelimb primordia is visible but sections show that the ectoderm is thickening in the forelimb regions. A pseudostratified layer about four cells thick and about 7 somites long protrudes directly out from the somatopleure on each side. Mesenchyme accumulates within the budlike outgrowths, and by 10 days they have begun to bend in a ventral direction. The mesenchyme cells are covered first by a layer of cuboidal ectoderm and then by a layer of flattened squamous ectoderm. The cuboidal layer forms a well-defined apical ectodermal ridge along the ventrolateral margin of each bud. At least one more row of cells is added, and a blood vessel appears below the ridge. The segmental arteries may be used to locate specifically the forelimb buds. Often the right bud is slightly anterior to the left one. As the embryo grows, the buds grow anteroposteriorly, but they grow directly outward at a faster rate. Since they do not grow anteroposteriorly as rapidly as the crown-rump length increases, the number of somites related to a bud is gradually reduced to about 5. At 11 days the footplate begins to develop in each bud and blastema condensations appear for the scapula, humerus, and ulna. At 12 days the footplate is polygonal (pentagonal) and the ectoderm and marginal sinus are complete. The sinus empties through the postaxial and preaxial veins into the postaxial part of the limb. The blastemas for all five metacarpals may be seen, joined to the common carpal blastema, and there is some chondrification in the radius and metacarpals 2, 3, and 4. At 13 days the footplate is deeply indented, and the apical ectoderm has begun to regress. Nerves supply the first phalanges, muscles are present throughout the limbs and ossification has started in the humerus, radius, and ulna. At 14 days the phalanges are separated distally, and necrosis begins in the intervening webbing. The marginal sinus breaks up and no apical ectoderm remains. Muscles are well developed throughout. Chondrification occurs in all the first phalanges, as well as in the second phalanges of digits 2, 3, and 4 and, in the scapula, radius, and ulna. Capillary invasion indicates the impending transformation of perichondrium into periosteum. At 15 days the digits are entirely separate, and ossification is proceeding rapidly everywhere, with the scapula and humerus ahead of the radius and ulna. The shoulder and elbow joints are well advanced. By 16 days all skeletal parts are chondrified. Ossification is well advanced in the scapula and in the shafts of the humerus, radius, and ulna. By birth (19 days) all skeletal parts are ossified except the carpus and pollex. (Note from table on page 212 that forelimb development precedes that of the hindlimb by about one day.)

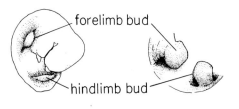
10 days   11 days

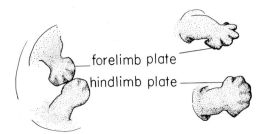
12 days   13 days

14 days

15 days

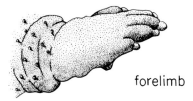
forelimb

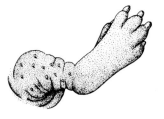
16 days

hindlimb

17 days

DEVELOPMENT OF APPENDAGES OF MOUSE EMBRYO

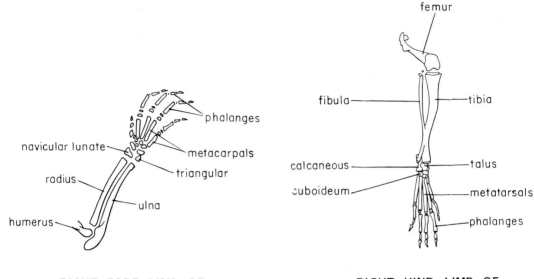

**RIGHT FORE LIMB OF
NORMAL ADULT MOUSE**

(Freye 1954, Grüneberg 1963)

**RIGHT HIND LIMB OF
NORMAL ADULT MOUSE**

(Searle 1963, Grüneberg 1963)

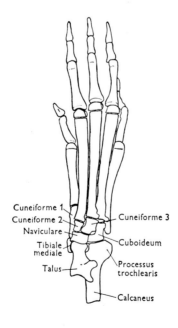

**RIGHT HIND FOOT OF
NORMAL ADULT MOUSE**

(Grüneberg 1956, 1963)

LIMB DEVELOPMENT*

AGE IN DAYS	FORELIMB	HINDLIMB
9	Limb buds as low ridges	
10	Limb buds/semicircular	Limb buds as low ridges
11	Separation into leg and circular footplate	Limb buds semicircular
12	Footplate pentagonal	Separation into leg and circular footplate
13	Footplate indented	Footplate pentagonal
14	Digits separated distally; still webbed proximally	Footplate deeply indented
15	Digits separated throughout, and very divergent; end phalanges beginning to show.	
16	Digits 2 to 5 nearly parallel and proximally webbed end phalanges clear, hindlimbs somewhat less advanced than forelimbs.	
17	Fingers and toes completely webbed as in newborn.	

* After Grüneberg, 1943 and 1963.

(Note: 15, 16, 17 day descriptions apply to both forelimb and hindlimb.)

The hindlimb buds appear at 10 days as crescent-shaped laterally directed projections largely mesenchyme. They lag behind the forelimbs in growth and differentiation, not exhibiting their characteristic elongation at 15 and 16 days. At 15 days cleared specimens show the tibia and fibula, and the tarsals have ossification centers. At the base of each appendage, the skin is folded, in preparation for further extension of the limb skeleton. Hair follicles are visible on the skin. The folds persist at 17 days but are partially flattened through stretching. The hindfeet are still webbed but have clawed digits.

SENSE ORGANS:[*] The large paired otic vesicles are apparent as early as 9 days. Anteriorly the paired telencephalic vesicles form bulges, just posterior to which the paired optic protuberances from the diencephalon arise at 10 days. At this time the first signs of olfactory pits also appear. Therefore, soon after the untwisting of the embryo, indications of sense organ development may be seen through the skin.

At 12 days the eyes are large oval swellings. Just posterior to them are the remnants of the otic vesicle openings, or endolymphatic ducts. During the next 3 days some eye parts become visible through the temporarily exposed cornea, only to be covered completely by the dorsal and ventral eyelids, which grow together by 16 days. The mouse is born blind because of this fusion of the eyelids.

The pinna, a double layered flat fold of skin, begins to grow forward over the external auditory meatus, on each side at 13 days. It almost completely covers the meatus by 17 days.

*External changes only. See page 245 for internal changes.

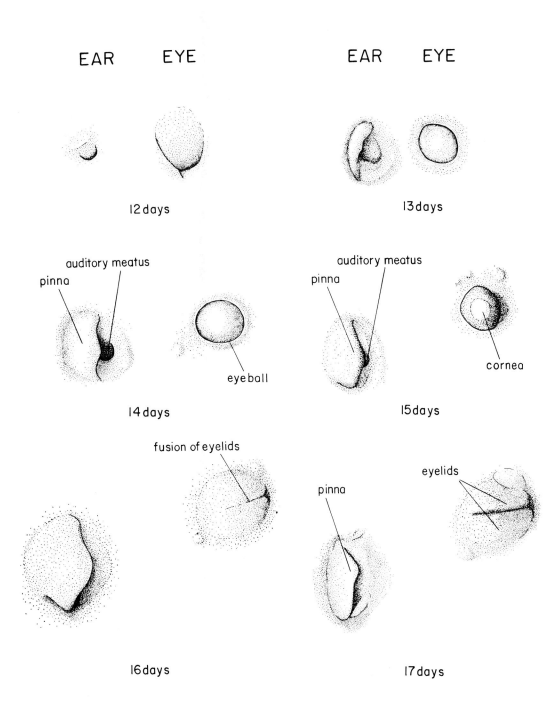

EXTERNAL VIEWS OF RIGHT EYE AND EAR
IN MOUSE EMBRYOS

The olfactory pit develops into external nares connected internally with ramifying tubes of the nasal chambers and ultimately with the pharynx just dorsal to the epiglottis. From the surface only facial changes around the external nares are evident.

SKIN, HAIR FOLLICLES, AND VIBRISSAE: Cup-shaped mammary welts may be seen at $10\frac{1}{2}$ days. The primary vibrissae are elevated on the face at 13 days. The earliest differentiation of the skin and related parts occurs at about 14 days, when occasional aggregations of cells, forerunners of the hair

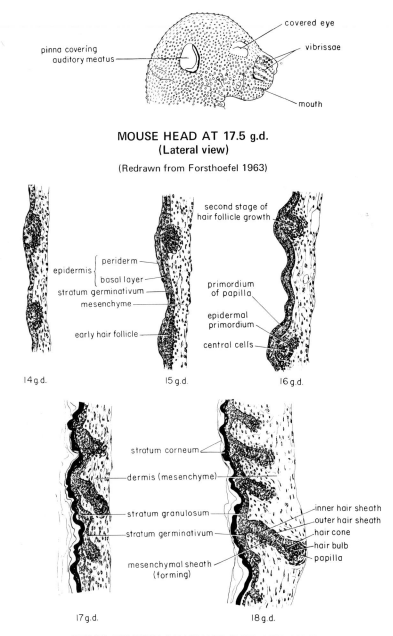

MOUSE HEAD AT 17.5 g.d.
(Lateral view)

(Redrawn from Forsthoefel 1963)

DEVELOPMENT OF MOUSE SKIN AND HAIR

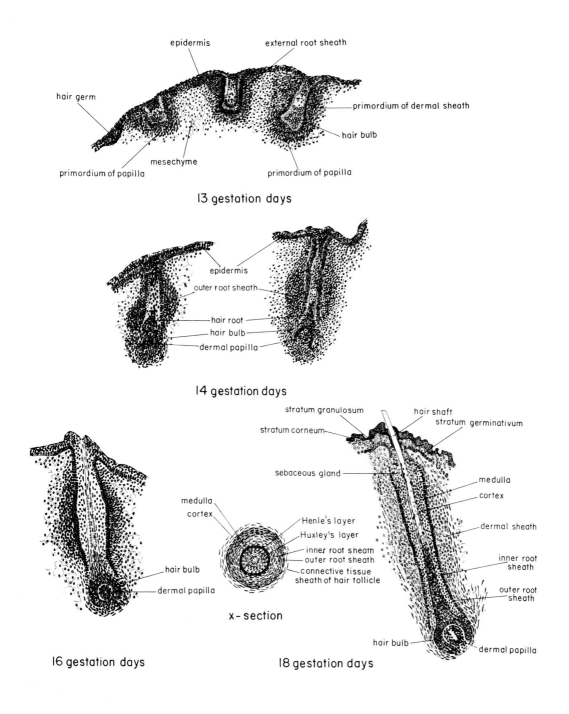

epidermis   external root sheath

hair germ

primordium of dermal sheath

hair bulb

primordium of papilla   mesechyme   primordium of papilla

13 gestation days

epidermis

outer root sheath

hair root
hair bulb
dermal papilla

14 gestation days

stratum granulosum   hair shaft
stratum corneum   stratum germinativum

sebaceous gland

medulla
cortex

medulla
cortex   Henle's layer
Huxley's layer
inner root sheath
outer root sheath
connective tissue
sheath of hair follicle

hair bulb

dermal sheath

inner root
sheath

outer root
sheath

hair bulb   dermal papilla

x-section

hair bulb

dermal papilla

16 gestation days   18 gestation days

DEVELOPMENT OF MOUSE VIBRISSAE

follicles, appear in a single layer of ectoderm. At $14\frac{1}{2}$ days the vibrissae papillae are invaginated, and ectoderm (periderm) is proliferating in the external
nares. By 15 days a second skin layer, the stratum germanitivum, arises and
between it and the first layer a basal layer develops. Much loose mesenchyme
is present below these layers, when the hair germs are forming. At 16 days
the periderm is thick in the nares, on the eyelids and in the external meatuses;
the vibrissae are cornified and papillae for the body hair follicles are visible.
Some of these follicles are in the second stage of growth. For the first time the
stratum corneum appears below a sloughing off layer of superficial squamous
epithelium. The hair follicles differentiate rapidly and are abundant, but not
until 18 days do hair bulb, core, and inner and outer sheaths all project deeply
into the skin (dermis). The periderm of the nares sloughs off. By 19 days the
vibrissae can be charted easily on the face and jaws. All their layers are apparent in cross section at this time. At birth the surface layer of epidermis is
cornified, but hair has not yet emerged.

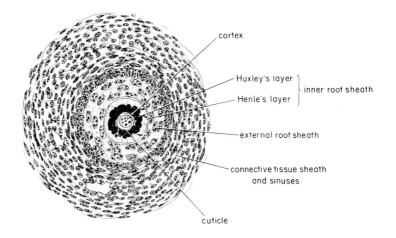

NEWBORN MOUSE VIBRISSA
(x-section under high power)

INTERNAL CHANGES

SKELETON: Chondrification centers may be seen first in sections as early as
11 days gestation. At $11\frac{1}{2}$ days there are blastema condensations for
the scapulna, humerus, and ulna, which begin to chondrify at $12\frac{1}{2}$ days. At this
time the blastemas for the radius and ulna are joined to the metacarpals by the
common carpal blastema. The interrelation of these various centers is an integral part of the developmental sequence. Nerves are also forming, some of
which reach the ends of the radius and ulna and vertebral condensation of mesenchyme is taking place. The precartilaginous nasal capsule, cartilaginous otic
capsule, and primordia of the otic ossicles appear by 14 days. Chondrification
is active, and many ossification centers are visible at $14\frac{1}{2}$ days and ossification
is widespread by 16 days.

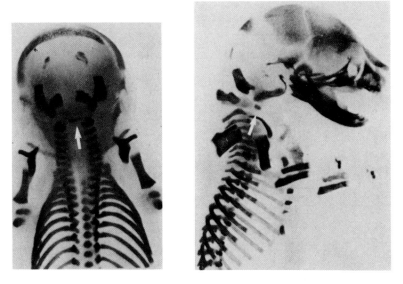

CFI mouse fetuses at 16 days gestation showing ossification center in the anterior arch of the atlas.

(Courtesy K. Hoshino, Congen. Anom. Japan 1967, Vol. 7:32–38.)

TIME OF ORIGIN OF VARIOUS ORGANS*

CFI MICE

ORGAN	GESTATION DAY	ORGAN	GESTATION DAY
Blood island	7+	Maxillary process	9
Foregut	7+	Forelimb bud	9
Heart mesoderm	7+	Lens placode	10
Head fold	7+	Nasal placode	10
Neural groove	8	Olfactory pit	10
Visceral arch I	8	Hindlimb bud	10
Neural tube anterior	9	Lens vesicle	11
Neural tube posterior	9	Lens fibers	11
Brain vesicle	9	Sclerotome	11
Eye cup	9	Eyelid	12
Otic vesicle	9		

* Data assembled from T. Ogawa 1967, Congenital Anomalies (Japan) 7:27-31.

GESTATIONAL AGE AND SOMITE NUMBER

GESTATIONAL AGE	SOMITE NUMBER
7 days	1 – 5
8 days	8 – 13
9 days	13 – 20
10 days	22 – 29
11 days	36 – 42
12 days	48 – 53
13 days	60+

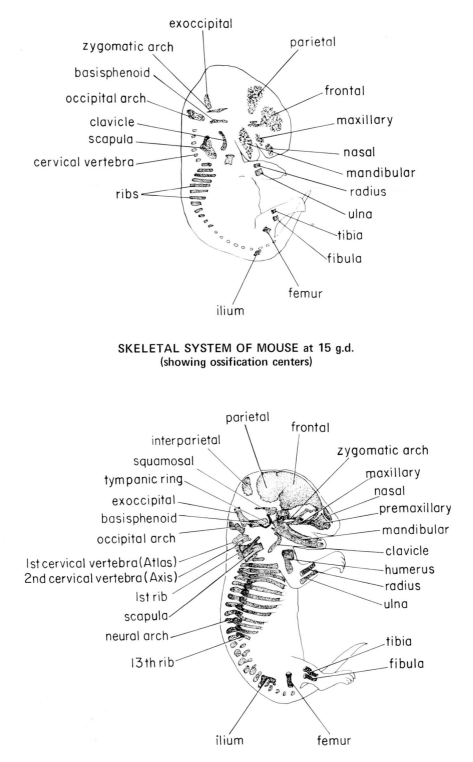

SKELETAL SYSTEM OF MOUSE at 15 g.d.
(showing ossification centers)

SKELETAL SYSTEM OF MOUSE at 16 g.d.

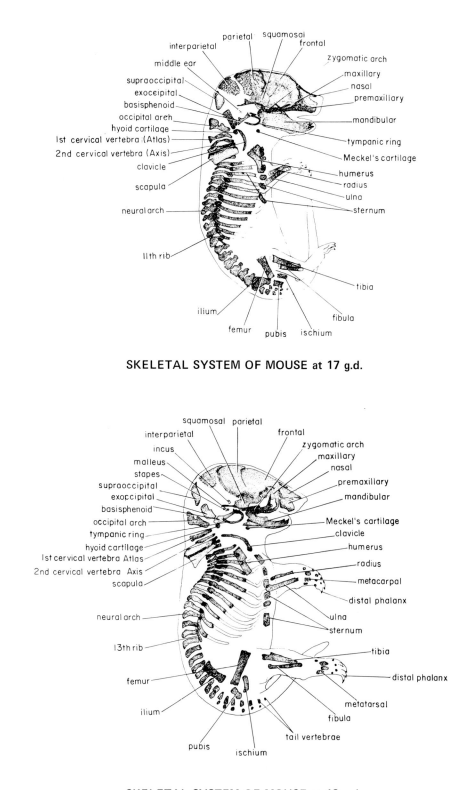

SKELETAL SYSTEM OF MOUSE at 17 g.d.

SKELETAL SYSTEM OF MOUSE at 18 g.d.

MOUSE FETUS FROM 15 to 18 days GESTATION
(showing extent of development of the skeleton)

(Spalteholz technique)

The chondrification centers themselves apparently are not targets of gene action. The osseous skeleton that develops from them tends to reflect conditions affecting the cartilaginous precursors. Therefore, most abnormalities of the skeleton can be traced to the cartilage and in turn to the membraneous blastema in which it arises. The mesenchymal stage is most vulnerable to environmental trauma. Trauma as early as $7\frac{1}{2}$ days may cause stunting, through damage to the mesenchyme, and hence to chrondrifications which affect the osteoblasts arising later in these centers.

The rib centers for the anterior (thoracic) regions appear early, by 12 days, and develop ahead of those for more posterior regions. Some of the membrane bones of the skull and jaw can be identified at this time. At 14 days frontal and zygomatic ossifications start to form. At 15 days the basic ossification centers are sufficiently developed that alizarin red S stains them specifically. Not only the appendage centers but also the cranial, pectoral, and pelvic centers can be recognized. The centras chondrify at 16 days. By 17 days the sternum is formed, the cranial cavity is almost encased, the tympanic rings can be identified, and the mandibular articulations at the zygomatic arches are established. At 18 days the appendage parts become stainable, and the bones contain hematopoietic marrow. The cranial parts have not yet fused but are expanding as the brain within enlarges. The scapulae are the largest bones in the body at this time.

The vertebrae originate from the sclerotomes. Each sclerotome consists of an anterior or cranial zone of low density and a posterior or caudal zone of high density, with a sharp demarcation line known as the sclerotomal fissure (fissure of von Ebner) between the two zones. The fissure does not reach the mid-line in the vicinity of the notochord, and it is soon invaded by cells from the more caudal zone. The somites are bounded by segmental blood vessels. The mesenchymal (sclerotomal) material for the vertebrae is bounded by sclerotomal fissures. The fissures separate this material into segments alternating with the myotomal and dermatomal portions of the somites, and so the vessels that were intersegmental with respect to the somites become midsegmental with respect to the vertebrae. The zones of lesser density chondrify to give rise to the centra of the vertebra, and the zones of higher density (perichordal discs) form the annuli fibrosi of the intervertebral discs.

The notochord is at first a cylindrical rod of cells of uniform consistency. As chondrification proceeds, it thins inside the centra but shows some swelling at the levels of the intervertebral discs. Eventually it is completely ossified, and only an acellular notochordal sheath remains in the vertebrae.

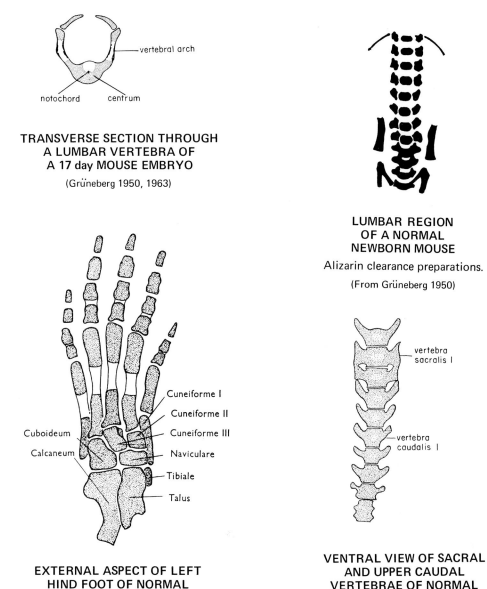

TRANSVERSE SECTION THROUGH
A LUMBAR VERTEBRA OF
A 17 day MOUSE EMBRYO

(Grüneberg 1950, 1963)

LUMBAR REGION
OF A NORMAL
NEWBORN MOUSE

Alizarin clearance preparations.

(From Grüneberg 1950)

EXTERNAL ASPECT OF LEFT
HIND FOOT OF NORMAL
NEWBORN MOUSE

(Grüneberg 1953, 1963)

VENTRAL VIEW OF SACRAL
AND UPPER CAUDAL
VERTEBRAE OF NORMAL
NEWBORN MOUSE

(Grüneberg 1953, 1963)

CHONDRIFICATION AND OSSIFICATION CENTERS
IN SKELETAL DEVELOPMENT:CFI MICE*

BONES **CRANIAL BONES**	CHONDR. GESTATION DAYS	OSSIF.	BONES **FACIAL BONES**	CHONDR. GESTATION DAYS	OSSIF.
Basioccipital	13	14	Maxilla	—	—
Exoccipital	13	14	Lateral plate	—	15
Supraoccipital	16	16	Palatine process	—	15
Parietal	—	15	Zygomatic process	—	15
Interparietal	—	15	Premaxilla	—	—
Frontal	—	14	Median plate	—	15
Sphenoidal	—	—	Horizontal plate	—	15
Presphenoid body	17(?)	18	Perpendicular plate	—	16
Lateral process	13	15	Zygomatic	—	16
Inter pterygoid	13	14	Lacrimal	—	16
Alar process	14	15	Palatine	—	15
Squamosal	—	14	Nasal	14	16
Tympanic ring	—	16	Vomer	—	15
			Mandible body	—	13
			Coronoid process	—	16
			Condyloid process	—	15
			Angular process	—	16
			Hyoid process	14	17

TRUNK & GIRDLE BONES			**BONES OF EXTREMITIES**		
Clavicle	—	13	Humerus	13	14
Scapula	13	14	Radius	13	14
Ribs	13	14	Ulna	13	14
Sternebrae	15	16	Carpals	14	—
Ilium	13	15	Metacarpals	13	16
Ischium	14	16	Digits (forelimb)		
Pubis	14	16	Proximal	14	18
Vertebrae			Middle	14	18
Cervical arch	13	14	Distal	15	17
body	13	18	Femur	13	14
Thoracic arch	13	15	Tibia	13	14
body	13	15	Fibula	13	14
Lumbar arch	13	15	Tarsals	13	18
body	13	15	Metatarsals	13	16
Sacral arch	13	16	Digits (hindlimb)		
body	13	16	Proximal	14	18
Caudal arch	13	18	Middle	15	18
body	13	17	Distal	15	17

*Data assembled from K. Hoshino 1967, Congenital Anomalies (Japan) 7:32–38. (The data of this table do not coincide with data of the following table, derived 34 years earlier. Author)

APPEARANCE OF OSSIFICATION CENTERS IN THE MOUSE*

SKULL	Prenatal Time Day	Hr.	Postnatal Time Days	SKULL	Prenatal Time Day	Hr.	Postnatal Time Days
Alisphenoid	15	11		Orbitosphenoid	18	6	
Basihyal	17	1		Palatine	15	11	
Basioccipital	15	11		Parietal	15	11	
Basisphenoid	15	11		Periotic	17	1	
Ceratohyal			17	Premaxilla	15	11	
Epihyal			21	Presphenoid	18(17)	6	
Ethmoid			2(1)	Pterygoid	15	11	
Exoccipital	15	11		Squamosal	15	11	
Frontal	15	11		Stylohyal			20
Interparietal	15	11		Supraoccipital	15(17)	11	
Jugal	15	11		Thyrohyal			4
Lacrimal	17(16)	1		Tympanic	16		
Mandibular	15	11		Tympanohyal			21
Maxilla	15	11		Vomers	15	11	
Nasal	17(16)	1					

ANTERIOR APPENDAGES	Prenatal Time Day	Hr.	Postnatal Time Day	Hr.	POSTERIOR APPENDAGES Bone	Prenatal Time Day	Hr.	Postnatal Time Day	Hr.
Clavicle:					Ilium:				
Primary Center	15(14)	0			Primary center	15	11		
Sternal Ossicle			35		Iliac crest			14 weeks	
					Ischium:				
Scapula:					Primary center	17(16)	1		
Primary center	15	11			Pubis:				
Coracoid			5(2)	11	Primary center	17(16)	1		
Subcoracoid			8(7)		Pectineal eminence	17	1		
Acromion			35		Os Acetabulum			12 days	
					Femur:				
Humerus: :					Primary center	15	11		
Primary center	15	11			Greater trochanter			14(11)	
Greater tuberosity			5(4)	11	Lesser trochanter			14(11)	
Head			7(7)	10	Head			15(14)	
Capitellum			5(3)	11	Medial condyle			7	
Trochlea			5(3)	11	Lateral condyle			9	
Medial condyle			9(7)		Tibia:				
Lateral condyle			19(7)		Primary center	15	11		
					Distal epiphysis			9(6)	
Ulna:					Proximal epiphysis			18(7)	
Primary center	15	11			Greater tuberosity			18(16)	
Distal epiphysis			5(3)	11	Fibula:				
Proximal epiphysis			5(5)	11	Primary center	15	11		
					Distal epiphysis			7(6)	
Radius:					Proximal epiphysis			21(17)	
Primary center	15	11			Patella			18	
Distal epiphysis			5(3)	11	Fabella			18	
Proximal epiphysis			9(7)		Semilunar cartilage) medial			19(15)	
					) lateral			(18)	

* Adapted from M. L. Johnson, Am. Anat. 52 241-271 (1933). Dates in parenthesis are different, given by Z. T. Wirt-schafter in "The Genesis of the Mouse Skeleton: A Laboratory Atlas." 1960, C. C. Thomas, Springfield, Illinois.

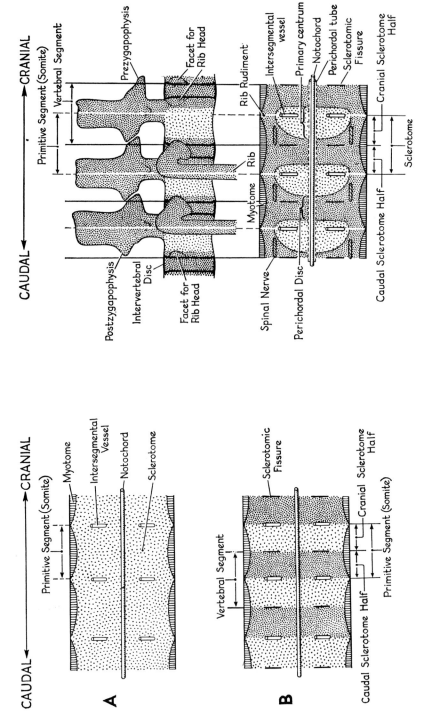

CRANIAL

CAUDAL

Prezygapophysis

Facet for Rib Head

Vertebral Segment

Primitive Segment (Somite)

Postzygapophysis

Intervertebral Disc

Facet for Rib Head

Rib

Myotome

Rib Rudiment

Intersegmental vessel

Primary centrum

Notochord

Perichordal tube

Sclerotomic Fissure

Cranial Sclerotome Half

Sclerotome

Caudal Sclerotome Half

Spinal Nerve

Perichordal Disc

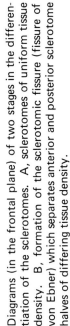

Diagrams illustrating the participation of the sclerotome-halves in the formation of the definitive vertebrae and intervertebral discs. In the upper half, vertebrae and ribs are shown as seen from the right side; in the lower half of the diagram, the sclerotomic origin of the various parts is shown in a frontal section.

(Sensenig 1949, Grüneberg 1963)

A

CAUDAL

CRANIAL

Primitive Segment (Somite)

Myotome

Intersegmental Vessel

Notochord

Sclerotome

B

Sclerotomic Fissure

Vertebral Segment

Caudal Sclerotome Half

Cranial Sclerotome Half

Primitive Segment (Somite)

Diagrams (in the frontal plane) of two stages in the differentiation of the sclerotomes. A, sclerotomes of uniform tissue density. B, formation of the sclerotomic fissure (fissure of von Ebner) which separates anterior and posterior sclerotome halves of differing tissue density.

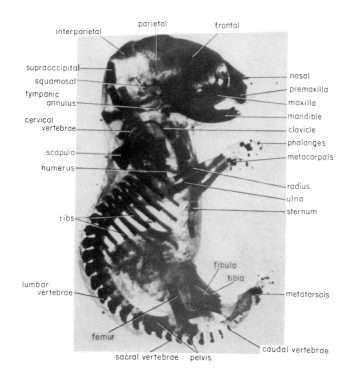

SKELETON OF MOUSE EMBRYO at 18.5 days

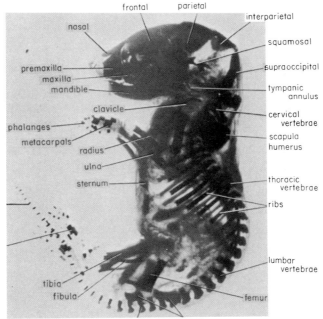

SKELETON OF MOUSE EMBRYO
Newborn

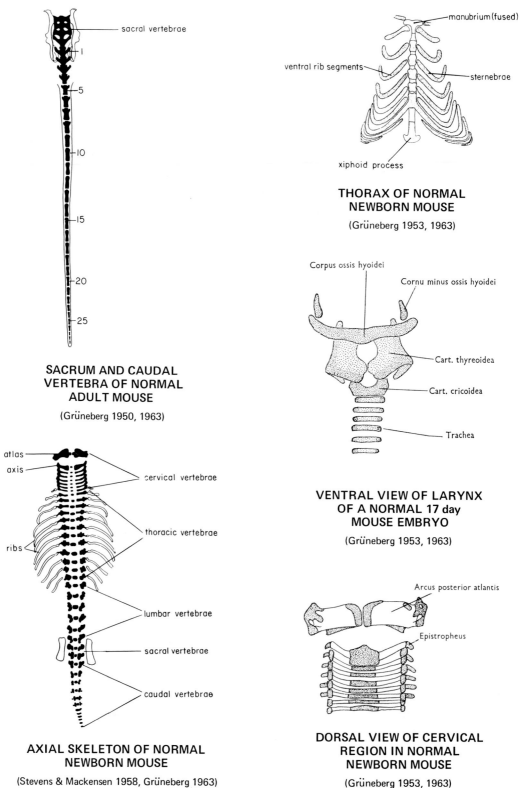

SACRUM AND CAUDAL VERTEBRA OF NORMAL ADULT MOUSE

(Grüneberg 1950, 1963)

sacral vertebrae

THORAX OF NORMAL NEWBORN MOUSE

(Grüneberg 1953, 1963)

manubrium (fused)

ventral rib segments

sternebrae

xiphoid process

Corpus ossis hyoidei

Cornu minus ossis hyoidei

Cart. thyreoidea

Cart. cricoidea

Trachea

VENTRAL VIEW OF LARYNX OF A NORMAL 17 day MOUSE EMBRYO

(Grüneberg 1953, 1963)

atlas
axis
cervical vertebrae
thoracic vertebrae
ribs
lumbar vertebrae
sacral vertebrae
caudal vertebrae

AXIAL SKELETON OF NORMAL NEWBORN MOUSE

(Stevens & Mackensen 1958, Grüneberg 1963)

Arcus posterior atlantis

Epistropheus

DORSAL VIEW OF CERVICAL REGION IN NORMAL NEWBORN MOUSE

(Grüneberg 1953, 1963)

TOOTH DEVELOPMENT: The embryonic and maternal genotypes are both relevant here. Prenatal (genetic and nutritional) factors are of prime importance, as is also lactation, in determining the growth and ultimate size of the teeth. Inadequate mesenchyme building material may result in failure to achieve a minimum level of growth and, therefore, in a lack of teeth. Lactation is genetically controlled, but can be influenced by the environment.

The mouse has a monophyodont dentition and a dental formula of: incisors 1/1; cuspids 0/0; premolars 0/0; and molars 3/3, making a total of 16. The first molars arise first, followed by the second and then the third, the latter at 4 to 6 days after birth. The first molars are the largest and the third the smallest. The development of the molars in the upper jaw lags behind those in the lower jaw by 12 to 24 hours, and the third molars may be missing in the first litter of any female.

Before gestation day 12 there is no indication of the differentiation of the dental lamina, which is the first of normal dental structures to appear. Molar development will be described in gestational sequence. *

Day 12: The first indication of odontogenesis is the stratification of the oral epithelium along the free margin of the jaws in the vicinity of the future dental arches. The dental lamina consists of 5 layers of cuboidal cells at the most. The remaining oral epithelium has one or two layers of cuboidal cells. There are numerous mitotic figures in both the oral ectoderm and the underlying mesoderm.

* Based upon studies by Dr. S. A. Cohn, 1957, Am. Jour. Anat. 101:295-320.

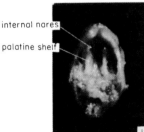
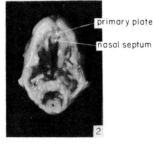
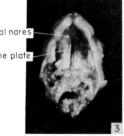

internal nares
palatine shelf

primary plate
nasal septum

internal nares
palatine plate

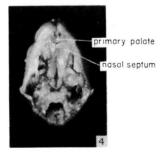

primary palate
nasal septum

Fig. 1 — Normal palate, stage #1, inbred C57BL and A/Jax mice.
Fig. 2 — Normal palate, stage #3.
Fig. 3 — Normal palate, stage #4.
Fig. 4 — Normal palate, stage #6. The palatine shelves have already begun to fuse.

Courtesy G. Callas and B. F. Walker, 1963, Anat. Rec. 145:61-71. Chronological age of all four mice was between 14 days and 7 hours and 15 days and 6 hours, with staging based upon actual development of the palate.

PALATE DEVELOPMENT IN THE MOUSE

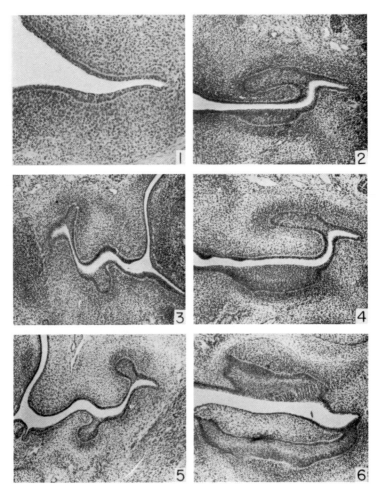

PRENATAL DEVELOPMENT OF MOUSE MOLARS — 12 to 15 days

Fig. 1 — Day 12. Sagittal section. Thickening of oral epithelium in region of lower jaw denotes beginning of dental lamina.

Fig. 2 — Day 13. Sagittal section. Note thickening of dental lamina in both jaws. Adjacent mesenchyme cells stain darkly.

Fig. 3 — Day 13. Frontal section of dental lamina.

Fig. 4 — Day 14. Sagittal section. Note the growth in length and thickness of the dental lamina as compared to Fig. 2.

Fig. 5 — Day] 4. Frontal section. Tooth buds for upper and lower first molars are present.

Fig. 6 — Day 15. Sagittal section. Note the marked increase in length of the dental lamina. The shallow invagination on its deep surface marks the site of the enamel organs of the first molars.

(Courtesy Dr. S. A. Cohn, Am. Jour. Anat. 101:295-320, 1957.)

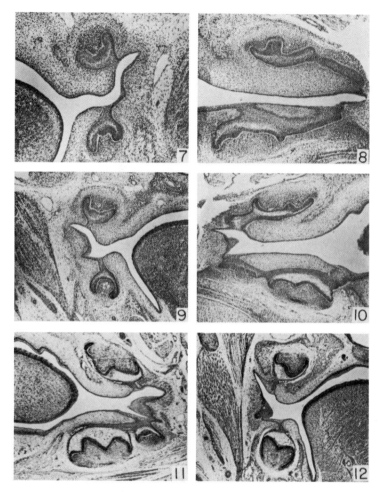

PRENATAL DEVELOPMENT OF MOUSE MOLARS — 15 to 18 days

Fig. 7 — Day 15. Frontal section. Cap stage of enamel organ of upper and lower first molars.

Fig. 8 — Day 16. Sagittal section. Enamel organs of the first molars can be seen on each dental lamina.

Fig. 9 — Day 16. Frontal section. Note the deepening of the invagination in the enamel organ of the lower first molar and the thinning out of its connection to the dental lamina.

Fig. 10 — Day 17. Sagittal section. Observe the increase in size of the enamel organs of the first molars. Note epithelial island in lower molar.

Fig. 11 — Day 18. Sagittal section. Enamel organs of the first molars are in the bell stage. Note the increase in stellate reticulum. The second lower molar is also visible.

Fig. 12 — Day 18. Frontal section. Enamel organs of the first molars are nearly separated from the dental lamina. Note how the outer enamel epithelium has thinned out.

(Courtesy Dr. S. A. Cohn, Am. Jour. Anat. 101:295-320, 1957.)

Day 13: Cell proliferation causes the thickening and elongation of the dental lamina. Growth in length parallels growth of the jaws. The dental lamina consists largely of cuboidal cells, while its basal layer consists of low columnar cells whose nuclei are more or less oriented toward the oral cavity. The relatively clear cytoplasm of these cells lies next to the underlying mesoderm. The originally cuboidal cells on the oral surface of the dental lamina are now more flattened.

Day 14: The length of the dental lamina is increased by anterior-posterior growth. Both upper and lower first molars now show enamel organs. A club-shaped tooth bud is formed by rapid cell proliferation on the basal surface of the dental lamina. Around the base of the bud are deeply staining mesodermal cells. Bone is beginning to form.

Day 15: The lengthening of the dental lamina continues. The tooth buds of the first molars show invaginations on their deep surfaces. This invagination is filled with deeply staining mesodermal cells which initiate the formation of the dental papilla and vascular elements. Between the central cells of the enamel organ a stellate reticulum appears, which defines an outer and inner enamel epithelium. The primordium of the dental sac may be seen as two or three layers of flattened mesenchymal cells surrounding the enamel organ.

Day 16: The size of the enamel organs of the first molars is now markedly increased by proliferation of cells from the inner and outer enamel epithelium. The intercellular spaces of the stellate reticulum is increased. The outer enamel epithelium is reduced to a thin layer of low columnar cells lying over a few layers of squamous cells. The inner enamel epithelium is but a single layer of columnar cells whose nuclei show no uniformity in position. These cells increase in height during proliferation. The future dentino-enamel junction is seen as a narrow eosinophilic zone between the inner enamel epithelium and the adjacent cells of the dental papilla. The expanding invagination of the enamel organ continues to be filled by the proliferation of cells of the dental papilla. The well-defined dental sac consists of several layers of flattened cells oriented closely around the enamel organ. Cells within the dental papilla and adjacent to the dental sac are bathed in increasing vascularity. The cell stalk connecting the enamel organ and the dental lamina becomes tenuous. Bone appears around the tooth germs of the first molars. Tooth buds of the second molars now make their appearance, to which the dental lamina extend posteriorly.

Day 17: There is accentuation of growth of the enamel organs of the first molars, especially in the cervical region, increasing the size of the pattern of the future crown. The stellate reticulum is also expanded, both by cell proliferation and enlargement of the interstices. The stratum intermedium first appears as a layer of flattened cells between the stellate reticulum and the inner enamel epithelium. An epithelial island is first visible, and will participate in root formation during post-natal development. There is cell proliferation in both the inner enamel epithelium and the dental papilla. Enamel organs of the second molars are now in the cap stage, similar to those of the first molars on day 15.

Day 18: The enamel organs of the first molars are now in the bell stage, beginning to form the pattern of the cusps and the outline of the future dentino-enamel junction. There are now no more than three layers of cuboidal cells in the outer enamel epithelium. Vascular sprouts and fibroblasts invade the outer enamel epithelium and the stellate reticulum. The inner enamel epithelial cells continue to proliferate, and capillaries of the dental papilla are seen. The connection between the enamel organ and the dental lamina is all but lost. The enamel organs of the second molars are comparable to those of the first molars on gestation day 16.

Days 19 and 20: The crown pattern of the enamel organs of the first molars is almost completed. Proliferation in the cervical region persists for some time after birth. Certain cells of the inner enamel epithelium are transformed into fully differentiated ameloblasts which more than double their height and their nuclei become oriented farthest from the future dentino-enamel junction. Ameloblast development begins at the high points which correspond to the growth centers on the cusps. Some of the cells of the inner enamel epithelium remain in the low columnar state, and may be seen extending along an entire margin of each cusp near its apex, invaginating pulpward to form one or more shallow depressions which disrupt the even contour of the future dentino-enamel junction, leaving the pulpal surface convex. The invagination appears to carry inward the cells of the stratum intermedium and stellate reticulum. Cells from the dental papilla differentiate on day 20 into odontoblasts, aligning themselves opposite the ameloblasts, and begin to form predentin. Odontoblasts differentiate next to the convex surface of the cord of undifferentiated cells of the inner enamel epithelium and a concavity of predentin forms along an entire margin of each cusp to be enamel-free during subsequent development. These changes occur in all molars. The squamous cells of the stratum intermedium become more cuboidal opposite ameloblasts undergoing differentiation but remain unchanged opposite the cord of undifferentiated cells. The second molars reach this stage of development by day 18, as the bony encapsulation of the first molars is almost completed and extends to the second molars.

TABLE 6

TABULAR SUMMARY OF MOLAR DEVELOPMENT IN THE MOUSE*

	MOLAR 1	MOLAR 2	MOLAR 3
Tooth bud appearance	Gestation day 14	Gestation day 16	Post-natal day 4 to 6
Initial dentinogenesis	Gestation day 20	Post-natal 2	Post-natal 10
Initial amelogenesis	Post-natal 1, 2	Post-natal 3, 4	Post-natal 11, 12
Eruption to oral cavity	Post-natal 16, 17	Post-natal 18, 19	Post-natal 28, 29
Functional occlusion	Post-natal 24	Post-natal 25	Post-natal 35
Cusps: upper/lower	3/5	3/4	3/3
Roots: upper/lower	3/2	3/2	3/1

* Modified from Cohn: Am. Jour. Anat. 101:295-320, 1957.

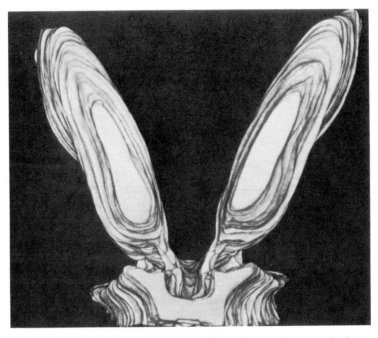

Photograph of wax plate reconstruction of normal upper incisor
germs, epithelial and oral epithelium. Labial view.
(Courtesy P. A. Knudsen, Aarhus from Acta Odontologica 23:391–409, 1965)

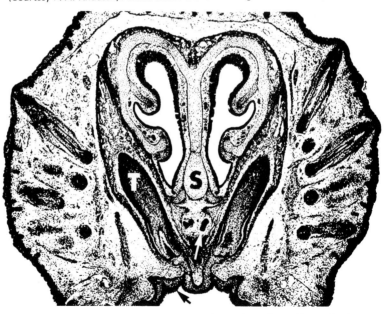

Normal mouse embryo. 18 days old. Frontal section. Upper incisor
germs.
T = tooth germ
S = septal cartilage
White arrow = infraseptal nasal glands
Black arrow = oral epithelium
(H.E.).
(Courtesy P. A. Knudsen, Aarhus from Acta Odontologica 24:55–77, 1966)

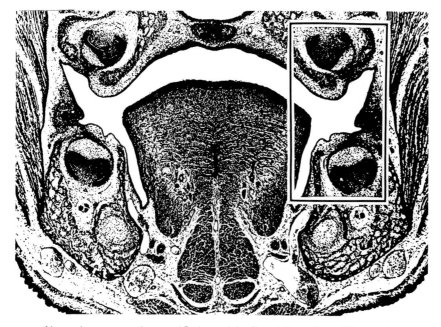

Normal mouse embryo. 18 days old. Frontal section. First molar germs in the upper and lower jaws. Note the position of the tooth germs and the size of the vestibule of the oral cavity (in the frame)
(Courtesy P. A. Knudsen, Aarhus from Acta Odontologica 24:55–77, 1966)

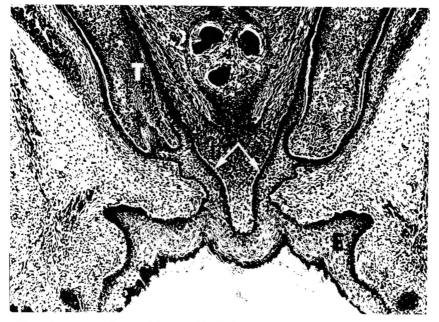

Normal mouse embryo. 18 days old. Epithelial laminae connecting the incisor germs and the oral epithelium. The portion near the tooth germ comprises stellate reticulum and outer dental epithelium, whilst the remainder has the same appearance as the epithelium in the oral cavity (arrows show the transition).
T = tooth germ
E = oral epithelium
(Courtesy P. A. Knudsen, Aarhus, from Acta Odontologica 23:71–89, 1965)

In summary, molar development in the mouse is not completed until about day 35 after birth. The number of cusps (upper/lower) for the three molars is: 3/5, 3/4, and 3/3 and the number of roots is 3/2, 3/2, and 3/1. The enamel-free areas typical of rodent molars are initiated before birth because of the failure of certain ameloblasts to undergo cyto-differentiation. The squamous cells of the stratum intermedium appear to be related to the cyto-differentiation of the ameloblasts under which condition they become cuboidal.

For completion of molar development during post-natal days 0-35 see the original reference. Dr. Cohn writes: "The mouse molar is an excellent and readily available source of didactic material of all periods of tooth development for use by students."

The incisor primordia arise at 13 days, just posteriorly to the lip furrow band. At 14 days the inner and outer enamel epithelia are differentiated, and nearby is Meckel's cartilage. Within each primordium is a stellate reticulum. By 15 days the dental papillae appear, as the inner and outer enamel epithelia are expanded, more particularly in the ventral incisors. By 16 days all of the aforementioned structures have proliferated, the surrounding mesenchyme is being organized into bone, and ameloblasts and odontoblasts ring the papillae. At 18 days the alveolar bone formation is prominent, incorporating the abundant surrounding mesenchyme. Complete differentiation of the distal and intermediate portions of the incisors depends upon association with the proximal portions. This results in complete and integrated parts for the upper and lower incisors in the mouse.

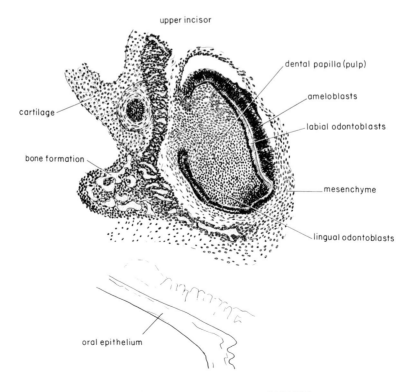

INCISOR DEVELOPMENT — MOUSE

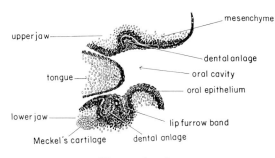

13 gestation days

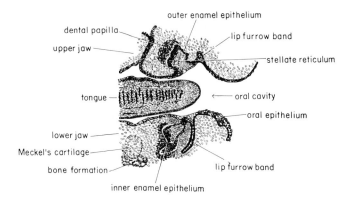

14 gestation days

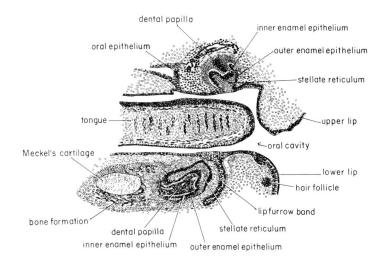

15 gestation days

INCISOR DEVELOPMENT — MOUSE

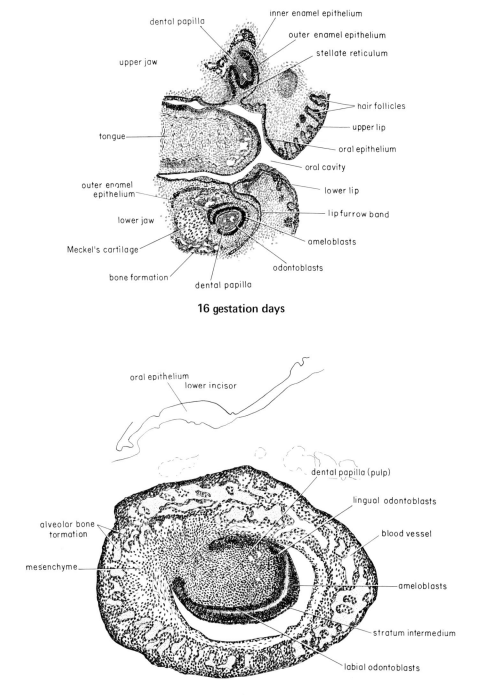

dental papilla

inner enamel epithelium

outer enamel epithelium

stellate reticulum

upper jaw

hair follicles

upper lip

oral epithelium

tongue

oral cavity

outer enamel
epithelium

lower lip

lip furrow band

lower jaw

ameloblasts

Meckel's cartilage

odontoblasts

bone formation

dental papilla

16 gestation days

oral epithelium

lower incisor

dental papilla (pulp)

lingual odontoblasts

blood vessel

alveolar bone
formation

mesenchyme

ameloblasts

stratum intermedium

labial odontoblasts

18 gestation days

INCISOR DEVELOPMENT — MOUSE

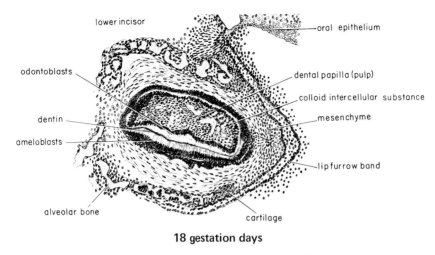

lower incisor — oral epithelium

odontoblasts — dental papilla (pulp)

colloid intercellular substance

dentin — mesenchyme

ameloblasts

lip furrow band

alveolar bone

cartilage

18 gestation days

INCISOR DEVELOPMENT — MOUSE

BRAIN: The central nervous system is the first system to develop and to differentiate, and one of the last to be completed. The cerebellum, for instance, is not entirely differentiated until some days after birth. Grossly, however the primary parts of the brain can be identified soon after the neural groove, neural plate, and head process stage at $7\frac{1}{2}$ days and by 14 days the brain is typically that of a mammal. Development is very rapid. At 7 3/4 days the neural plate has its limiting folds, and the otic disc and preotic transverse sulcus are visible. At 8 days the neural tube is closed in part, neural crests have spread ventrolaterally and cranial ganglia V, VII, VIII, IX, and X are apparent. At $8\frac{1}{2}$ days the anterior neuropore begins to close and the optic sulcus appears. Then the three primary brain vesicles and the cranial flexure form; the rhombencephalon acquires neuromeres; the cranial ganglia separate from the neural tube, and the trigeminal ganglion (V) becomes very large with an ophthalmic branch. At 9 days the otic vesicle is closed; Rathke's pocket can be seen; the roof of the rhombencephalon is thin; the neural crests are independent units; the lamina terminalis is present; the anterior neuropore is closed; and the posterior neuropore is closing. At $9\frac{1}{2}$ days the posterior neuropore closes, and at 10 days the telencephalic evaginations are evident. The brain is now subdivided into five major sections, the trigeminal ganglion has its three main branches (ophthalmic, maxillary, and mandibular), the cranial ganglia are distinct, and the acoustic (VIII) nerve is in direct contact with the otic vesicle. At $10\frac{1}{2}$ days the infundibulum is an outpocketing of the diencephalon, the lamina terminalis is well defined, the facial ganglion (VII) sends its nerve to the brain, the cervical spinal ganglia are metameric, and cranial nerves V to X all have established their connections with the brain.

At 11 days the infundibulum begins to pinch off from the diencephalon, the roof of the fourth ventricle is very thin, and the cervical flexure appears. The first cervical ganglion degenerates, and the second cervical ganglion is recognizable at the level of the sixth somite. The olfactory nerve (I) fibers reach to the telencephalon, on the underside of which are the short, anteriorly projecting olfactory bulbs, really extensions of the brain. The diencephalon is obscured from above, except for a slight portion on each side, by the large and bulbous single-lobed mesencephalon. In lateral view the diencephalon can be identified by the optic chiasma and infundibulum. The metencephalon (cerebellum) forms a transverse fold posterior to the mesencephalon, and just posterior to its center is the triangular and lucid (because it is so thin) roof of the myelencephalon, which tapers rapidly into the spinal cord. At $12\frac{1}{2}$ days the cerebellum is much thickened, the posterior choroid plexus is apparent in the roof of the myelencephalon, the pontine flexure is present, the telencephalic vesicles are enlarging, and the spinal ganglia are visible to about the level of somite 40. At 13 days oculomotor nuclei appear; the cerebral hemispheres partly cover the diencephalon; the anterior choroid plexi project into the first and second ventricles; and the corpora striata, epiphysis, optic thalamus, and foramen of Monro can be seen.

By 15 days gestation all these parts are further developed, especially by enlargement, but the basic relations remain unchanged. In lateral view the cerebellum is tucked among mesencephalon, myelencephalon, and spinal cord. Ventrally both the optic chiasma and infundibulum are clearly visible. The optic nerves to the brain are distinct, the foramen of Monro is reduced, there is a choroid plexus in the third ventricle, and the pontine flexure has straightened the body. At 17 days the olfactory nerve (I) and the optic nerve (II) emerge from telencephalon and diencephalon respectively, the cerebellum is enlarged but not yet folded; and the infundibulum persists. The greatest growth has occurred in the telencephalon which will give rise to the cerebral hemispheres.

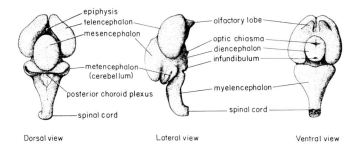

14 g.d. — 6 mm.

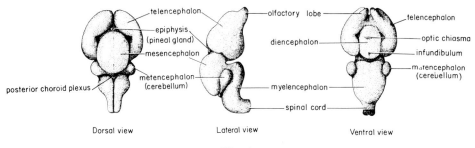

15 g.d.

MOUSE BRAIN

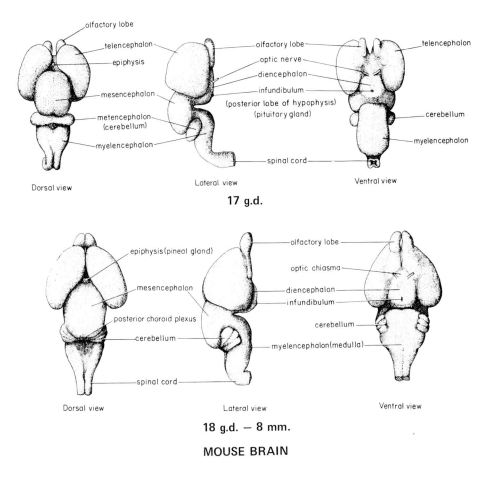

Dorsal view   Lateral view   Ventral view

17 g.d.

Dorsal view   Lateral view   Ventral view

18 g.d. — 8 mm.

MOUSE BRAIN

At 18 days the brain is essentially complete in general topography. The telencephalon is separated into two cerebral lobes, or hemispheres, the olfactory lobes are prominent, and the cerebellum is beginning its characteristic transverse folding. At 19 days it has both shallow and deep transverse folds; although it is not fully differentiated until some time after birth.

Through recent studies (Nimi et al '61) it is possible to describe in detail the development of one part of the brain, the diencephalon of the mouse.

The structural changes of the diencephalon from gestation day 10 onward can be divided principally into the development of the epithalamus, dorsal thalamus, ventral thalamus, and hypothalamus all by the ventricular sulci. Representative mid-transverse sections from this original source have been redrawn for reproduction here. These four zones can be distinguished from each other by the ventricular sulci or eminences as well as by other structural changes that appear during development (see opposite page).

There are three stages of nuclei development in the diencephalic wall. The second stage, from gestation day 13 to 15, shows differentiation of the various layers: germinal, mantle, and marginal. These layers increase in volume while they show cellular differentiation into distinct nuclear groups.

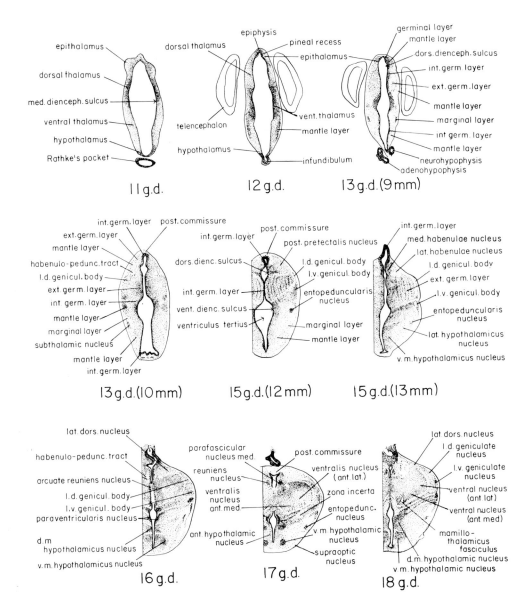

**ONTOGENETIC DEVELOPMENT OF THE
DIENCEPHALON OF THE MOUSE**

(Courtesy of Niimi, Harada, Kusada, and Kishi 1961.)

This differentiation seems to progress from caudal to cephalic levels of the diencephalon. During the third stage, from gestation day 16 until after birth, the orderly structure within the original three layers is lost in favor of differentiation into special central nervous system nuclei. The dorsal thalamus shows this differentiation last.

The *epithalamus* is large on gestation day 13, but decreases in size and at day 16 the median habenular nucleus arises from the external germinal layer, and the lateral habenular nucleus arises from the mantle layer. It appears that the pretectalis and posterior nuclei may arise from the mantle layer also, while the posterior commissure nucleus arises from the external germinal layer.

The *dorsal thalamus* is rather poorly developed at gestation days 11 and 12, but shortly increases in volume. By the 16th day the internal germinal layer differentiates into the median group of nuclei known as the reuniens arcuatus and reuniens medianus. The massa intermedia is formed by day 17 by the fusion of the internal germinal layer from both sides. During days 17 and 18 the external germinal layer differentiates into the anterior and mesial groups of nuclei. In the anterior group are the parataenialis and both the anterior dorsal and ventral nuclei, but the anterior medialis nucleus has not yet formed. The mesial group differentiates into the paramedianus, parafascicularis, laminaris, and median dorsal nuclei. From the mantle layer there develop the latero-dorsal geniculate body and on days 15 and 16 the median dorsal geniculate body. During days 16 to 18 the latero-dorsal and ventral nuclei are gradually developed within the mantle layer.

The *ventral thalamus* is well developed by 12 days gestation, but then it begins to decrease in volume. All of the nuclei of this ventral thalamus develop from the mantle layer. By days 15 and 16 there can be distinguished the reticular nucleus, the zona incerta, the latero-ventral geniculate body, and the median ventral geniculate body. Before this, on day 13, can be seen the massa cellularis reuniens thalami.

By 11 days gestation the *hypothalamus* is not well developed, but shortly increases in volume. By the 16th day the internal germinal layer differentiates into the periventricular group of nuclei; the preopticus, the hypothalamicus, the suprachiasmaticus, and the supraopticus nuclei. The mantle layer differentiates on days 15 and 16 into the mesial group of nuclei: the massa cellularis reuniens hypothalami, the median preopticus, the anterior hypothalamicus, the ventro-medial hypothalamicus, the dorso-medial hypothalamicus, the dorsal hypothalamicus, the posterior hypothalamicus, as well as nuclei for the mammillary body. The mammillary body includes the magnocellularis, parocellularis, intercalatus, supramamillaris, and the premamillaris dorsal and ventral nuclei. The anlage of the mamillary body is first seen on gestation day 13. Most of the paraventricular nucleus arises from the mantle layer. During days 15 and 16 the marginal layer differentiates into the lateral group of nuclei: the lateral preoptic, the lateral hypothalamicus, the mamilloinfundibularis, the subthalamicus, and the entopeduncularis nuclei.

In general, the development of the diencephalic nuclei of the mouse is quite similar to that of other vertebrates studied.

SPINAL CORD: The earliest indication of the spinal cord is the neural plate seen at $8\frac{1}{2}$ days. The posterior neuropore closes at 9 days. At 10 days the neural groove is closed into a tube in limited regions, forming a large neural canal and the relatively thick layer of surrounding neuroblasts. The neural tube extends temporarily into the tail but decreases in diameter and disappears there by 13 days. It occurs as a completed structure in the tail, not developing from neural plate as in the body. Neural crests appear in the body simultaneously with the closing of the neural tube. More laterally are the paired somites, the most anterior of which may show early differentiation. By 11 days the spinal cord has thickened by the proliferation of its cells, and the ependymal layer, ventral horns, spinal ganglia, and spinal nerves are visible. Abundant sclerotomal mesenchyme is distributed around the cord, except dorsally, and encloses the ventrally located remnant of the notochord. By 12 days the texture of the cord is quite typical in transverse sections with marginal white and more central gray mantle layers and an ependymal layer that is thicker dorsally than elsewhere. Spinal ganglia are prominent with both dorsal and ventral roots. Development proceeds with histological and cytological refinement of parts. Neuroblasts fill the expanding mantle layer, spinal arteries arise both dorsally and ventrally, the invading centra entirely obliterate the notochord and cartilaginous neural arches appear. The cord is depressed from the dorsal skin, and mesenchyme completely surrounds it. At 17 days it could satisfy the requirements of any adult vertebrate cord, even to the presence of sympathetic ganglia. The original neural tube persists as a very small subcentral hole surrounded by a thin layer of ependymal cells. At 18 days the spinal cord is structurally and functionally complete.

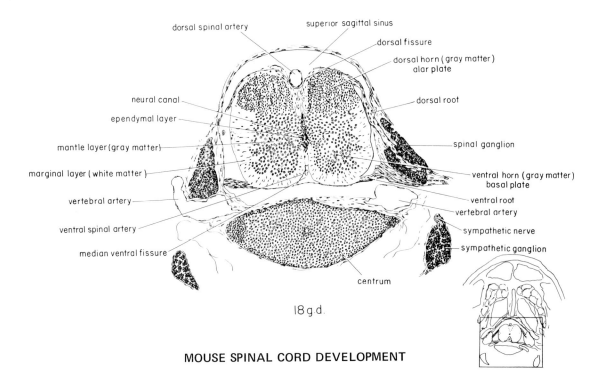

MOUSE SPINAL CORD DEVELOPMENT

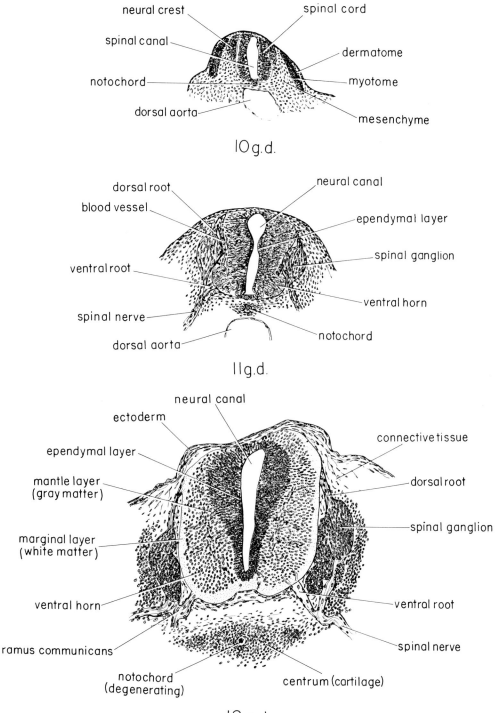

neural crest — spinal cord

spinal canal

notochord

dorsal aorta

dermatome

myotome

mesenchyme

10 g.d.

dorsal root — neural canal

blood vessel

ventral root

spinal nerve

dorsal aorta

ependymal layer

spinal ganglion

ventral horn

notochord

11 g.d.

neural canal

ectoderm

ependymal layer

mantle layer
(gray matter)

marginal layer
(white matter)

ventral horn

ramus communicans

notochord
(degenerating)

connective tissue

dorsal root

spinal ganglion

ventral root

spinal nerve

centrum (cartilage)

12 g.d.

MOUSE SPINAL CORD DEVELOPMENT

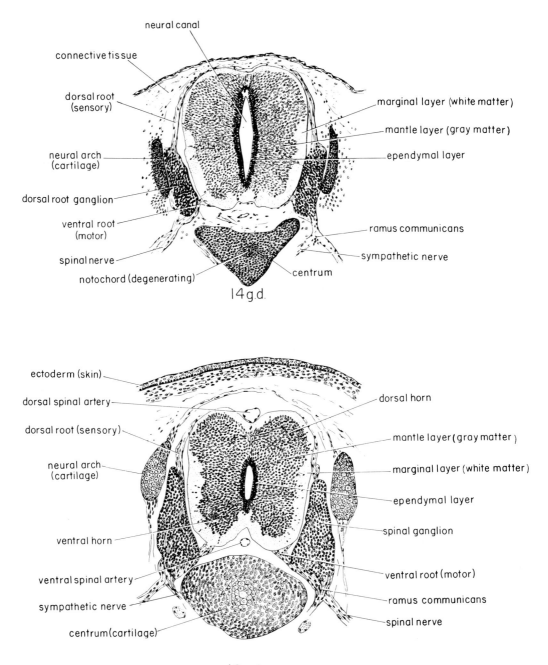

neural canal

connective tissue

dorsal root (sensory)

neural arch (cartilage)

dorsal root ganglion

ventral root (motor)

spinal nerve

notochord (degenerating)

marginal layer (white matter)

mantle layer (gray matter)

ependymal layer

ramus communicans

sympathetic nerve

centrum

14 g.d.

ectoderm (skin)

dorsal spinal artery

dorsal root (sensory)

neural arch (cartilage)

ventral horn

ventral spinal artery

sympathetic nerve

centrum (cartilage)

dorsal horn

mantle layer (gray matter)

marginal layer (white matter)

ependymal layer

spinal ganglion

ventral root (motor)

ramus communicans

spinal nerve

16 g.d.

MOUSE SPINAL CORD DEVELOPMENT

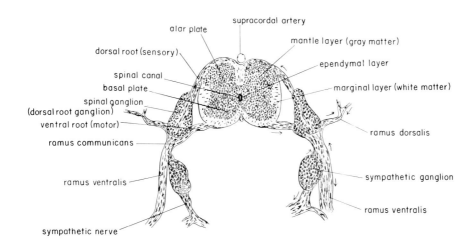

SPINAL CORD AND NERVES OF MOUSE
17 gestation days (transverse section)

SENSE ORGANS:

The Eyes (Optic Apparatus): At 8 days gestation the optic vesicles appear, the optic stalk and double-layered cup begin to close on day $9\frac{1}{2}$, the lens vesicle first invaginates on day 10. By day $10\frac{1}{2}$ the optic cup first comes into contact with the ectoderm of the head, the optic stalk shortens, the choroid fissure forms and the nervous layer of the retina is prominent. On day 11 the lens vesicle closes off from the head ectoderm. By $11\frac{1}{2}$ days the lens shows outer epithelial and inner fibrous parts, the inverted choroid fissure is deep, the hyaloid vessels and plexi are formed, the vitreous mesenchyme accumulates and the nervous layer of the retina is at least 4 times the thickness of the pigmented layer. Pigment appears in the retinal region and the posterior lens cells elongate. On day 12 the choroid fissure closes, and by $12\frac{1}{2}$ days the lens becomes separate from the epidermis and drops below the surface. Its fibrous part is expanding at this time. The vitreous humor forms and mesenchyme intercepts the lens and the epidermis. On day 13 lens fibers invade the lens vesicle, the primary lid fold begins to form, nerve fibers appear in the optic stalk, and the oculomotor nuclei form. By $13\frac{1}{2}$ days the lens expands vertically more than laterally, the anterior chamber of the eye forms, the nervous layer of the retina is at least 8 times as thick as the pigmented layer, the optic (II) nerves can be identified, the iris forms, and the upper eyelids and the ocular muscles arise.

By 14 days the lens epithelium, facing the anterior chamber, is cellular and becomes thinner as the lens enlarges. The posterior chamber contains scattered cells and the optic nerve (II) is clearly defined. The lower eyelid also forms, growing toward the upper eyelid, with which it will fuse to cover the eye before birth. By $14\frac{1}{2}$ days both corneal and scleral mesenchyme form around the eyeballs.

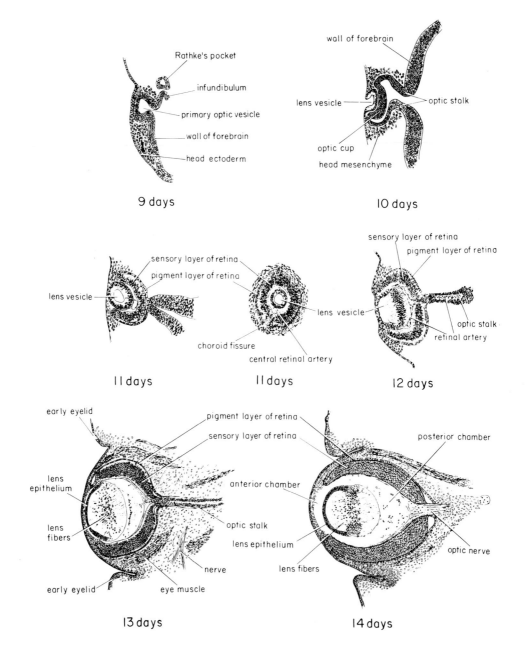

SECTIONS OF THE EYES OF THE EMBRYONIC MOUSE

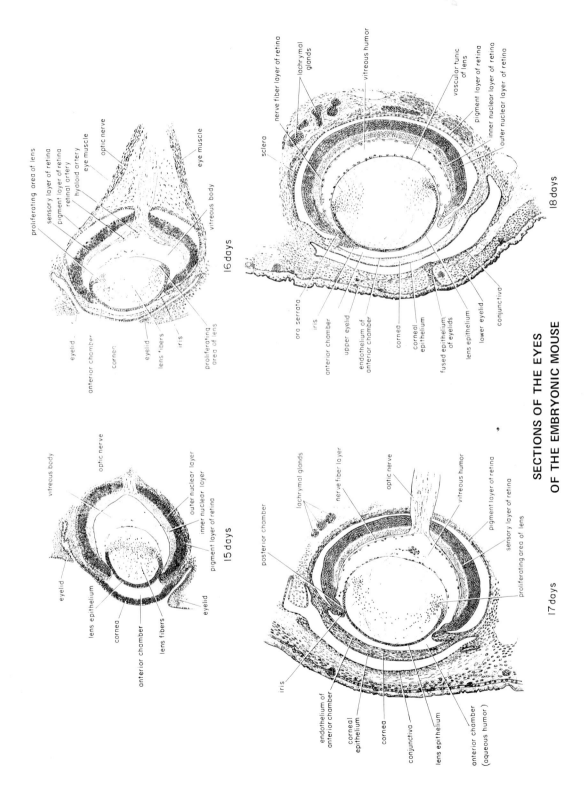

16 days

proliferating area of lens
sensory layer of retina
pigment layer of retina
retinal artery
hyaloid artery
eye muscle
optic nerve
eye muscle
vitreous body
proliferating area of lens
iris
lens fibers
eyelid
cornea
anterior chamber
eyelid

18 days

sclera
nerve fiber layer of retina
lachrymal glands
vitreous humor
vascular tunic of lens
pigment layer of retina
inner nuclear layer of retina
outer nuclear layer of retina
ora serrata
iris
anterior chamber
upper eyelid
endothelium of anterior chamber
cornea
corneal epithelium
fused epithelium of eyelids
lens epithelium
lower eyelid
conjunctiva

15 days

vitreous body
optic nerve
outer nuclear layer
inner nuclear layer
pigment layer of retina
eyelid
eyelid
lens epithelium
cornea
anterior chamber
lens fibers

17 days

posterior chamber
lachrymal glands
nerve fiber layer
optic nerve
vitreous humor
pigment layer of retina
sensory layer of retina
proliferating area of lens
iris
endothelium of anterior chamber
corneal epithelium
cornea
conjunctiva
lens epithelium
anterior chamber (aqueous humor)

SECTIONS OF THE EYES
OF THE EMBRYONIC MOUSE

By 15 days the retina begins to show cellular differentiation into the inner and outer nuclear layers, as the lens fibers further elongate and the optic nerve increases in diameter. The epidermis covering the lens begins to differentiate into the cornea. The pigmented layer is distinct but not yet highly pigmented. By 16 days the pigmented and retinal layers are separated, the iris forms, and the lens becomes elongate-oval in a dorso-ventral direction. The lens vesicle has disappeared except at the margins of the lens epithelium. The vitreous humor accumulates and a large blood vessel appears near the entrance of the hyaloid artery. The eyelids fuse, the cornea and sclera show stratified epithelia, and neuroblasts appear in the retina.

Day 17 sees the first appearance of the lachrymal glands in the mesenchyme near the eye, the iris is well advanced and the ora serrata can be identified. Beneath the closed eyelids there is differentiation of the layers into the endothelium of the anterior chamber, the cornea, and the outermost corneal epithelium. The innermost layer of the closed eyelids becomes the conjunctiva. The retina is further differentiated into its several layers (the mouse has no cones) and a nerve fiber layer lines the retina from the optic nerve. A ciliary ring is formed. By 18 days the eye is essentially complete, with a vascular tunic around the lens, the inner and outer nuclear layers of the retina clearly distinct, and the sclera encapsulating the eye. The cornea is about 3 times as thick as the sclera, and the iris is fully formed.

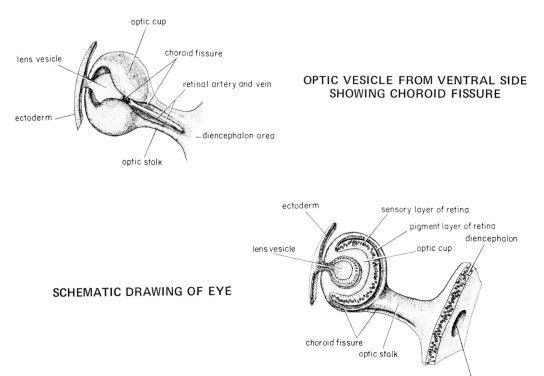

OPTIC VESICLE FROM VENTRAL SIDE SHOWING CHOROID FISSURE

SCHEMATIC DRAWING OF EYE

neatus is partly covered by the pinna forming from the third visceral arch. The tympanic cavities form from the remnants of the first visceral clefts. True cartilage does not appear in the otic capsule until about day $14\frac{1}{2}$, the utrico-saccular duct is reduced, the cochlear duct is further twisted and the pinna almost completely overlaps the outer auditory meatus. By 16 days the endolymphatic duct has expanded, the cochlea has 1 3/4 turns, the otic capsule is extensively chondrified, and the meatus is filled with periderm and covered by the pinna. By 17 days the ear parts are essentially complete with tympanic cavity, auditory tube, nasopharyngeal ostium, cartilaginous ossicles in mesenchyme, and a plugged meatus. By 19 days the otic capsules are ossified, and the pinna seals over the meatus.

The Nose (Olfactory Apparatus): By gestation day $8\frac{1}{2}$ the olfactory placodes form, the olfactory vesicles develop by day $10\frac{1}{2}$ and olfactory pits by day 11. By day $11\frac{1}{2}$ deep olfactory pouches (vomeronasal organs) are formed while mitral cells originate in the olfactory epithelium and corresponding triangular (Cajal) cells of the accessory bulb appear, sometimes even earlier. A few superficial tufted cells begin to appear, but not many are seen until later (about day 18). By day 12 short axon cells of Cajal are found and continue to appear for about 6 days more. On the 11th day there is rapid proliferation of granule cells of both the olfactory and accessory bulbs, and this proliferation continues for several weeks after birth. By day 12 the olfactory pouches narrow to the tube-like vomeronasal organ and by day $12\frac{1}{2}$ olfactory (I) nerves form. By day $13\frac{1}{2}$ the bucconasal partition is ruptured and the nasal capsule and septum become pre-cartilaginous. By $14\frac{1}{2}$ days choanae are slit-like, there is a long vomeronasal organ, thick walled tubes in the nasal septum, and a nasolachrymal duct. By day 15 the palatine processes meet mesially, and the nasopalatine duct is separated from the posterior choanae. By day 16 the palate is completed, there is a long nasopharyngeal meatus, secondary choanae, and the nares are plugged temporarily. On day 17 lateral nasal mucous glands appear and the nares open on day 19.

THE ENDOCRINE ORGANS:

The Thyroid, Parathyroid, and Ultimobranchial Bodies: At $8\frac{1}{2}$ days gestation the first indication of the median thyroid thickening in the floor of the foregut occurs with a distinct primordium of evaginating endoderm appearing a few hours later. By 10 days the duct forms. There develop paired lateral thyroid diverticulae and by 11 days there remains only a strand of tissue where the original thyroid evagination had formed, connecting the thyroid with the pharynx. Portions of the laryngotracheal groove from which originated the thyroid duct may remain in the floor of the foregut. The parathyroid primordia from the dorsal part of the 3rd visceral pouch, and the ultimobranchial bodies from the epithelium of the 4th to the 6th visceral pouches, appear by this time. By 12 days, the thyroid diverticulae lose their cavities and become cellular aggregates, as the parathyroids and ultimobranchial bodies become detached. By 13 days the paired thyroid gland becomes lobular and forms a crescent shaped collar around the lower pharynx, comprising the thin isthmus which is ventral to the 3rd tracheal ring. The parathyroids are now attached to the thyroid and the regressing ultimobranchial bodies are embedded in the lateral wings of the thyroid substance. By 17 days there are numerous follicles in the bi-lobed thyroid well supplied with blood vessels, and this endocrine gland is encapsulated. Since the thyroid follicles begin to show colloid the follicle stimulating hormone (FSH) must be elaborated by the pituitary gland (hypophysis). Tracheal cartilage intercepts the thyroid and the trachea. By 15 days the beginning of organization of the parathyroids may be seen. These are all but surrounded by thyroid follicles, but easily distinguished from them. The non-glandular parathyroids are dorso-laterally located and are separated from the thyroid substance by a layer of connective tissue. (See page 186 for illustrations of earlier stages of thyroid development.)

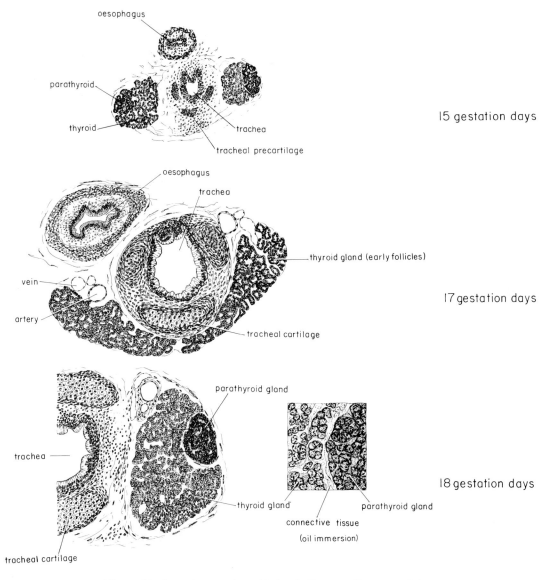

oesophagus

parathyroid

thyroid

trachea

tracheal precartilage

15 gestation days

oesophagus

trachea

thyroid gland (early follicles)

vein

artery

tracheal cartilage

17 gestation days

parathyroid gland

trachea

thyroid gland

parathyroid gland

connective tissue

(oil immersion)

tracheal cartilage

18 gestation days

(Note: To show the parathyroid glands these sections were taken anterior to the isthmus which therefore does not show.)

*The Thymus**: At 12 days gestation the first indication of thymus development appears when the subscapular primordium is covered by a basement membrane separating the epithelial bud from the surrounding mesenchyme. It arises from the epithelium of the 3rd and 4th visceral pouches, hence its original lining is endodermal, as is that of the thyroid. The cells actually resemble those of the pharyngeal endodermal epithelium and carry cilia and microvilli on their free surfaces. The adjoining cells are connected by desmosomes, and within the plasma membranes of the cells are oval structures, honey-combed by thin walled tubes. Stellate mesenchymal cells of the branchial area become flattened and spindle-shaped as they are compressed by the outgrowing endodermal thymic bud, and differentiate to form a capsule of fibroblastic cells. Some of these appear to be elaborating collagen. The free surfaces of the outermost cells of the primitive thymic bud are covered by a fine fibrillar membrane which separates the endodermal cells from the surrounding mesenchyme. One of three prominent and large nuceoli are seen as fenestrated spheres in dense granular opaque substance of the cytoplasm of these cells, closely adherent to the nuclear membrane whose outer lamella is beaded with ribosomes. There is a paucity of organelles in the cytoplasm, except for some free ribosomes and small oval mitochondria. Cristae are few, short, and unoriented. Golgi are present but not significant. The vacuoles are large, round, have translucent matrices, and are reduced as development proceeds. The plasma membranes of contiguous cells are lined intermittently by desmosomes, and terminal bars attach them at the luminal surface. Large epitheloid cells may also be seen interspersed among the thymic cells at their periphery. These possibly degenerating cells have dark globular inclusions. The thymus as an endocrine organ appears to be proliferating much faster than is the nearby thyroid.

By 13 days some of the endodermal cords differentiate into stromal elements with pseudopodia which insinuate themselves between the adjacent cells and encircle others. These free cells differentiate along the divergent lines of lymphopoiesis. The thymus primordium is now well encapsulated by fibroblasts which have derived much connective tissue and extra-cellular collagenous material. There is no evidence yet of any lobule formation or vascularization in the thymic matrix. In the peripheral area there appear many mitoses, and there is an outermost double layer of cells just beneath the capsule, distinguishable from those within, which begins to show heterogenity. Some enlarged cells invade the thymic mass and cause the appearance of lacunae, which spaces are soon filled with smoother, and more rounded cells that probably are stem cells for lymphopoiesis. The lymphocyte precursors have rather opaque hyaloplasm, with increased ribosome population and polygonal aggregates. Granules of ribonucleoprotein and mitochondria are also evident. Cristae become more regular and numerous. There are membranous filaments which attach immature thymic lymphocytes to the circumscribing epithelial cells.

By 14 and 15 days the mouse thymus shows the most rapid histogenetic and organogenetic changes, assuming the structure of the adult and a position close to the pericardial cavity. The surface of the thymic primordium has become scalloped and undulating, and trabeculae begin to invade and divide the organ

---

* This description combines the excellent original study by Dr. F. T. Sanel, yet to be published, and our direct observations.

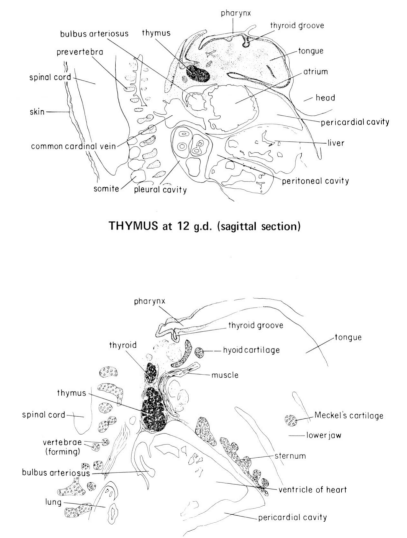

THYMUS at 12 g.d. (sagittal section)

THYMUS at 14 g.d. (sagittal section)

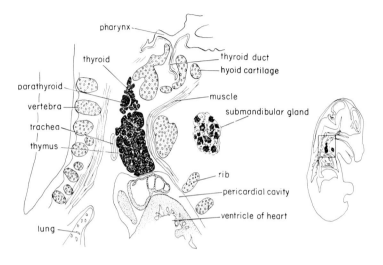

THYMUS at 15 g.d. (parasagittal section)

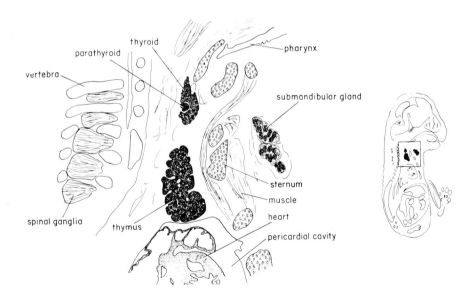

THYMUS at 16 g.d. (parasagittal section)

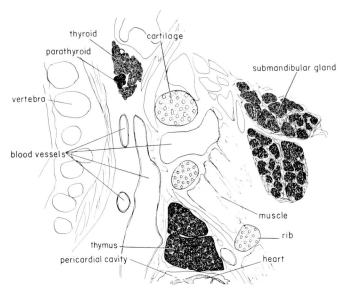

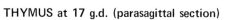

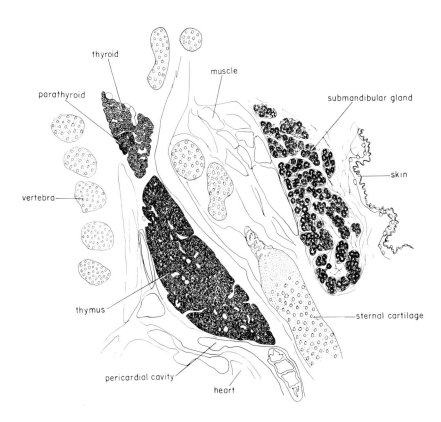

THYMUS at 17 g.d. (parasagittal section)

THYMUS at 18 g.d. (parasagittal section)

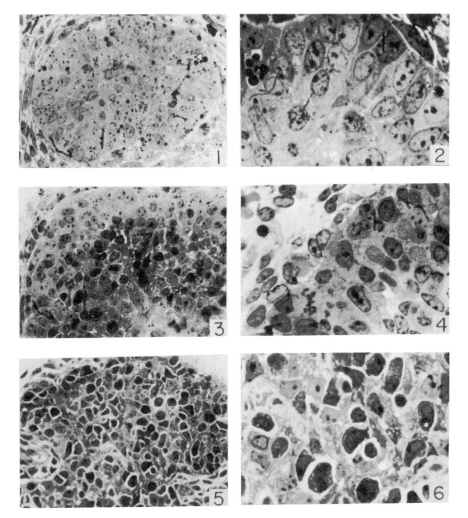

DEVELOPMENT OF MOUSE (C57 black) THYMUS GLAND

Fig. 1 — The mesenchymal cells of the branchial area are compressed by the out-growing endodermal bud. Notable are the numerous cells in mitosis, the pycnotic inclusions and the relative homogeneity of the cell population. (12 days)

Fig. 2 — The epithelium of the anlagé, predominantly columnar, is primitive and undifferentiated. In a few cells, slight condensation of nuclear chroma-tin and increasing basophilia are seen. (12 days)

Fig. 3 — The mesenchymal cells, now fibroblastic, encapsulate the still predomi-nately epithelial primordium. Transforming columnar cells in the sub-cortical area disrupt the endodermal cords. (13 days)

Fig. 4 — A slightly higher proportion of epithelial cells in division are seen in peripheral areas. Lymphoid precursors are sometimes identifiable. Note the erythroblast in the pericapsular space. (13 days)

Fig. 5 — Invasive septa are beginning to shape the anlagé into lobules. Stromal cells have developed long cytoplasmic processes that encircle the differ-entiating thymocytes. (14 days)

Fig. 6 — The vacuolated cytoplasm of the epithelial cells is characteristic of early stages. Many large lymphocytes are developing within the thymic reticulum. (14 days)   (Courtesy of F. T. Sanel)

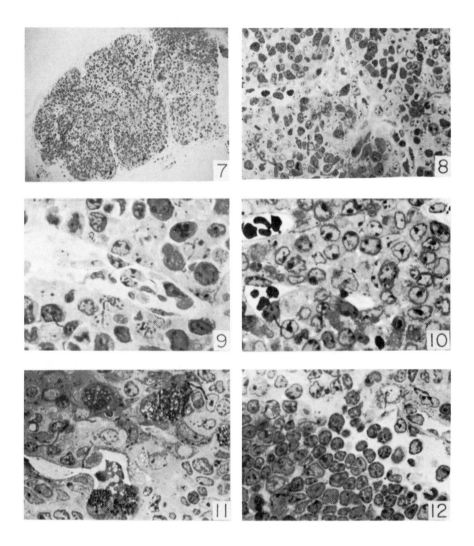

DEVELOPMENT OF MOUSE (C57 black) THYMUS GLAND

Fig. 7 — A low power view of a definitely lymphoid organ now recognizable as the thymus. Blood vessels are limited for the most part to the trabeculae although a rare capillary has penetrated the parenchyma. (15 days)

Fig. 8 — Some polarization of like cell species has occurred, but division into cortex and medulla is as yet not apparent. Note the similarity of the mononuclear cells within the blood vessels to those of the matrix. (15 days)

Fig. 9 — Cuboidal epithelium frequently underlies the capsule. Free lymphoid cells predominate in subcortical areas. (15 days)

Fig. 10 — Large ovate reticulo-epithelial cells form a compact stroma in the medulla and are numerous in this quite vascular region. (18 days)

Fig. 11 — Granule containing cells in the vicinity of the medullary blood vessels may be mast cells, although these did not stain metachromatically. Basophilic inclusions within stromal cells nearby are not uncommon. (19 days)

Fig. 12 — Thymic lymphocytes at various stages of development crowd into perivascular channels at the cortico-medullary junction and appear to be streaming into the systemic circulation via these pathways. (19 days)

(Courtesy of F. T. Sanel)

into lobules. Blood vessels (parenchymal) extend into the septa carrying ery-throid elements. Some lobules may even begin to show cortical and medullary differentiation. The thymic cells are rounded and basophilic, lodged within a reticulum or stroma the cells of which have cytoplasmic elongations often linked by desmosomes. It is within such a matrix that the early lymphoid cells divide and mature. The small lymphocytes measure from 12 to 18 microns and the nucleocytoplasmic ratio is now 2:1. Chromatin masses form an irregular ring around the inner wall of the nuclear membrane, and the nucleoli are less con-spicuous. Ribosomes are being concentrated. There is evidence that the epi-thelial cells are being transformed into stellate reticulo-epithelial cells with long processes buttressing the tissue. These are the cytoreticulum or skeleton of the gland. There are also columnar cells in the acinar formation containing aggregates of particulate glycogen; and hyperthrophied cells similar to those in the medullary zone of the adult thymus; all of epithelial origin.

At 16 to 19 days the largest proportion of thymocytes are medium sized, but there are also large thymocytes, and some small cells indigenous to the mature organ. There is no evidence at any time of a germinal center in the development of the thymus. The epithelial cell derivatives are more or less segregated into similar groups, the stromal cells with cytoplasmic prolonga-tions polarized in the cortical regions. Some cells are compressed into colum-nar or cuboidal shapes in an acinar configuration, often contiguous cells joined by desmosomes at the luminal surface of which are microvilli bearing hairy or bristle-like material, and conventional cilia. There is evidence of glycogen storage. The hypertrophied cells occasionally seen in the adult thymic medulla are first seen at 18 days. These cells may contain odd-shaped vacuoles, myelin figures, and flocculent material. By this time (18 days) the thymus has reached almost its natal size and is bounded by vertebrae, pericardium, and the sternal cartilage. There is some evidence that the thymic epithelial cells are secre-tory, even during the last phases of development. The capillary walls of the thymus contain fenestrae, thus supporting the contention that the gland has an endocrine function.

The Pituitary Gland: Rathke's pocket (hypophyseal pouch) is first defined on gestation day $8\frac{1}{2}$ but the infundibular evagination of the diencephalon, which is to be intimately associated with Rathke's pocket, does not occur until about day $11\frac{1}{2}$. Rathke's pocket becomes detached by 12 days and shows cellular pro-liferation ventro-mesially in the ultimate pars distalis (anterior lobe). By 14 days there is a solid hypophyseal stalk and the infundibulum becomes pinched off and connected only by an infundibular recess to the diencephalon. The lumen of the original Rathke's pocket is compressed. That portion of the roof of the original Rathke's pocket adjacent to the ectoderm of the infundibulum will be-come the pars intermedia, and below all will be the mesenchymal precursors of the sphenoid cartilage. The arachnoid space which partially encircles the de-veloping pituitary gland from below is the beginning of the sella turcica. Thus, by 14 days the pars distalis (anterior lobe) and intermedia can be identified and distinguished, and by 16 days the pars neuralis (posterior lobe) also develops. Thus all three of the major components of the pituitary gland are differentiating by 16 days gestation. The pars neuralis is really dorsal to the pars intermedia but does not extend quite as far anteriorly as does the more ventrally located pars distalis. By 17 days the pituitary gland seems to elaborate some of its normal secretions. By 18 days the cell aggregates of the anterior lobe become lobular and occlude the remnants of Rathke's pocket. Lumina are trapped in the anterior lobe cell proliferation, and by this time it is quite obvious that an endocrine gland is forming.   The cell configurations of the three parts of the pituitary are quite distinct, the most glandular-looking portion being the anterior lobe. Its lining epithelium develops from brain ectoderm and Rathke's pocket ectoderm, but it is quickly surrounded by mesenchyme.   While it is a single gland of three distinct functioning parts, the pituitary arises primarily through the cooperative development of ectoderm from two originally unrelated regions.

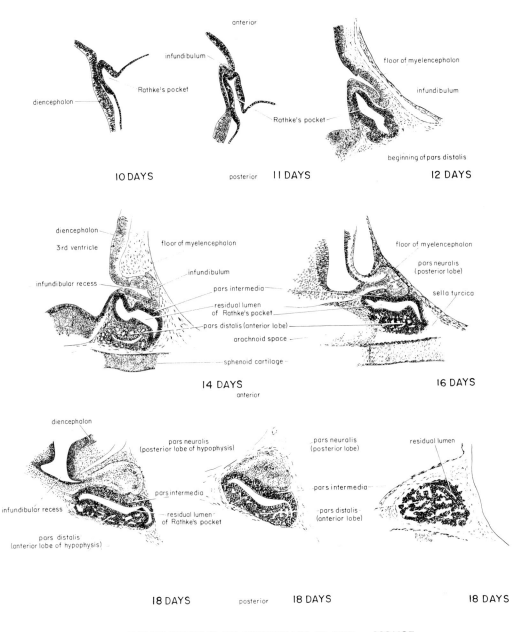

DEVELOPMENT OF PITUITARY GLAND — MOUSE
(Sagittal sections)

The Pineal Gland (Epiphysis): By $11\frac{1}{2}$ days there is the first indication of the epiphyseal evagination from the roof of the diencephalon. It occurs just posterior-dorsal to a line dividing the diencephalon from the telencephalon. By 12 days this becomes a distinct evaginating pocket which is pinched off from the forebrain by 13 days, and by 16 days takes on a lobular and glandular structure. A tubular connection from the 3rd ventricle (diocoel) persists as the pineal differentiates. At about 18 days it becomes highly glandular and the original tubular connection with the brain becomes the pineal stalk. It is surrounded by the pia mater.

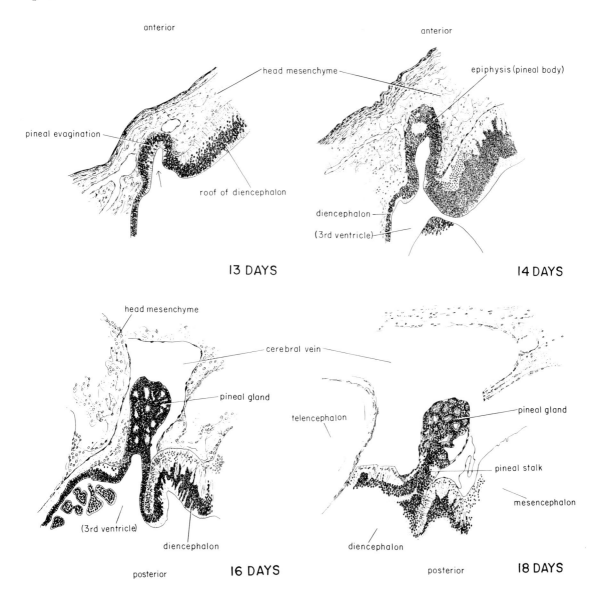

DEVELOPMENT OF PINEAL GLAND – MOUSE
(Sagittal sections)

The Adrenal Glands: On day 11 the adrenal blastemas are seen and the sympathetic neuroblasts to the adrenal cortex by $12\frac{1}{2}$ days. The consolidation of the cortex and medulla (but no capsule) occurs by 14 days; strands are found in the medulla and cells in nests, which begin to control glycogenesis, by 17 days. By 19 days or just before birth the adrenal medulla is distinguishable.

THE DIGESTIVE SYSTEM: The foregut forms as early as $7\frac{1}{2}$ days, the oral (stomodeal) membrane ruptures on day 8, the liver primordium forms and the yolk stalk becomes a narrow tube with differentiation of the hindgut at 9 days. By $9\frac{1}{2}$ days the laryngotracheal groove, liver, and dorsal pancreas begin to form, the vitelline duct closes, and the cloaca develops. By 10 days the stomach begins to differentiate, the liver acquires trabeculae, the endoderm of the vitelline duct is resorbed and the proctodeal membrane is formed. At this time the 1st pharyngeal pouch reaches the outer head ectoderm, the 2nd forms the visceral groove, the 3rd touches the head ectoderm. By $10\frac{1}{2}$ days the ultimo-branchial bodies form from the epithelium of visceral pouches 4 to 6; the stomach is well differentiated, both the duodenum and the hepato-pancreatic ducts are formed, the much enlarged dorsal pancreas has a duct, and the ventral pancreas can be identified. On day 11, the tongue primordium is developed, the trachea separates from the esophagus, the stomach changes its position to oblique, the liver becomes lobular, the cloacal cavity is divided into urogenital and rectal portions, the post anal gut regresses, and an umbilical hernia is first formed. By day 12, the dorsal and ventral pancreatic rudiments fuse, and the umbilical hernia begins to be incorporated into the intestinal loop. By 13 days, the dental lamina- appear as placodes, epithelial cords form the salivary glands, the distal end of the tongue is free, and the cloaca has a urogenital sinus and rectum. On day 14, the dental lamina grow deeply into gaps between the incisor and molar parts, the salivary glands have secretory buds, there is an enormous umbilical hernia, the pancreas develops islets and tubules and is well vascularized, and the proctodeal membrane begins to rupture. By day 15 the molar lamina have a trough-shaped rim, and the incisor germs are long. By day 16 the umbilical hernia is usually withdrawn. By day 17 the molar lamina with two concentrations of the enamel organ and papillary blastema are present; the upper and lower incisors lengthen; tall ameloblastic epithelia and papillae are vascular; liver glycogen is accumulating; the pancreas is compact with islets, ducts, acini and less connective tissue; the gut is withdrawn from the umbilicus. By day 18, the incisors are about to erupt with their dentine and enamel layers, the molars have transverse ridges, the intestinal derivatives become functionally active with secretory granules in the acini cells of the pancreas. Regarding the further development of the teeth, the post-natal sequence of eruption involves first the incisors, then the 1st, 2nd, and 3rd molars - a process which may take up to 5 weeks.

The Liver: At 9 days the liver diverticulum first forms, and proliferates cells rapidly beginning at $9\frac{1}{2}$ days gestation. Shortly thereafter epithelial cords appear in the liver primordia. Nearby is a dorsal pancreatic constriction and by $10\frac{1}{2}$ days the liver acquires large blood vessels which later become the vitelline veins. Beginning at 11 days, the ventral pancreatic rudiments appear and

---

* Tooth development is described in more detail under the heading of Skeleton.

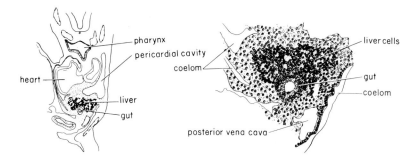

LIVER at 10 g.d.

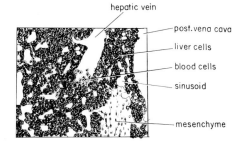

LIVER at 11 g.d.

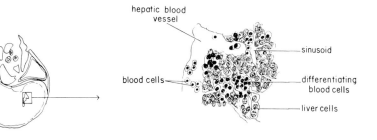

LIVER at 12 g.d.

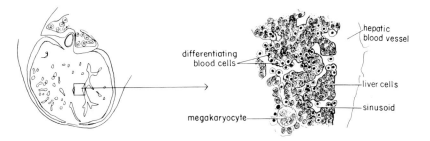

LIVER at 13 g.d.

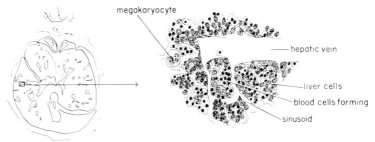

LIVER at 14 g.d.

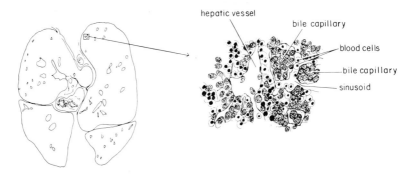

LIVER at 15 g.d.

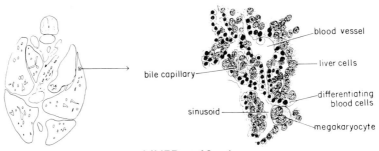

LIVER at 16 g.d.

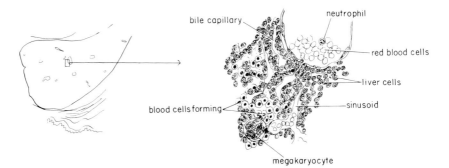

LIVER at 18 g.d.

soon thereafter the stomach expands. By $11\frac{1}{2}$ days the dorsal and ventral pancreatic rudiments fuse near the duodenum, and by $13\frac{1}{2}$ days vacuoles appear in the stomach epithelium suggesting its glandular function. It is not until day 16 that there is proliferation of the gastric glands.

In sections, liver cells may be identified as early as day 10, closely associated with the foregut and somewhat surrounded by trapped coelomic spaces. Close by is the large posterior vena cava. By day 11 the liver cells surround the hepatic vein, and are close to the posterior vena cava. Blood cells appear in the forming blood vessels, the liver mass is full of sinusoids and is invaded by mesenchyme. By day 12 there appear many early hematopoietic cells intermingled with the liver cells, some lying freely within the hepatic blood vessels. These include hemocytoblasts, myeloblasts, promyelocytes, metamyelocytes, polymorphonuclears, proerythroblasts, basophilic erythroblasts, polychromatiophilic erythroblasts, and orthochromatic erythroblasts. The last three are found in substantial numbers while the others are less than 1%. By day 13 some megakaryocytes are present, indicating the hematopoietic function of the liver. There will also appear many nucleated red cells. By day 15 bile capillaries can be distinguished from blood capillaries since they are devoid of blood cells of any kind, and are lined with true liver cells. By 16 days all of these elements have increased in number and volume, with the quantity of differentiating blood cells appreciably increased. By 18 days the liver appears to be more hematopoietic than endocrine, with numerous sinusoids and blood cells in various stages of formation.

The Spleen: At 13 days, the primordium of the spleen, located just dorsal to the stomach and near the level of the gonad, may first be seen. It does not enlarge as rapidly as does the thymus, nor does it change its location. It may be found at 15 and 16 days approximately where it originated. It is suspended by the dorsal mesogastrium into the peritoneal cavity. By 17 days it is elongating and by 18 days is a long and slender organ tucked in between the metanephros, stomach, and outer body wall. It is not yet differentiated into the typical pulp areas.

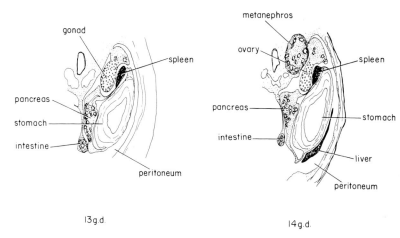

13 g.d.   14 g.d.

DEVELOPMENT OF THE SPLEEN
(Drawn to scale)

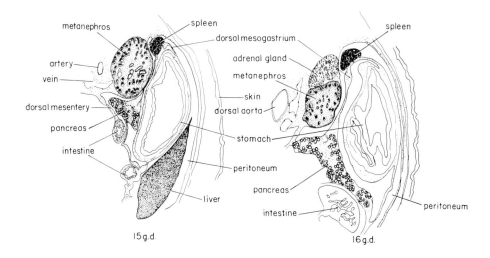

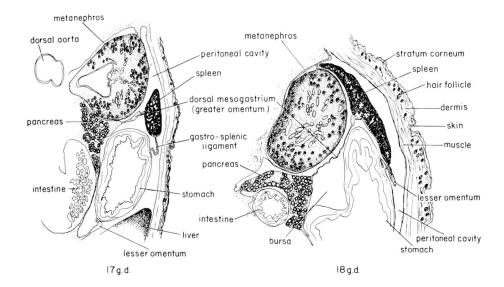

DEVELOPMENT OF THE SPLEEN

RESPIRATORY SYSTEM: By gestation day $8\frac{1}{2}$ the first pharyngeal pouch is formed; the second pouch and indications of the laryngo-tracheal groove appear on day $9\frac{1}{2}$; the third pouch develops by day 10 as the laryngo-tracheal groove extends into a pair of lung buds. At this time the trachea separates from the esophagus, the 4th and 6th pharyngeal pouches are formed, and larynx and trachea are distinguishable. On day $10\frac{1}{2}$, the right lung bud is slightly in advance of the left. By day 11, the larynx acquires arytenoid lateral swellings and the connection with the trachea is a slit, the upper part of the larynx is completely compressed, and bronchi are formed. By day $11\frac{1}{2}$, the pharyngeal pouches close off, there are arytenoid and epiglottis swellings, the tracheal-pharyngeal connection is very narrow, the bronchi have 2 or 3 buds, and the pleuro-peritoneal septa develop. On day $12\frac{1}{2}$, the cervical sinus closes, the trachea is occluded, the right bronchus has both secondary and tertiary buds (with fewer in the left) and the pleuro-peritoneal cavity persists. At 13 days, the epiglottis is separated from the arytenoid bodies by a transverse groove, the epiglottis grows separately from the lower larynx (which is occluded) the bronchial tree has many branches and buds, and the pleuro-peritoneal connection is narrowing. By $14\frac{1}{2}$ days, the larynx is enlarged, the bronchial tree branched further, and lungs are lobed and vascularized. There is a hyoid but no laryngeal or tracheal cartilage, and the diaphragm now separates the peritoneal and pleural cavities. By day 16, the epiglottis projects into the nasopharyngeal meatus, the laryngeal and tracheal cartilages appear, and the larynx acquires a ventricle with secondary grooves. By 17 days the epiglottis remains as a nasopharyngeal duct with pre-cartilage, the laryngeal ventricle expands, and the lung tissue differentiates. By day 19, just before birth, the cavity of the lower larynx and trachea becomes continuous with the ventricle of the upper part, the cartilaginous epiglottis at the root of the tongue is covered by the soft palate, the nares are open so the air passage to the lung is now complete and the lungs can become inflated with air. The mouse fetus can be delivered by caeserian section at this time.

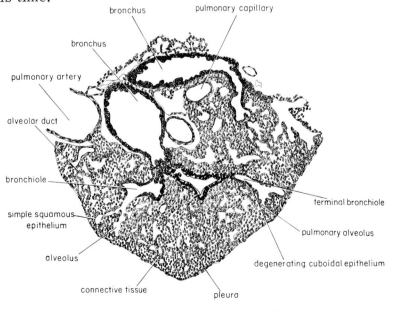

**DEVELOPING PULMONARY LOBULE
IN LUNG OF MOUSE at 17 g.d.**

CIRCULATORY SYSTEM: As early as 7 days there are paired cardiac mesenchymal primordia which are fusing. By $7\frac{1}{2}$ days there is a pleuro-
peritoneal cavity that is crescent-shaped, and endothelial cells between the
endoderm and the splanchnic mesoderm. By 8 days, a tubular heart with the
beginning of myocardial contractions appears; the tubular portion of the heart
elongating and being bent into an S-shape, the atrial and bulbo-ventricular areas
clearly defined with the latter contracting. By $8\frac{1}{2}$ days the first pair of aortic
arches appear; and by 9 days aortic arch #2, cardiac veins, the sinus venosus,
umbilical veins, the atrium, ventricular loop, arterial trunk all develop and the
circulation is established. By $9\frac{1}{2}$ days, the first three aortic arches are prominent, and the interventricular sulcus is formed; aortic arch #1 begins to regress,
#4 is forming, and by 10 days, paired heart chambers may be seen. At $10\frac{1}{2}$ days
of the four pairs of aortic arches #1 and #2 regress as #3 and #4 enlarge and become well developed, #5 and #6 begin to form, the atrial septum appears as
well as the atrio-ventricular canal. By day 11, aortic arch #7 (pulmonary)
forms, the interventricular septum appears, the atrio-ventricular cushions and
septa form, small pulmonary veins grow into the left atrium, aortic arch #5
regresses, common cardinal veins appear, and the hepatic vein collects blood
from the vitelline and umbilical veins through the liver. The atrial septum is
not complete at this time but the dorsal and ventral atrio-ventricular cushions
form. Aortic arches #3 and #4 are very well developed and symmetrical, #5
disappears, #6 is asymmetrical with the right part narrowing, #7 is branching
off of #6. By 12 days, the foramen ovale forms, the atrial septum is almost
complete, the cervical sinus closes, the aortic trunk is partially divided into
systemic and pulmonary stems, the internal carotids and pulmonaries enlarge,
and the auricle forms. By $12\frac{1}{2}$ days, the heart and aortic trunk are longitudinally divided and the inferior cardinal veins are formed. Aortic arch #3 is
symmetrical, #4 has only the left portion which is the large systemic arch with
right #4 being small and leading to the vertebrae and subclavian arteries of the

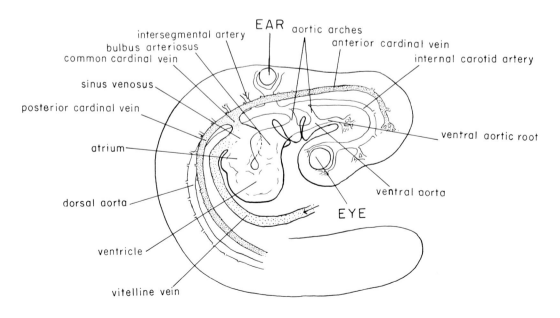

**DIAGRAM OF HEART AND CIRCULATORY SYSTEM OF RIGHT SIDE
9 DAY MOUSE EMBRYO**

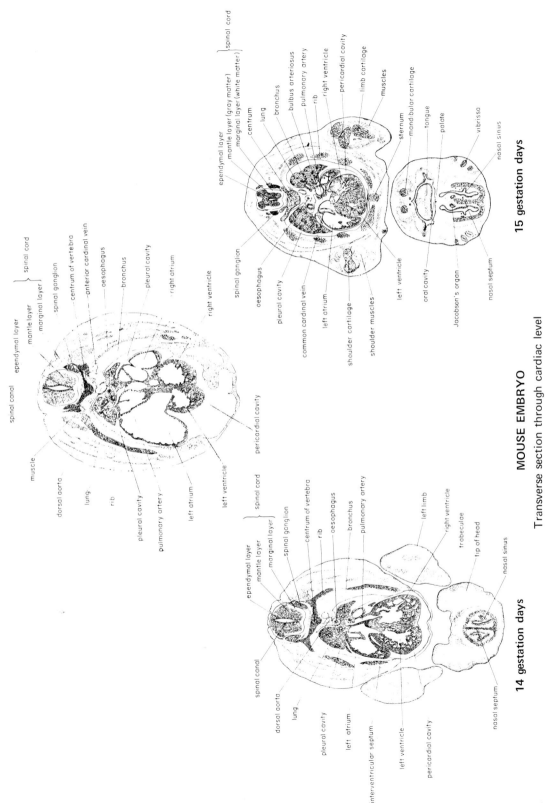

MOUSE EMBRYO

Transverse section through cardiac level

14 gestation days

15 gestation days

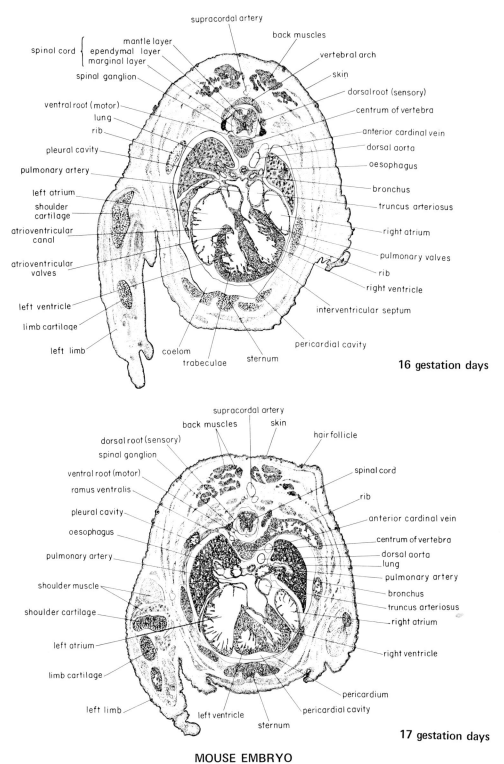

supracordal artery
back muscles
spinal cord { mantle layer
ependymal layer
marginal layer
spinal ganglion
vertebral arch
skin
dorsal root (sensory)
centrum of vertebra
ventral root (motor)
lung
rib
anterior cardinal vein
dorsal aorta
pleural cavity
oesophagus
pulmonary artery
bronchus
left atrium
truncus arteriosus
shoulder cartilage
atrioventricular canal
right atrium
pulmonary valves
atrioventricular valves
rib
right ventricle
left ventricle
limb cartilage
interventricular septum
left limb
coelom
trabeculae
sternum
pericardial cavity

16 gestation days

supracordal artery
back muscles
skin
dorsal root (sensory)
hair follicle
spinal ganglion
spinal cord
ventral root (motor)
ramus ventralis
rib
anterior cardinal vein
pleural cavity
centrum of vertebra
oesophagus
dorsal aorta
lung
pulmonary artery
pulmonary artery
bronchus
shoulder muscle
truncus arteriosus
shoulder cartilage
right atrium
left atrium
limb cartilage
right ventricle
left limb
left ventricle
pericardium
sternum
pericardial cavity

17 gestation days

MOUSE EMBRYO

Transverse section through cardiac level

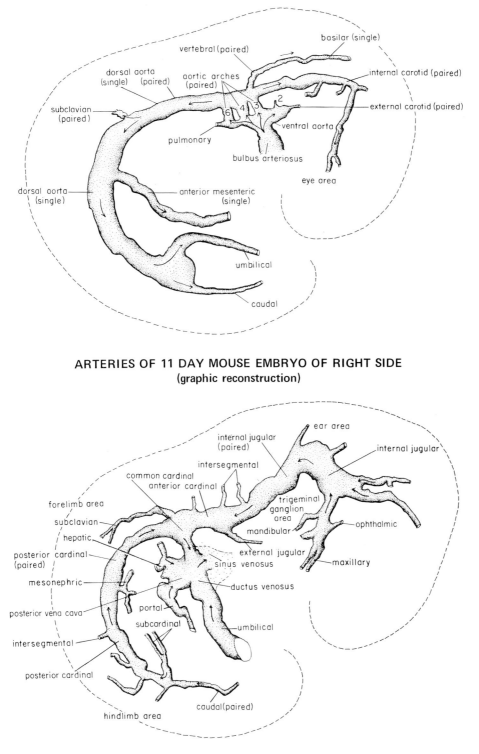

ARTERIES OF 11 DAY MOUSE EMBRYO OF RIGHT SIDE
(graphic reconstruction)

VEINS OF 11 DAY MOUSE EMBRYO OF RIGHT SIDE
(graphic reconstruction)

right side. The left portion of #6 goes to the dorsal aorta and the right portion of #6 regresses except for connections with the pulmonary arteries. By 13 days, the dorsal aorta is reduced between aortic arches #3 and #4, the aortic trunks separate, the inferior vena cava collects blood from the renal, vitelline, and umbilical veins and includes the hepatic vein, the right inferior cardinal vein is interrupted and the left is reduced. The aortic arch #3 is the only symmetric pair left, having lingual and carotid branches. By 14 days, the right portion of #4 loses its connection with the dorsal aorta and supplies the right subclavian and vertebral arteries. The left portion of aortic arch #4 further enlarges as the systemic arch and receives a duct from #6 left. After giving off the left subclavian and vertebral arteries it becomes the descending aorta while the interatrial foramen persists. By $14\frac{1}{2}$ days the fetal circulatory system is complete and functioning, with the large vitelline circuit in addition to the allantoic (umbilical) vessels. At birth the vitelline and umbilical (allantoic) veins collapse and degenerate, the oval and interatrial foramina close, and the arterial duct (#6 left) regresses and disappears.

BLOOD OF THE MOUSE FETUS: Blood islands may be seen in the splanchnic mesoderm of the 7 and 8 day mouse embryo and are the only source of red cells for the embryo through day 11. This yolk sac hematopoeisis provides the primitive generation of erythrocytes which are exceptionally large nucleated cells. They may have volumes some 4 or 5 times that of erythrocytes of the adult mouse, and may be compared with pre-erythrocytes and normoblasts of bone marrow. During days 11 and 12 many may be seen in mitosis, but by day 14 their nuclei are more likely to be pycnotic. After day 15 these cells tend to diminish rapidly. Active hemopoiesis in liver, spleen and bone marrow occurs only after the circulation is established on day 9. From day 12 through 16 the liver is almost exclusively the fetal hematopoietic organ, and may contain hematopoietic foci even after birth. This is true to a lesser degree for the spleen and marrow, with the spleen showing erythropoiesis and myelopoiesis by day 15 and the bone marrow by day 16. The spleen is the first site for myelopoiesis, followed by bone marrow myeloid hematopoiesis. Within the liver at 15 days may be found some myoblasts, and neutrophilic and basophilic leukocytes. The erythroid precursors are quite similar to those found in bone marrow, and are of several distinct sizes. Hepatically derived erythrocytes are enucleated and measure about 8 microns in diameter, being slightly larger than the 6 micron erythrocytes of the adult. Lymphocytes are first seen in the developing thymus. It is difficult to obtain sufficient fetal blood for proper analysis before day 17, so that the following data are from 15 males and 15 females of 17 days gestation.

Fetal mice were removed from the uterus and their amniotic sacs, and the mucous was removed from their noses to allow normal respiration. The mice were kept alive without anesthesia for the few minutes required until the blood was collected. It was usually difficult to obtain from a single mouse enough blood for the entire count, so that two litter-mate mice were used. The hemoglobin, red cell count, and platelets were determined from one mouse while the white cell count and the differentials were taken from the other. The data below are taken from 15 fetuses of each sex, so represent an average. For comparison similar counts are given for adults of both sexes at 9 weeks of age, all of the same strain.

TABLE 7

BLOOD ANALYSIS OF FETAL AND ADULT MICE*

| | 17 DAY FETUSES | | 9 WEEK OLD ADULTS | |
	MALES (15)	FEMALES (15)	MALES (25)	FEMALES (25)
Hemoglobin	8.9	8.3	14.1	15.1
RBC (Erythrocytes)	3,438,666	3,244,000	9,140,800	9,700,300
WBC (Leukocytes)	1,720	1,533	10,520	9,976
Platelets:	453,333	454,000	1,361,600	1,342,000
Neutrophils:Segmented	21.9	25.1	27.1	23.1
Neutrophils:Stabophils	10.3	7.1	1.1	0.4
Neutrophils:Metamyelocytes	1.5	0.8	0.0	0.0
Neutrophils:Myelocytes	0.1	0.1	0.0	0.0
Eosinophils:	2.5	1.0	3.5	4.1
Eosinophils:Meta	0.5	0.0	0.0	0.0
Monocytes:	18.7	19.7	5.8	4.3
Monocytes:Young	0.9	1.1	0.0	0.0
Lymphocytes:	31.8	30.4	61.5	67.0
Lymphocytes:Young	2.1	2.4	0.2	0.1
Blasts:	0.5	0.8	0.0	0.0
Histiocytes:	0.3	0.1	0.0	0.0
Phagocytes:	0.4	0.7	0.0	0.0
Granular-Vacuolated Cells:	0.9	4.1	0.0	0.0
Early Type Ring Nuclear Cell:	5.3	4.6	0.04	0.1
Double Nucleated Lymphocytes:	0.1	0.1	0.5	0.5
Nucleated Erythrocytes:	48.6	53.7	0.0	0.0
Unidentifiable Cells:	2.0	0.4	0.0	0.0

* Blood analyses made by Csilla Somogyi.

  Differential values are in percentages.

The blood was generally taken from the jugular vein which is always plainly visible through the thin and translucent skin. The mouse was held in the left hand, the index finger pushing its head back and the thumb on the lower body. The jugular vein was pierced with especially fine scissors. The three suction pipettes were prepared in advance, complete with rubber tubing which allowed single hand manipulation. A single drop of blood was found to be enough for the differential count smear, after which the fetuses were disposed of by over-anesthetization.

Due to the fragility of the blood cells, differential counts were difficult to make, many of the cells being damaged by the smearing process. This was especially true of day 16, the cells being younger. Generally two smears were counted on day 17 to get the 100 cells. The erythrocytes were partially of the fetal type, appearing larger in the smears. Much stippling, polychromatophilia, and nucleated erythrocytes were seen (see accompanying photographs).

Counts at birth (day 19) showed a marked increase in hemoglobin, erythrocytes, a slight increase in the leukocytes, a great increase in neutrophils, a decrease in lymphocytes and, of course, a decrease in nucleated reds and the younger leukocyte forms.

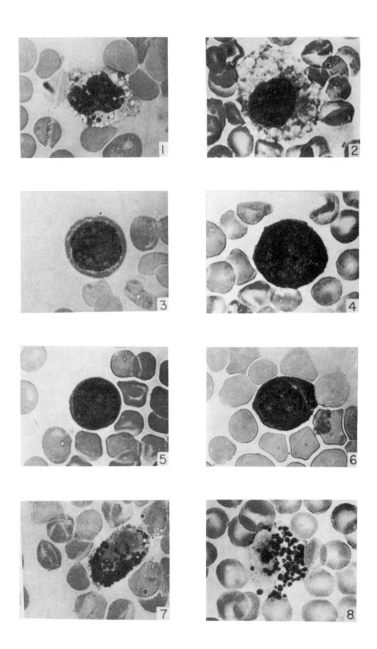

BLOOD TYPES OF THE 17 DAY FETAL MOUSE

Figs. 1, 2   — Granulated-vacuolated cells
Figs. 3 to 6 — Blasts: note dark cytoplasm and nucleoli
Figs. 7, 8   — Mast cells; tissue basophils

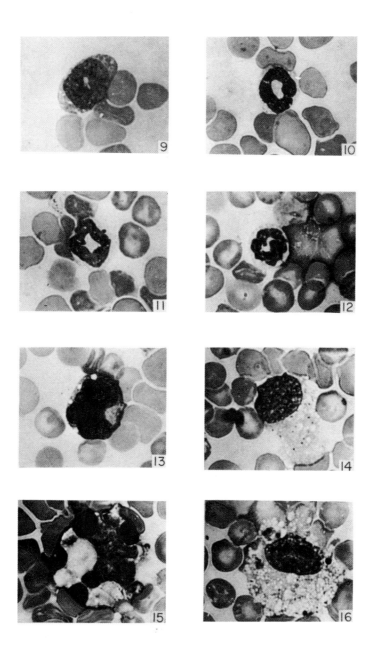

BLOOD TYPES OF THE 17 DAY FETAL MOUSE

Figs. 9 to 12 — Neutrophils; showing partial developmental
Fig. 13   — Young monocyte with vacuoles
Fig. 14   — Granular histiocyte, normally found only in adult bone marrow
Fig. 15   — Phagocyte with inclusions
Fig. 16   — Phagocytic histiocyte, not often found in mature blood

THE EXCRETORY SYSTEM: By 8 days, the nephrogenic cords appear, and by 10 days (in frontal sections) one may see the pronephric tubules and collecting pronephric ducts, suspended within the coelom. The tubules are enmeshed in mesenchyme, and by $10\frac{1}{2}$ days the nephric ducts make contact with the cloaca, and the mesonephroi begin to form tubules. By 11 days nephrotomes can be found, some with apparent nephrostomes opening into the coelom (peritoneal cavity). Each mesonephros consists of crowded nephric units not yet differentiated. The anterior tubules appear to be attached to nephric ducts, there are no renal corpuscles, and the metanephric urethral buds grow out from the posterior part of the nephric ducts. The cloaca begins to divide and some anterior nephrostomes may persist. By 12 days the transient mesonephric tubules will be seen forming and some will be degenerating, just dorsal to the gonad primordium. The pronephroi begin to degenerate but their ducts remain in the female as oviducts. The mesonephric (Wolffian) ducts can be seen, clearly lined with cuboidal epithelium and lying posterior to the tubules. They connect with the uro-genital sinus. The metanephroi will also be seen more posteriorly without any association with the mesonephroi. The metanephric ducts leading to the cloaca arise from the pelvis of each kidney with two branches (calyces) and are known as the future ureters. The cloaca is partially separated into the intestinal and uro-genital parts. Thus by 12 days the mouse embryo has portions of all three of its embryonic excretory systems. By $12\frac{1}{2}$ days the mesonephric tubules are scattered and only the anterior few connect with the mesonephric ducts. Posteriorly they are the vestigial renal corpuscles. The mesonephroi are shorter than the associated gonad primordia, and their ducts are short, and open into the urogenital portion of the cloaca. The external urethral papilla and perineum form. By 13 days there remain a few mesonephric tubules, and in the male each Wolffian (or mesonephric) duct is transformed into a vas

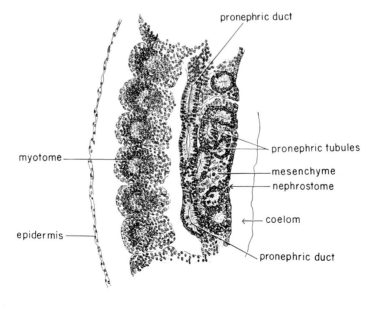

MOUSE EXCRETORY SYSTEM at 10 g.d.
(Sagittal section)

deferens. Each metanephros becomes highly tubular, and its ureter is a single, large tube clearly lined with cuboidal epithelium. By 13 days, the anterior and the cloacal membrane form as the metanephroi differentiate. By $13\frac{1}{2}$ days the mesonephroi regress, except for their anterior parts which become the epididymi of the male. The metanephroi move anteriorly behind the gonads and make contact with the adrenals, forming tubules and ampullae. The mesonephric and metanephric ducts open separately into the urogenital sinus. At 14 days the very close proximity of each metanephros and the gonad is obvious, and within the metanephros are collecting tubules and glomeruli with their Bowman's capsules. Thus relatively early in the development of the mouse, the true kidney is actively differentiating. Much of the kidney is mesenchyme and closely associated with it now is the adrenal gland. The ureters connect with the bladder and the cloacal membrane is beneath the urethral papilla. By 15 days the metanephros is encapsulated and its internal organization is that of a vertebrate kidney. By 16 days the two ureters can be seen expanding into the pelves of the kidneys and the medullary and cortical regions become distinguishable. The urogenital sinus and rectum are separated and the perineum develops further. By 17 and 18 days the kidney is fully differentiated and able to function, but excretion by the fetus continues to be primarily via the placenta.

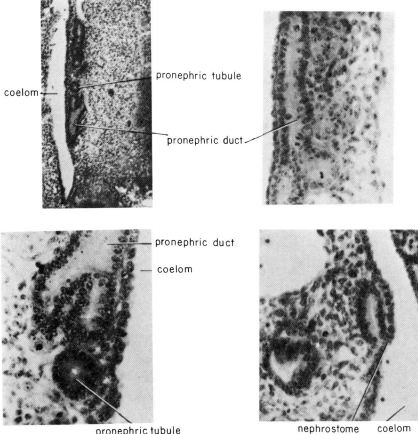

coelom —
pronephric tubule
pronephric duct

— pronephric duct
— coelom

pronephric tubule
nephrostome   coelom

MOUSE PRONEPHRIC KIDNEY at 11 g.d.

(Photographs)

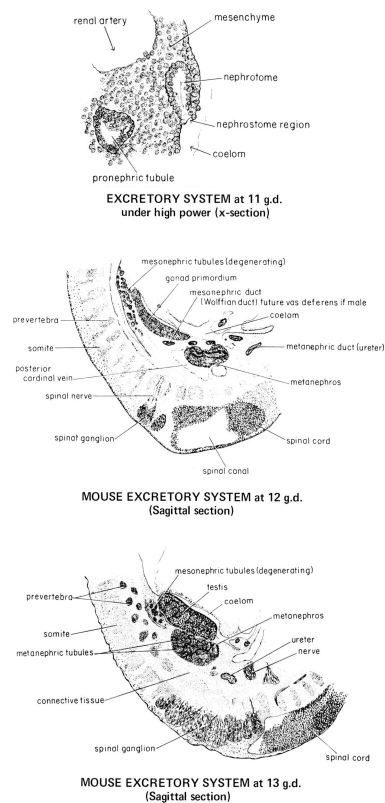

renal artery

mesenchyme

nephrotome

nephrostome region

coelom

pronephric tubule

**EXCRETORY SYSTEM at 11 g.d.
under high power (x-section)**

mesonephric tubules (degenerating)

gonad primordium

mesonephric duct
(Wolffian duct) future vas deferens if male

coelom

prevertebra

somite

metanephric duct (ureter)

posterior
cardinal vein

metanephros

spinal nerve

spinal ganglion

spinal cord

spinal canal

**MOUSE EXCRETORY SYSTEM at 12 g.d.
(Sagittal section)**

mesonephric tubules (degenerating)

prevertebra

testis

coelom

metanephros

somite

metanephric tubules

ureter

nerve

connective tissue

spinal ganglion

spinal cord

**MOUSE EXCRETORY SYSTEM at 13 g.d.
(Sagittal section)**

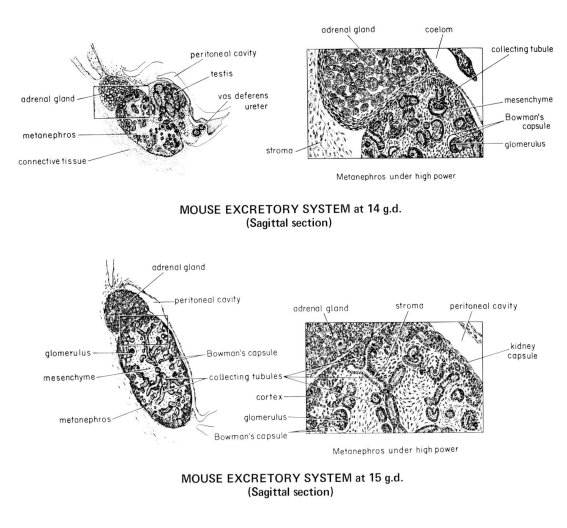

MOUSE EXCRETORY SYSTEM at 14 g.d.
(Sagittal section)

MOUSE EXCRETORY SYSTEM at 15 g.d.
(Sagittal section)

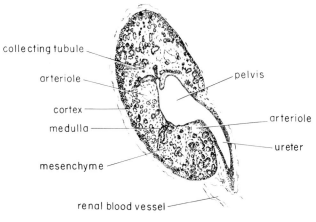

MOUSE EXCRETORY SYSTEM at 16 g.d.
(Sagittal section)
no mesonephros remaining

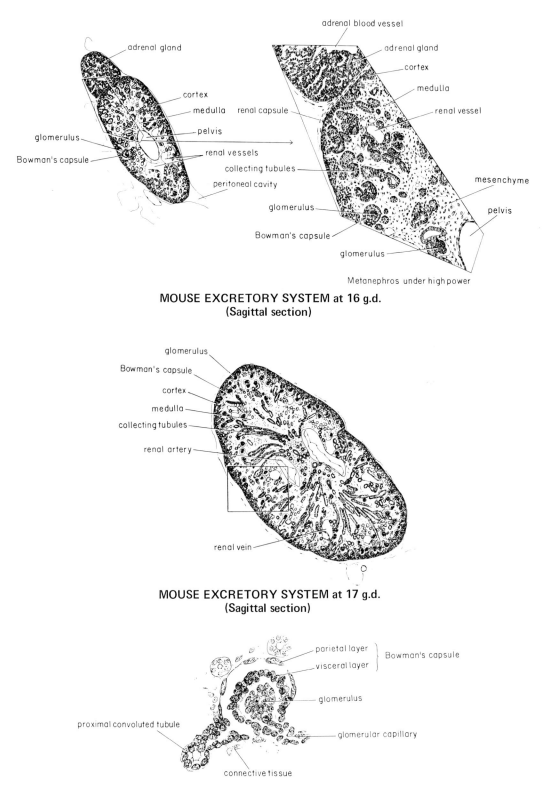

MOUSE EXCRETORY SYSTEM at 16 g.d.
(Sagittal section)

MOUSE EXCRETORY SYSTEM at 17 g.d.
(Sagittal section)

MOUSE EXCRETORY SYSTEM at 18 g.d.

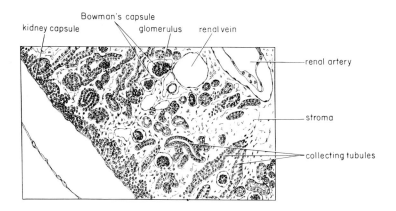

KIDNEY under high power

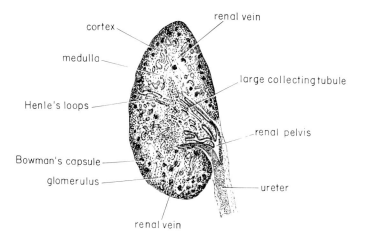

MOUSE KIDNEY at 18 g.d.
(x-section)

THE MUSCULAR SYSTEM: By as early as $5\frac{1}{2}$ to 6 days paired mesocardial primordia appear and cardial myoblasts by $7\frac{1}{2}$ days, with myocardial contractions as early as 8 days. By $7\frac{1}{2}$ days myotomes arise from somites in the cervical region and by 11 days myotomes appear with longitudinally directed myoblasts with ventral processes in some. By 13 days, the myoblasts show active proliferation and somite muscle fibrils can be found on day $14\frac{1}{2}$. Skeletal muscles become contractile, myosin synthesis is accelerated by day 15, and cross striated myofibrils are seen in the development of all of the musculature by 16 days. PAS - positive, diatase digestible material can be found uniformly distributed in the myo-tubes between days 15 and 16. Secondary fibers increase rapidly on days 17 and 18 but the diameter remains less than in the primary fibers in most muscles. Spindles may be seen developing in many muscles, containing 2 to 4 fibers. By 19 days the secondary fibers are well developed in most muscles, and both types of fibers are in the myotube stage. Another or tertiary fiber now appears, developing between the primary and secondary fibers. Muscle fibers are changing their shape to that found in the post-natal mouse. At birth some secondary fibers exceed the diameter of the primary fibers, the intercellular space is reduced, myotubes are still found even in tertiary fibers. The spindles usually consist of from three to five fibers. (See Wirsen & Larsson '64.)

THE GONADS:

The Testes: From 10 to 100 primordial germ cells may be found in the presumptive male mouse embryo in its yolk sac endoderm at $7\frac{1}{2}$ to 8 days gestation. These cells continue to undergo mitoses but while doing so they migrate via the gut mesentery and their own motility at about $8\frac{1}{2}$-9.0 days to reach the dorsal mesenteries and the left and right coelomic angles by 10 days. Immediately thereafter, by $10\frac{1}{2}$ days, the gonad primordia form close to the mesonephroi, in anticipation of the arrival of the migrating germ cells. At 11 days

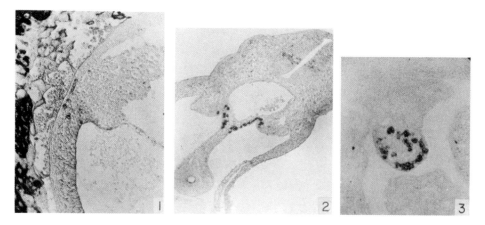

GERM CELL MIGRATION

Fig. 1 — 8 day mouse — stained for alkaline phosphatase
Fig. 2 — 10 day mouse — stained for alkaline phosphatase
Fig. 3 — 13 day mouse — stained for alkaline phosphatase

(from Mintz)

the gonad primordia are indistinguishable as to sex and all of the germ cells have reached their destination by $11\frac{1}{2}$ days. During transit their number increases to well over 5,000 cells, and these are believed to be the sole progenitors of the Type A stem cell spermatogonia of the adult testis. At 12 days the genital (germinal) ridge and newly arrived primordial germ cells appear as condensations of cells ventral to and apparently closely associated with the mesonephric masses. They project into the coelom and consist of blastemas containing all the primordial germ cells, often in the process of dividing. In both male and female the origins and migration of these primordial germ cells are identical, but now their organization and distribution begin to give the clue to the presumptive sex. In the males the primordial germ cells tend to become centrally located almost immediately. By 13 days the genital ridges enlarge considerably with interstitial tissue beginning to be distinguishable from the sex cords of the testis. At this time, the epithelium covering the genital ridge of the presumptive testis develops a slight photophatase-positive reaction. The genital ridge remains closely associated with the mesonephros and is surrounded by germinal epithelium. By 14 days each testis cord will show early spermatogenesis, particularly in the division of Type A spermatogonia. By $14\frac{1}{2}$ days gestation the interstitial tissue and the germinal epithelium are clearly defined. By 15 days the testis has prominent sex cords, each of which will become a seminiferous tubule as spermatogenesis begins. The cords now seem to contain rather uniform but indifferent cells, some of which are precursors of spermatogenetic cells and others of Sertoli cells. Gonocytes increase in size and in

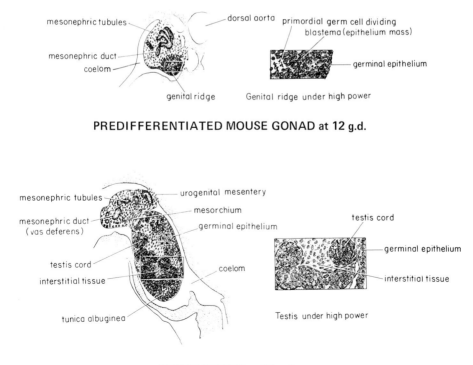

PREDIFFERENTIATED MOUSE GONAD at 12 g.d.

MOUSE TESTIS at 13 g.d.

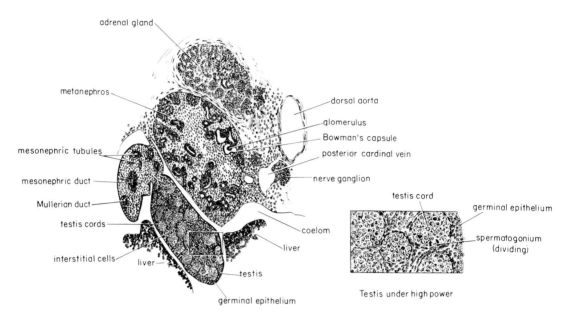

MOUSE TESTIS at 14 g.d.

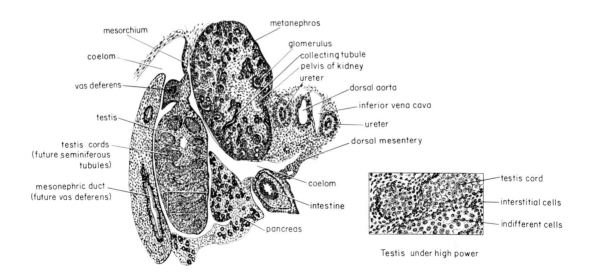

MOUSE TESTIS at 15 g.d.

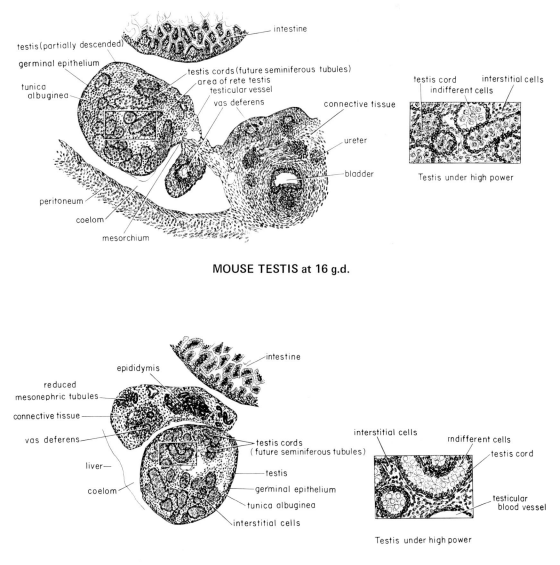

intestine

testis(partially descended)

germinal epithelium

tunica albuginea

testis cords (future seminiferous tubules)
area of rete testis
testicular vessel
vas deferens

connective tissue

ureter

bladder

peritoneum

coelom

mesorchium

testis cord   interstitial cells
indifferent cells

Testis under high power

MOUSE TESTIS at 16 g.d.

intestine

epididymis

reduced
mesonephric tubules

connective tissue

vas deferens

liver

coelom

testis cords
(future seminiferous tubules)

testis

germinal epithelium

tunica albuginea

interstitial cells

interstitial cells   indifferent cells

testis cord

testicular
blood vessel

Testis under high power

MOUSE TESTIS at 17 g.d.

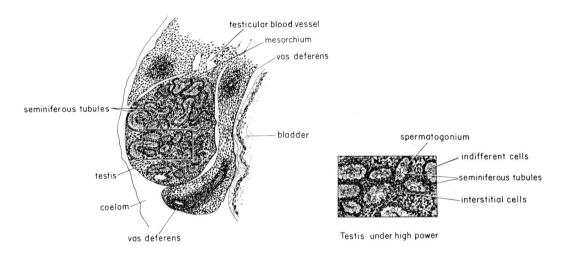

MOUSE TESTIS at 18 g.d.

their cytoplasmic inclusions, are centrally placed in the sex cords, and divide frequently in early fetal life. They divide in the fetal testis until day 16 or 17 when all divisions cease, to be resumed after birth. They continue maturation during the first post-natal weeks and become immature type A spermatogonia which then enlarge and initiate the cycle of spermatogenesis. The indifferent cells are mononucleate, continually divide, and after spermatogenesis begins, the remaining indifferent cells become Sertoli cells. Spermatogenesis in the mouse actively begins by 9 days after birth. Each vas deferens, and remnants of the associated Wolffian body may be seen close by the testes in sections of the fetus at this time. The Mullerian (pronephric) ducts persist in both sexes but function only in the female (as oviducts). The secondary sex characters begin differentiation, the males acquiring seminal vesicles and prostates with the regression of other and vestigial gonoducts by day 17. There is not much change during the next two days except in the enlargement of the seminiferous tubules, and vascularization of the tunica albuginea. Sex at birth can be determined by the distance between the anus and urethral papilla which is greater in the male than in the female.

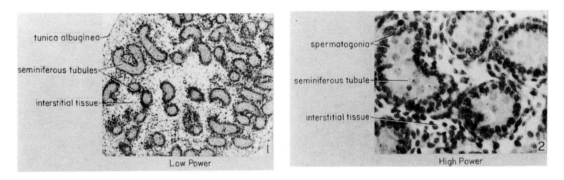

MOUSE TESTIS AT BIRTH

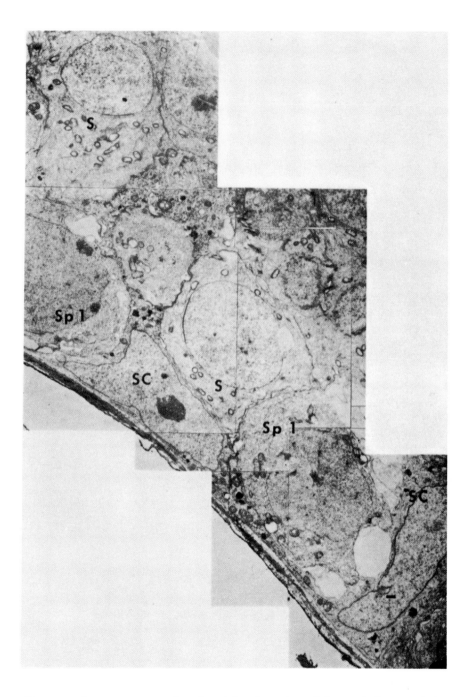

Electron microscope composite of peripheral portion of the seminiferous tubule of the mouse. The primary spermatocytes (Sp 1), adjacent to the external limiting membrane, are identified by the presence of synaptinemal complexes within the nuclei. They are embedded in the Sertoli cells (SC) and bounded centrally by the spermatids (S).

(Courtesy Dr. P. J. Gardner, from Gardner & Holyoke, Anat. Rec. 130:391-401, 1964.)

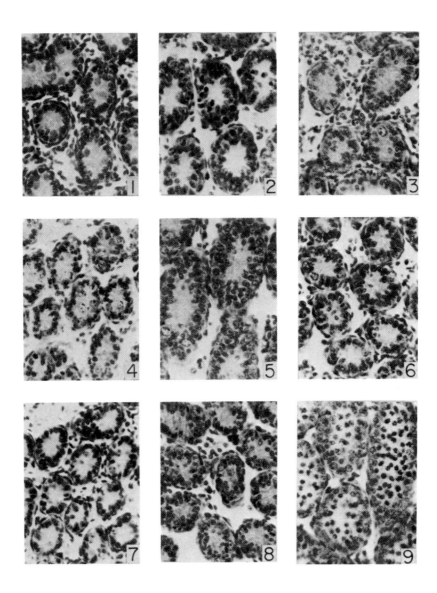

NEO-NATAL DEVELOPMENT OF MOUSE TESTIS

Fig. 1 — Section of testis of newborn mouse. Note small tubules, uniformity of spermatogenetic cells, and scattered interstitial tissue. No maturation.

Fig. 2 — Mouse testis at one day of age. Note beginning of lumen of seminiferous tubules.

Fig. 3 — Mouse testis at two days of age.

Fig. 4 — Mouse testis at 3 days of age.

Fig. 5 — Mouse testis at 4 days of age. Still no active differentiation of spermatogenetic cells.

Fig. 6 — Mouse testis at 6 days of age. Still relatively little interstitial tissue.

Fig. 7 — Mouse testis at 7 days of age.

Fig. 8 — Mouse testis at 8 days of age.

Fig. 9 — Mouse testis at 9 days of age and the beginning of active spermatogenesis. Note most cells with distinct chromosomes.

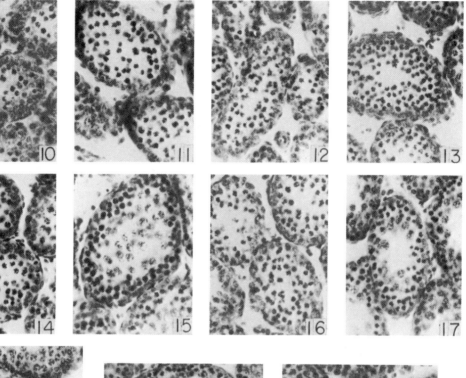

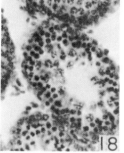

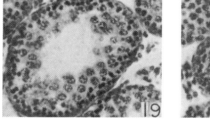

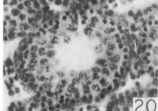

NEO-NATAL DEVELOPMENT OF MOUSE TESTIS

Fig. 10 — Mouse testis at 10 days of age. Note enlarging tubule with many cells in mitosis.

Fig. 11 — Further enlargement of tubules, and spermatogenetic cells rather scattered.

Fig. 12 — Mouse at 12 days of age, Sertoli cells first become obvious and most cells in spermatogonial stages.

Fig. 13 — Mouse testis at 13 days of age.

Fig. 14 — Mouse testis at 14 days of age.

Fig. 15 — Mouse testis at 15 days of age. Note many large primary spermatocytes loose within the seminiferous tubule, and spermatagonial stages at periphery.

Fig. 16 — Various stages of spermatogenesis seen at 16 days of age.

Fig. 17 — Secondary spermatocytes appear at about 17 days of age.

Fig. 18 — Spermatogonia and spermatocytes are seen at 18 days, and the tubule is much enlarged.

Figs. 19 & 20 — Still no spermatids or spermatozoa, but many primary and secondary spermatocytes and very much enlarged seminiferous tubules.

The Ovaries: The origin and migration of the primordial germ cells to their final site is the same in both sexes. In the female some 10 to 100 primordial germ cells arise at $7\frac{1}{2}$ to 8 days in the yolk sac splanchnopleure and migrate, via the gut mesentery, to the genital ridges during which process they increase manifold by mitosis. By day 13 the genital ridge is much enlarged by the aggregation of the migrating and mitotically active primordial germ cells, some of which will be in leptotene stage. The degenerating mesonephroi are nearby. The ostia are formed but the temporarily solid oviducts extend toward

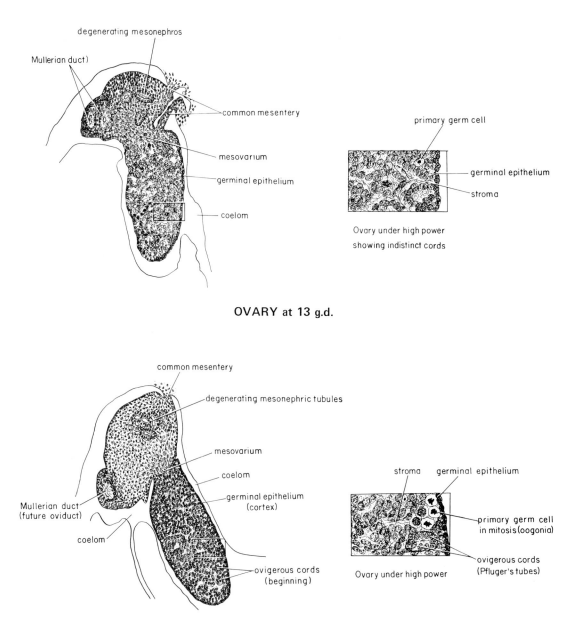

OVARY at 13 g.d.

OVARY at 14 g.d.

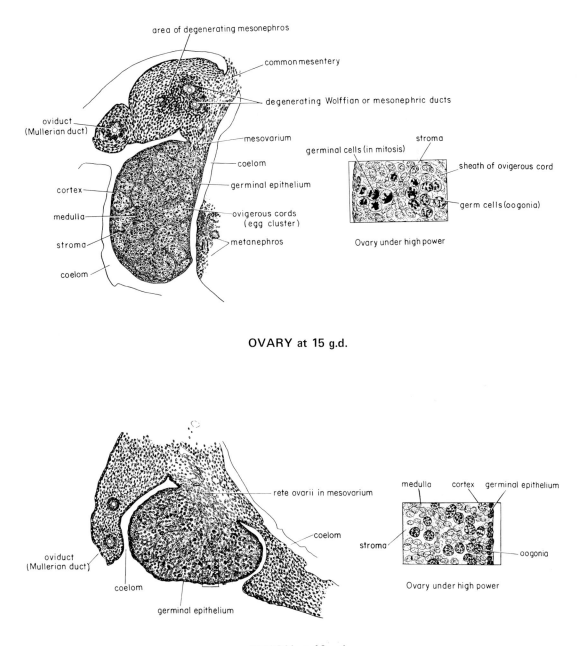

OVARY at 15 g.d.

OVARY at 16 g.d.

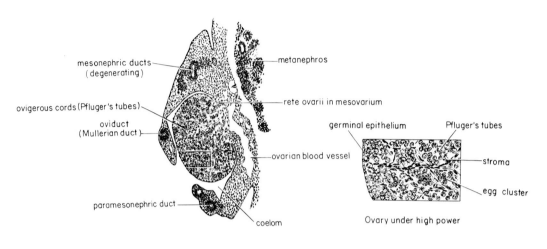

mesonephric ducts
(degenerating)

metanephros

ovigerous cords (Pfluger's tubes)

rete ovarii in mesovarium

oviduct
(Mullerian duct)

germinal epithelium

Pfluger's tubes

stroma

ovarian blood vessel

egg cluster

paramesonephric duct

coelom

Ovary under high power

OVARY at 17 g.d.

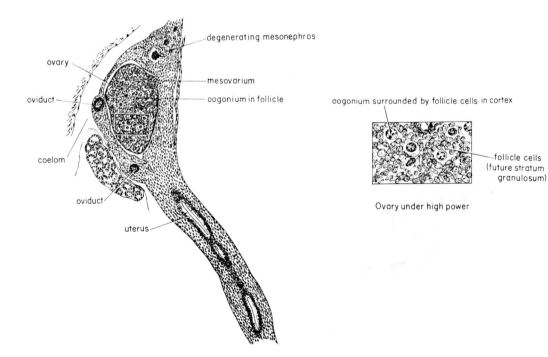

degenerating mesonephros

ovary

mesovarium

oviduct

oogonium in follicle

oogonium surrounded by follicle cells in cortex

coelom

follicle cells
(future stratum
granulosum)

oviduct

uterus

Ovary under high power

OVARY at 18 g.d.

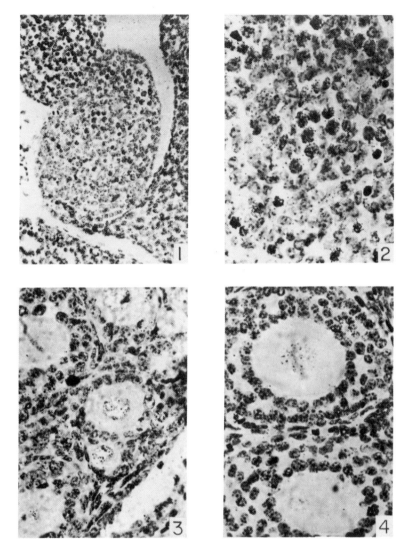

Fig. 1 — Ovary on 16th day of embryonic life.

Fig. 2 — 10 μCi tritiated thymidine was injected into the amniotic sac of a mouse embryo on the 15th day of pregnancy. The embryo was killed 24 hours later and an autoradiograph prepared of its ovary (Feulgen stain; K_2 liquid emulsion, Ilford; exposure 7 days). A group of labelled oöcytes in leptotene is seen in the center.

Fig. 3 — Tritiated thymidine injected on 14th day of embryonic life. The mouse was killed 7 days after birth. In the autoradiograph many labelled oöcytes are seen indicating that the oöcyte that synthesized its DNA on the 14th day of embryonic life persists in the post partum ovary.

Fig. 4 — Labelled oöcyte in a growing follicle of a 14 day old mouse. The radioactive DNA precursor had been injected on the 14th day of embryonic life.

(Courtesy Dr. Hannah Paters, Copenhagen, Denmark)

the cloaca. Germinal epithelium can be distinguished from the internal stroma (mesenchyme) and occasionally primary germ cells may be seen dividing. The early organization within the genital ridge is such that the germ cells are peripherally located in the presumptive ovary. At 14 days oögonia are in mitosis directly beneath the germinal epithelium. Stroma is abundant. The oviducts by-pass the nephric ducts to reach the pelvis and the urogenital sinus by 14 days. Soon (15 days) the oögonia appear in clusters, surrounded by a sheath of ovigerous cords, and loose mesenchymal stroma. By 16 days the cortex and medulla differentiate so that the presumptive sex of the female can be verified. Oögonia are abundant and many begin their maturation (synaptene) by day 16. By 17 days some will be in pachytene and others in diplotene and diakenesis. They cluster toward the cortex, as Pfluger's tubes, which are not really tubular at all. The females acquire their utero-vaginal canals. By 18 days follicle cells appear clustered around a selected oögonium (often in dictyate stage) that is destined to become an ovum. All presumptive female germ cells enter meiosis before birth, and there is no evidence of epithelial regeneration. Until recently the pre-natal preparation of mouse gametes was not clearly understood. It seems quite definite that such preparation is complex and crucial to the proper production of functional gametes during the later reproductive life of the mouse. Effects on these prenatal primordial germ cells may be mediated through their environment or through more complex and possibly unpredictable genetic factors. Some of these undetermined factors are the origin of the primordial germ cells in the yolk sac endoderm in the first place; the means of transit to the genital ridges, whether actively or passively; the factors which insure the synchronous onset of meiotic prophase in the oöcytes; the mechanism for meiotic pairing of allelomorphic chromosomes; and the relationship of the environment and genetic factors in gonad differentiation. The frequency in which ovo-testis combinations are found suggests that there is ontogenetically little difference in the production of sex dimorphism in the mouse. Finally, there is such rapid proliferation of the primordial germ cell during transit that the challenging question arises as to what controls the size of the ultimate gamete population. Thus there are many pages of normal embryology yet to be written.

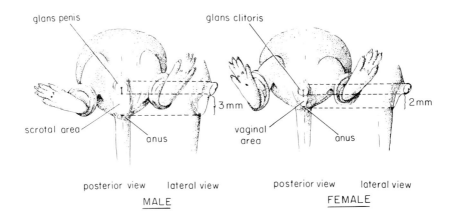

EXTERNAL GENITALIA OF THE NEW BORN MOUSE

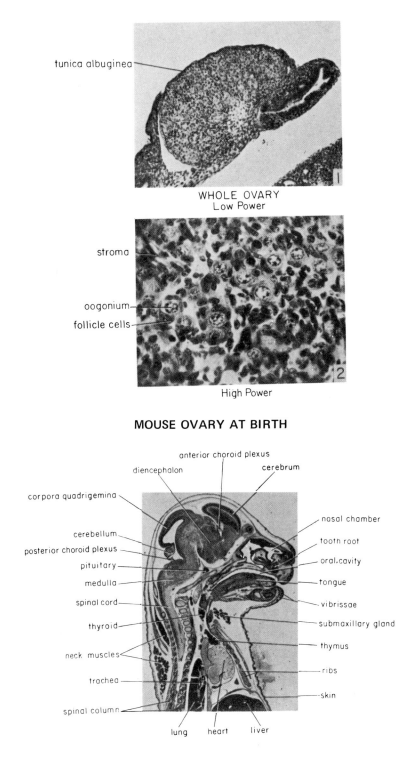

tunica albuginea

WHOLE OVARY
Low Power

stroma

oogonium
follicle cells

High Power

MOUSE OVARY AT BIRTH

anterior choroid plexus
diencephalon   cerebrum

corpora quadrigemina

cerebellum   nasal chamber

posterior choroid plexus   tooth root

pituitary   oral cavity

medulla   tongue

spinal cord   vibrissae

thyroid   submaxillary gland

neck muscles   thymus

trachea   ribs

spinal column   skin

lung   heart   liver

HEAD AND CHEST OF MOUSE AT BIRTH
(Sagittal section)

Addenda

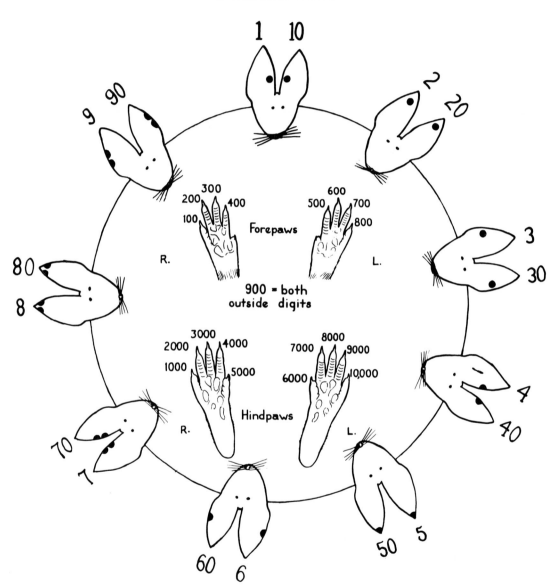

This plan shows how it is possible to mark permanently individual mice up to a total of 10,000. The black spots represent holes punched with an ear punch, and the toes are to be clipped. If this system is followed by all, and a mouse escapes its confines, it can be readily identified and returned to its proper place.

ANESTHESIA DOSAGES FOR ADULT MICE

Veterinarians Nembutal diluted to 10% with distilled water or saline may be injected intra-peritoneally for quick action or intra-muscularly for slower action. Duration about 1 hour.

WEIGHT OF MOUSE	CUBIC CENTIMETERS 10% SOLUTION	WEIGHT OF MOUSE	CUBIC CENTIMETERS 10% SOLUTION
10 grams	0.13	23 grams	0.30
11 ''	0.14	24 ''	0.31
12 ''	0.15	25 ''	0.32
13 ''	0.17	26 ''	0.34
14 ''	0.18	27 ''	0.35
15 ''	0.19	28 ''	0.36
16 ''	0.20	29 ''	0.38
17 ''	0.22	30 ''	0.39
18 ''	0.23	31 ''	0.40
19 ''	0.24	32 ''	0.42
20 ''	0.25	33 ''	0.43
21 ''	0.26	34 ''	0.44
22 ''	0.28	35 ''	0.46

COMPARATIVE AGES OF MOUSE AND HUMAN EMBRYOS*

MOUSE AGE, DAYS	HUMAN AGE, DAYS	MOUSE AGE, DAYS	HUMAN AGE, DAYS
0	1	8 4/5	24 1/3
1	2	9	25 1/2
2	3	9 1/2 – 9 2/3	26
4	4	10	27
5	5 – 6	10 1/2	28 1/2
5 1/2	7 – 8	11	30 3/4
6	9 – 10	11 1/2	33 1/2
6 1/2	11 – 13	12	36
7	14 – 17	12 1/3	36 1/2
7 1/2	18 – 20	13 – 13 1/2	38
8	20.5	14 1/2	47
8 1/3	21.0	15 1/2	65
8 1/2	22.0	16 1/2	84 1/2
8 2/3 – 8 3/4	23.0		

*In part from the author's "Vertebrate Embryology, The Dynamics of Development", Harcourt, Brace & World, Inc., 1964.

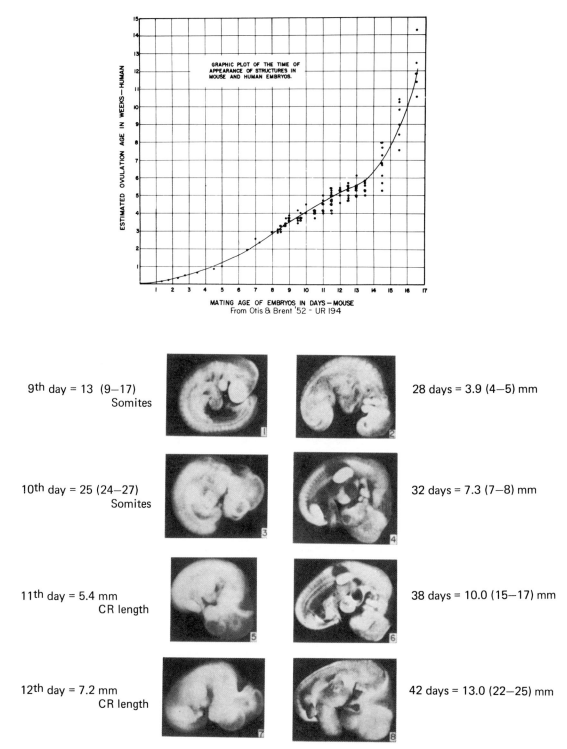

9th day = 13 (9—17)
Somites

28 days = 3.9 (4—5) mm

10th day = 25 (24—27)
Somites

32 days = 7.3 (7—8) mm

11th day = 5.4 mm
CR length

38 days = 10.0 (15—17) mm

12th day = 7.2 mm
CR length

42 days = 13.0 (22—25) mm

COMPARISON OF MOUSE AND HUMAN EMBRYOS
(Courtesy Alden "Laboratory Atlas of the Mouse Embryo" and
Carnegie Institute of Washington)

EXTRAPOLATION TABLE FOR MOUSE TO RAT EMBRYONIC AGES

MOUSE	RAT	MOUSE	RAT
1	2	8.5 – 9.0	10.5
2	3.25	9.5	11
3	4	10	11.5
4	5	10.25	11.75
4.5	6	10.50	12.125
5	6.75	11	12.5
5.5	7.25	12	13
6.0	7.50	12.5	13.5
6.5	7.75	13.0	14.5
7	8.5	14.5	15.5
7.5	9	15	16
7.75	9.5	16 – 16.5	17 – 18
8	10	17 – 19	19 – 22
Post Partem: 1 to 20	1 to 16	Post Partem: 21+	17+

(These are based upon similarities in embryological development but cannot apply to every minute detail.)

RELATION OF MOUSE EMBRYONIC AGE TO SOMITE NUMBER
AND TO CROWN-RUMP LENGTH*

AGE, DAYS	SOMITES	LENGTH, MM.	AGE, DAYS	SOMITES	LENGTH, MM.
8	1 – 4		12½	49 – 51	8.9
8½	5 – 12	2	13	52 – 60	9.4
9	13 – 20	2.2	13½		9.8
9½	21 – 25	3.3	14½	61 – 64	11.2
10	26 – 28	3.8	15	65	
10½	29 – 36	5.2	15½		13.7
11	37 – 42	6.2	16½		16.1
12	43 – 48	7.2			

*From the author's "Vertebrate Embryology, The Dynamics of Development", Harcourt, Brace & World, Inc., 1964.

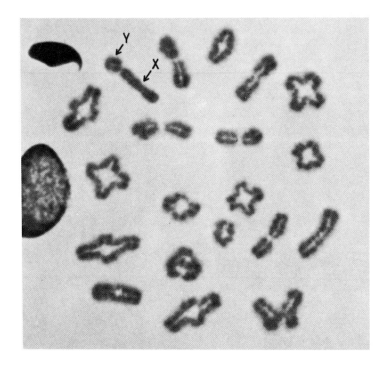

Bivalent chromosomes at late diakinesis in a primary spermatocyte from a CBA mouse. The heteromorphic XY bivalent is easily recognised. The chromosomes of seven autosomal bivalents are associated by single terminal chiasmata. In eleven others there is a single subterminal chiasma and in one bivalent, two diasmata. The appearance differs from cell to cell due to variation in the number and positions of chiasmata. Air-dried preparation.

(Courtesy Dr. C. E. Ford and Dr. E. P. Evans)

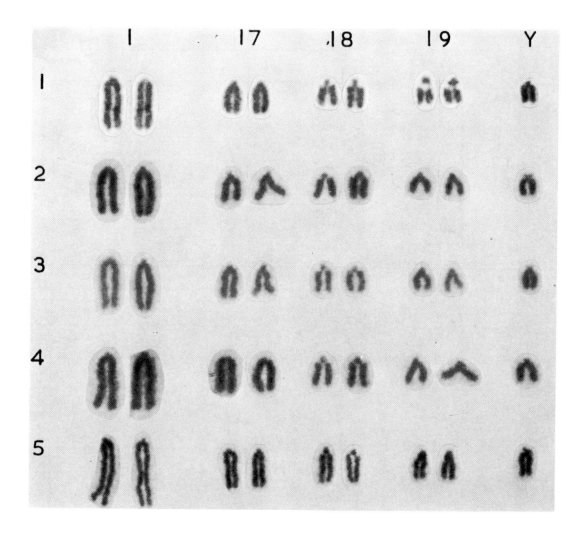

Selected chromosomes from five somatic cells of male CBA mice to demonstrate the distinctiveness of the two shortest pairs of autosomes (18 and 19) and the Y chromosome. Autosome "pairs" 1 and 17 were chosen arbitrarily to indicate the range in length of the remaining chromosomes. Air dried preparation of lymph node and spleen.

(Courtesy Dr. C. E. Ford, from TRANSPLANTATION 4:333, 1966)

WEIGHTS AND LENGTHS OF MOUSE EMBRYOS FROM 9½ GESTATION DAY TO 18½ GESTATION DAY AT HALF DAY INTERVALS

GESTATION AGE OF MOUSE EMBRYO: DAYS	TOTAL NUMBER OF YOUNG IN 5 LITTERS	AVERAGE WEIGHT IN GRAMS	RANGE IN INDIVIDUAL WEIGHTS: gms.	AVERAGE LENGTH IN mm.	RANGE OF INDIVIDUAL LENGTHS: mm.
9½	38	0.0027		2.8	2 – – – – 3.5
10	47	0.0051		3.5	2 – – – – 4.5
10½	42	0.015	0.01 – – – – – 0.025	4.34	3 – – – – 5
11	43	0.0195	0.01 – – – – – 0.025	5.1	3 – – – – 6.5
11½	53	0.029	0.02 – – – – 0.04	6.1	4 – – – – 6.5
12	37	0.0417	0.02 – – – – 0.06	7.0	5 – – – – 8
12½	39	0.0637	0.04 – – – – – 0.08	8.03	7 – – – – 9
13	44	0.0827	0.05 – – – – 0.09	9.09	7 – – – – 9.5
13½	46	0.0961	0.035 – – – – 0.014	9.31	6 – – – – 10.5
14	45	0.148	0.08 – – – – 0.18	10.4	9.5 – – – – 11
14½	40	0.161	0.10 – – – – 0.22	10.7	9.5 – – – – 11.5
15	34	0.256	0.23 – – – – 0.32	12.51	11 – – – – 14
15½	36	0.317	0.16 – – – – 0.43	13.31	11 – – – – 14.5
16	46	0.439	0.20 – – – – 0.59	15.18	11 – – – – 17
16½	41	0.563	0.41 – – – – 0.70	16.7	15 – – – – 18.5
17	46	0.661	0.33 – – – – 0.93	17.39	14 – – – – 20
17½	45	0.910	0.79 – – – – 1.08	19.83	18 – – – – 22
18	46	0.877	0.57 – – – – 1.125	19.25	15 – – – – 24.5
18½	41	1.124	0.64 – – – – 1.50	21.53	18 – – – – 25

Note: All data from Bouin-fixed embryos. To correct for live data add 24% to weights and 6.5% to lengths. (Data compiled by L. Skaredoff.)

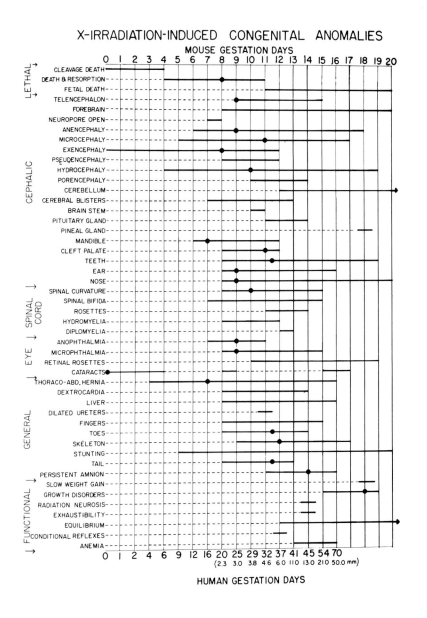

X-IRRADIATION-INDUCED CONGENITAL ANOMALIES

This chart shows the mouse gestation days (above) and the time during development when various x-ray-induced abnormalities can be produced. It will be noted that the majority of these anomalies occur following irradiation on days 8, 9, and 10 when the major activity is organogenesis. The dot represents the period of greatest sensitivity, and the solid line the extent of sensitivity leading to a specific congenital anomaly. It will also be noted that by day 13 the organs have been so well formed (differentiated) that congenital anomalies are more difficult to produce.

(From Rugh 1964 Radiology 82:917)

COMPARISON OF DEVELOPMENT: MOUSE AND HUMAN*

STAGE OF DEVELOPMENT	MOUSE DAYS	MOUSE HOURS	MOUSE SOMITES	HUMAN
2 cells		24—38	0	24—36 hours
4 cells		38—50	0	36—48 ''
5—8 cells		50—64	0	48—72 ''
9—16 cells		60—70	0	72—96 ''
Blastula		74—82	0	4.5 days
Implantation	4	12	0	6+ ''
Proamniotic cavity	5		0	7 ''
Primitive streak	6	12	0	13.5 ''
Head process	7		0	18 ''
Allantois	7	6	0	16.5 ''
Foregut pocket	7	22	0	18 ''
First somites	7	18	2	18 ''
4 somites	8	8	4	19 ''
Right & left heart primordia	8	8	4	19 ''
1st pharyngeal pouch	8	8	4	19 ''
Hindgut pocket	8	1	7	18 ''
Mediam thyroid primordium	8	12	6	19 ''
Optic sulcus	8	13	9	21 ''
1st aortic arch	8	13	9	19 ''
Anterior cardinals	8	14	9—10	21 ''
10 somites	8	14	10	20 ''
Thyroid primordia	8	16	12	20 ''
Otic invagination	8	18	13	21 ''
Anterior neuropore closing	8	18	13	21 ''
Both 1st & 2nd pharyngeal pouches	8	19	14	21 ''
Liver diverticulum	8	20	14—15	22 ''
Dors, mesocardium disappears	8	21	15	22 ''
Nephrogenic cord	8	21	15	20 ''
Ant. neuropore open in prosencephalon only	8	21	15	22 ''
Deep otic invagination	8	22	16	23 ''
Ant. neuropore closed	9	1	18+	23 ''
Otic cyst closed	9	2	19	28 ''
Oral membrane perforate	9	2	19	25 ''
1st & 2nd aortic arches	9	4	20	23 ''
Dorsal aorta fuses	9	4	20	23 ''
Dorsal flexure disappears	9	0		26 ''
Post. cardinal channel	9	4	20	26 ''
Rathke's pocket	9	9	23	29 ''

* Adapted from data in "Equivalent ages in mouse and human embryos" by E. M. Otis and R. Brent, 1952, UR-194.
Note: Mouse strains may differ as much as by 24 hours, and the human ages given are approximate and may be off by 1 week or more, particularly with the later stages.

STAGE OF DEVELOPMENT	MOUSE DAYS	MOUSE HOURS	MOUSE SOMITES	HUMAN
Anterior limb bud	9	9	23	26 days
Lat. thyroid diverticulum	9	12	24	23 ''
Proliferation of liver	9	12	24	22 ''
Post. neuropore closing	9	12	24	26 ''
Mesonephric tubules & duct	9	12	24	25 ''
Epithelial cords in liver	9	15	25	25 ''
Gall bladder separating	9	15	25	26 ''
Primary lung diverticulum	9	15	25	26 ''
Dors. pancreatic constriction	9	15	25	26 ''
Thickened lens disc	9	15	25	28 ''
Omental bursa	9	15	25	26 ''
3 pharyngeal pouches	9	15	25	24 ''
Endocardial cushions	9	18	26	25 ''
First 3 aortic arches	10	0	27	28 ''
Cerebral evagination	10	0	27	31 ''
Olfactory disc	10	0	27	28 ''
Vitelline veins in liver	10	12	28–36	25 ''
Rt. venous valve primordium	10	12	''	30 ''
Posterior limb bud	10	12	''	28 ''
Olfactory pit	10	12	''	29 ''
Lens vesicle	10	12	''	29 ''
Endolymphatic appendage	10	12	''	29 ''
Aortic arches # 3, 4, 6	10	12	''	
Primary intestinal loops	10	12	''	29 ''
Interventricular septum	10	12	''	29 ''
Mesonephric ducts to U.G. sinus	11	0	37–42	29 ''
Short uretric bud	11	0	''	29 ''
Trachea separate	11	0	''	29 ''
Thyroid remnant to pharynx	11	0	''	29 ''
Lens vesicle closed	11	0	''	31 ''
Ventral pancreatic rudiment	11	0	''	31 ''
Cochlear duct	11	0	''	33 ''
Endocardial cushion fused	11	0	''	35 ''
Subcardinals formed	11	12	''	30 ''
Cephalic umbil. veins atrophy	11	12		32 ''
Ant. limb curved	11	12		29 ''
Aortic-pulmonary septum	11	12		31 ''
Otic mesenchyme	11	12		
Expansion of stomach	11	12		31 ''
Hypophyseal evagination	11	12		33 ''
Pigment in retina	11	12		33 ''
Elongation post. lens cells	11	12		33 ''
Epiphyseal evagination	11	12		33 ''
Septum primum of heart	11	12		35 ''
Epithelium of semi-circ. canals	11	12		35 ''

STAGE OF DEVELOPMENT	MOUSE DAYS	MOUSE HOURS	MOUSE SOMITES	HUMAN
Fusion dors. & vent. pancreas	11	12		35 days
Bronchial areas of rt. lung	11	12		35 "
No lumen in Rathke's pocket	11	12		42 "
Pulmonary vein to left atrium	11	12		35–40 "
Olfactory nerves to brain	11	12		42 "
Vomeronasal organ	11	12		37 "
Mesenchyme of ribs	12	0		31+ "
Secondary bronchi	12	0		36+ "
Choroid fissure closed	12	0		40 "
Chondrification of neural process	12	0		37 "
Semicircular canals formed	12	0		37 "
Chondrification of centrum	12	0		37 "
Periderm present	12	0		45 "
Lens vesicle below surface	12	12		33 "
Subcardinal anastomose	12	12		40 "
Inf. vena cava enters heart	12	12		45 "
Sup. & Inf. colliculus separate	12	12		37+ "
Post. cardinals degenerate	12	12		45 "
Epithelial cords in testis	12	12		45 "
Choroid plexus in IV ventricle	12	12		49 "
Mesenchyme in ocular muscles	12	12		49 "
Rathke's pouch detached	12	12		50 "
Rt. dors. aorta between III & IV disappears	13	0		45 "
Aortic pulmonary septum complete	13	0		35 "
Esophageal submucosa thickens	13	0		35+ "
Atrio-ventricular valve	13	0		37+ "
Lens fibers inside vesicle	13	0		45 "
Otic capsule precartilaginous	13	0		45 "
Inter. ventricular septum complete	13	0		47 "
Primary lid fold	13	0		45 "
Inner neuroblasts of retina	13	0		49 "
Nerve fibers in optic stalk	13	0		49 "
Chondrification of ribs	13	0		45 "
Oculomotor nuclei	13	0		52 "
Initiation of aortic & pulmonary semilunars	13	12		35+ "
Vacuoles in stomach epithelium	13	12		45 "
Interdigital notches in hand plate	13	12		37+ "
Muscular primordia around esophagus	13	12		37 "
Cartilage advanced in humerus	13	12		49 "
Enucleate red cells 1%	13	12		51 "
Skin layers differentiated	13	12		52 "
Aortic pulmonary semilunars	14	12		37+ "
First intestinal villi	14	12		52 "
Cartilage in humerus	14	12		54 "

STAGE OF DEVELOPMENT	MOUSE DAYS	MOUSE HOURS	MOUSE SOMITES	HUMAN
Ossification of frontal & zygomatic arch	14	12		60 days
Saccule & utricle separated	14	12		60 ''
1st cartilage in otic capsule	14	12		60 ''
1st 7 ribs chondrified	14	12		60+ ''
Rudimentary periotic cistern	14	12		60+ ''
Enucleate red cells 25%	14	12		60+ ''
Scala tympani forming	14	12		63+ ''
Continuous muscle fibrils	15	12		63+ ''
Cerebellum fused at mid-line	15	12		68+ ''
Ossification of humerus	15	12		70+ ''
Stratum granulosum	15	12		75+ ''
Nucleate cells 5%	15	12		70 ''
Rib ossification	15	12		80 ''
Perichondrium — otic capsule	16	12		80 ''
Corpus callosum	16	12		90 ''
Ossification of centrum	16	12		90 ''
Alveoli in lungs	16	12		90 ''
Nucleate cells 1%	16	12		95 ''
Proliferation of gastric glands	16	12		105 ''

TECHNIQUE FOR THE STUDY OF EMBRYONIC AND FETAL MOUSE CHROMOSOMES

This technique is a slight modification of that found in Ford and Woollam 1963, Stain Technol. 38:271 with some changes suggested by Dr. C. E. Ford (personal communication).

The pregnant female is injected intraperitoneally with 0.3 ml of 0.025% colchicine 1½ hours before intended sacrifice. This period can be lengthened to 4 hours and might present more figures. Up to about 12 days gestation the fetuses are small enough to be used entirely. After that the fetuses may be removed, decapitated, and the liver excised and used alone. Whether the whole fetus or the excised liver is used, it is placed in 5 ml of 0.1% colchicine in Hank's balanced salt solution and broken up by slow aspiration, using, of course, a separate pipette for each sample. After 1.25 to 1.5 hour in suspension the whole is centrifuged for 5 minutes at 400 rpm and the supernatant removed. Add 5 ml of 1% sodium citrate and let suspension stand 15 to 20 minutes (no longer or the cells will cytolyze). Centrifuge 2 to 3 minutes slowly, to avoid damage to the cells. Add carefully 1 to 2 ml of fresh methanol; acetic acid (3:1), cells gently shaken and fixed at 4°C, for 30 minutes. Centrifuge, remove supernatant, and add sufficient 45% acetic acid to make suspension for slides (about 0.5 ml for 6 slides). Place suspension by drops on slides at 54°C. on hotplate with 2 rows of 4 drops each. Each drop will form a series of concentric rings. When dry, stain in LAO* (lactic-acid orcein) for at least 30 minutes, wash off excess stain with 3 changes of 45% acetic acid and air dry the slides.

Since large aggregates of cells are a problem, a very fine pointed pipette should be used, preferably one made from a 2 mm bore glass tubing drawn out in a flame to a fine point.

Abnormalities may include: 2N+ and 2 N- groups, fragmentation, stickiness and hence fusion of chromosomes, and grossly abnormal configurations. One must first become thoroughly acquainted with the normal compliment of 40 chromosomes, and identify as many pairs as possible before attempting to study anomalies.

*LAO (2 gm. synthetic orcein added to 50 ml glacial acetic acid, 42.5 ml of 85% lactic acid, and 7.5 ml of distilled water as per Welshons et al 1962, Stain Technol. 37:1-5).

TIME OF ORIGIN OF VARIOUS ORGAN PRIMORDIA IN THE MOUSE

ECTODERMAL DERIVATIVES:	7	8	9	10	11	12	13	14	15	16	17	18	19
NERVOUS SYSTEM:													
Neural folds		x											
Neural tube		x											
Neural crests			x										
Anterior neuropore		x											
Posterior neuropore		x											
BRAIN:													
Prosencephalon		x											
Mesencephalon		x											
Rhombencephalon		x											
Cephalic flexure		x											
Pontine flexure					x								
Telencephalon			x										
Olfactory lobes			x										
Lamina terminalis				x									
Diencephalon			x										
Optic vesicles			x										
Epiphysis					x								
Infundibulum			x										
Ant. choroid plexus								x					
Rathke's pocket			x										
Optic recess					x								
Mesencephalon :													
Cerebral peduncle							x						
Aqueduct of Sylvius					x								
Pons Varolli								x					
Isthmus				x									
Tuberculum posterius						x							
Metencephalon:			x										
Myelencephalon:			x										
Metatela (thin roof)				x									
Tela choroidea				x									
Post. choroid plexus						x							
NERVES:													
Cranial I				x									
Cranial II				x									
Cranial III							x						
Cranial IV					x								
Cranial V		x											
Cranial VI			x										
Cranial VII		x											
Cranial VIII		x											
Cranial IX		x											
Cranial X		x											
Cranial XI			x										
Cranial XII			x										

		GESTATION			DAYS							
ECTODERMAL DERIVATIVES:	7	8	9	10	11	12	13	14	15	16	17	18	19
SPINAL CORD:													
Marginal layer					x								
Mantle layer					x								
Ependymal layer					x								
Dorsal sensory roots			x										
Ventral motor roots					x								
Dorsal root ganglia			x										
Spinal nerves					x								
SENSE ORGANS:													
EYE:													
Optic vesicle		x											
Optic cup			x										
Optic stalk			x										
Optic chiasma							x						
Retinal layer							x						
Pigmented layer							x						
Lens				x									
Choroid fissure				x									
Vitreous humor						x							
Lachrymal gland											x		
Cornea											x		
Conjunctiva											x		
Iris											x		
EAR:													
Auditory vesicle		x											
Auditory cup			x										
Endolymphatic duct				x									
Sacculus					x								
Utriculus					x								
Cochlea						x							
Semicircular canals						x							
Otic capsule										x			
Pinna							x						
NOSE (Olfactory Organ):													
Placode			x										
Nasal pit				x									
Nasal sinus						x							
Naso-lachrymal duct								x					
Posterior choanae									x				
Nasal septum								x					
Nasopharyngeal duct									x				
Vomeronasal organ						x							
Palate										x			
Nasopharyngeal meatus										x			
Lat. nasal mucous glands											x		
Nares open													x
TAIL BUD:				x									
FORELIMB BUD:			x										

	G E S T A T I O N D A Y S											
ECTODERMAL DERIVATIVES:	7	8	9	10	11	12	13	14	15	16	17	18	19
HINDLIMB BUD:				x									
<u>SKIN</u>:													
Hair germ						x							
Vibrissae roots						x							
Periderm								x					
Stratum germanitivum									x				
Stratum corneum										x			
Mammary welts				x									
<u>STOMODEUM</u>:			x										
<u>PROCTODEUM</u>:			x										
ENDODERMAL DERIVATIVES:													
<u>DIGESTIVE SYSTEM</u>:													
Yolk sac	x												
Foregut (AIP)	x												
Hindgut (PIP)		x											
Oral cavity			x										
Palate						x							
Tongue					x								
Teeth								x					
Pharynx			x										
Pharyngeal pouches			x										
Thymus gland						x							
Thyroid gland			x										
Thyroid lobular						x							
Parathyroid gland					x								
Ultimobranchial body					x								
Salivary glands								x					
Larynx					x								
Esophagus					x								
Stomach					x								
Duodenum					x								
Pancreas					x								
Pancreatic duct						x							
Liver diverticulum				x									
Liver trabeculae					x								
Spleen							x						
Adrenal gland					x								
Small intestines					x								
Large intestines						x							
Umbilical hernia					x								
Umbilical hernia withdrawn										x			
Rectum							x						
Gall bladder					x								
Allantois	x												
Urinary bladder								x					
Cloaca			x										

| | GESTATION DAYS | | | | | | | | | | | |
ENDODERMAL DERIVATIVES:	7	8	9	10	11	12	13	14	15	16	17	18	19
RESPIRATORY SYSTEM:													
Laryngo-tracheal groove			x										
Trachea					x								
Epiglottis							x						
Bronchi					x								
Lung buds			x										
Bronchioles						x							
MESODERMAL DERIVATIVES:													
BODY CAVITIES													
& MESENTERIES													
Coelom (perit. cavity)		x											
Dorsal mesentery					x								
Ventral mesentery					x								
Dorsal mesocardium				x									
Ventral mesocardium					x								
Pericardial cavity		x											
Diaphragm									x				
Pleura cavity				x									
SKELETAL SYSTEM: (Chondrif.)													
Notochord (endodermal)	x												
Ribs - cartilage						x							
Ribs - bone							x						
Centrum (notochord)							x						
Basioccipital cartilage								x					
Nasal capsule								x					
Otic capsule								x					
Zygomatic arch											x		
Tracheal rings									x				
Sternum									x				
Shoulder girdle									x				
Scapula						x							
Humerus						x							
Radius						x							
Ulna						x							
Carpus						x							
Auditory ossicles											x		
Pre-vertebrae						x							
Meckel's cartilage								x					
Tooth development:													
Molar						x							
Dental lamina									x				
Molar enamel organ									x				
Incisors							x						
Enamel organ								x					
Ameloblasts											x		
Odontoblasts											x		

MESODERMAL DERIVATIVES:	GESTATION DAYS												
	7	8	9	10	11	12	13	14	15	16	17	18	19
CIRCULATORY SYSTEM:													
Heart	x												
Heart trabeculae					x								
Atria		x											
Ventricle		x											
Atrio-ventricular canal				x									
Sinus venosus			x										
Sino-auricular valve													
Bulbus arteriosus			x										
Ductus venosus					x								
Truncus arteriosus					x								
Inter-atrial septum				x									
Atrio-ventricular valve					x								
Arteries:													
Ventral aorta		x											
Aortic arch I			x		0								
Aortic arch II			x		0								
Aortic arch III					x								
Aortic arch IV				x									
Aortic arch V					x	0							
Aortic arch VI					x								
Aortic arch VII (Pulmon.)					x								
Dorsal aorta		x											
Internal carotid artery			x										
Intersegmental arteries					x								
Intercostal arteries					x								
Circle of Willis					x								
Iliac arteries					x								
Allantoic (umbilical) art.					x								
Vitelline (omphalomesent.)		x											
Caudal artery			x										
Basilar artery			x										
Pulmonary artery							x						
Cerebral artery			x										
Blood Islands:		x											
Veins:													
Allantoic					x								
Anterior cardinal			x										
Common cardinal (d. Cuvieri)					x								
Posterior cardinal						x							
Hepatic					x								
Hepatic portal					x								
Pulmonary	x												
Umbilical			x										
Sinus venosus			x										
Inferior vena cava				x									
Intervertebral							x						
Vitelline				x									

MESODERMAL DERIVATIVES:	7	8	9	10	11	12	13	14	15	16	17	18	19
UROGENITAL SYSTEM:													
Genital papilla								x					
Male external organs													
Female external organs													
EXCRETORY SYSTEM:													
Nephrogenic cord		x											
Pronephros			x										
Pronephric duct				x									
Mesonephros				x									
Mesonephric duct					x								
Metanephros						x							
Urogenital sinus					x								
Rectum							x						
Cloacal membrane							x						
Ureter						x							
Urethra										x			
GENITAL SYSTEM:													
Genital ridge primordium				x									
Testis					x								
Seminiferous tubules							x						
Vas deferens					x								
Wolffian duct					x								
Ovaries (genital ridge)			x										
Follicles, early oögonia							x						
Oviducts (Müllerian)			x										
Uterus											x		
MUSCULAR SYSTEM:													
Somites		x											
Myotomes		x											
Dermatome		x											
Scleratome		x											
Myocardia	x												
Myocardial contractions		x											
Myoblasts							x						
Muscle fibrils								x					

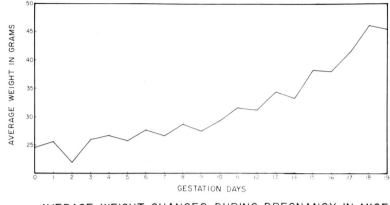

AVERAGE WEIGHT CHANGES DURING PREGNANCY IN MICE

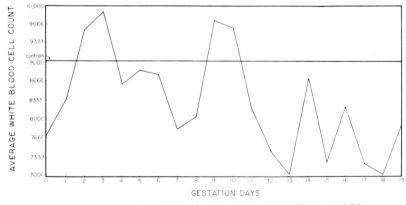

AVERAGE WHITE BLOOD CELL COUNT CHANGES
DURING PREGNANCY IN MICE

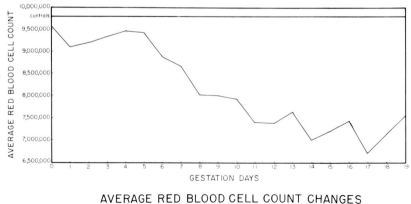

AVERAGE RED BLOOD CELL COUNT CHANGES
DURING PREGNANCY IN MICE

Glossary

ABORTION - Termination of pregnancy at a non-viable stage of the foetus.

ACHRONDROPLASTIC - Refers to miniature adult skeletal condition of some midgets.

ACIDOPHIL - Oxiphil; cell constituents stained with acid dyes, often used to designate an entire cell type. (*See* basophil.)

ACROBLAST - The association of idiosome and acrosome granule including the hypothetical acrosomic vacuole.

ACROSOME - A structure found in spermatids and spermatozoa arising from the acrosomic granule (which itself came from the idiosome of the Golgi body). Its shape is characteristic of the species and may be a rod, cone, or crescent, and there may also be an outer and inner zone. It stains well with PA-FA. It is not to be confused with the perforatorium. Functions in egg penetration.

ACROSOME PHASE - The period in spermiogenesis when the acrosome develops from the acrosomic granule, and the developing spermatid containing it takes the shape of the spermatozoa.

ACROSOME REACTION - Fusion of spermatozoon plasma membrane without a acrosome membrane opening the acrosome cavity and releasing the lytic agent hyaluronidase.

ACROSOMIC GRANULE - The single, round corpuscle which results from the fusion of the proacrosomic granules in the idiosome of young spermatids. (*Syn.*, das Korn, proacrosome, acrosomic bead, acrosome, archosome, mitosome, archoplasmic granule, idiospaerosome).

ACROSOMIC VACUOLE - Crescent-like space around the proacrosomic granules but may be a fixation artefact.

ACTIVATION - Stimulation of spermatozoon to accelerated activity, generally by chemical means (*e.g.*, fertilizin); process of initiating development in the egg; the liberation of naturally occurring evocators from an inactive combination.

ADAPTATION - Functional and correlative change, however brought about.

ADNEXA - Extra embryonic structures discarded before the adult condition is attained.

AER - Apical ectodermal ridge, the thickened outer margin of the mouse limb bud where the epidermis thickens for a short time at about 10.5 to 11.0 days gestation, before the mesenchyme condensations have begun to form for the digits. Disappears by day 13.

AFFINITY - Tendency of cells and tissues of the early embryo to cling together when removed from their normal environment. Equivalent to the cytarme of Roux.

AFTER-BIRTH - Extra-embryonic membranes which are delivered after the emergence of the fetus (mammal). Consists of placenta, amnion and yolk sac.

AGENESIS - Developmental failure of a primordium (*e.g.*, absence of arm or kidney).

AGGLUTINATION - Cluster formation; a spontaneously reversible reaction of spermatozoa to the fertilizin of egg-water. Active agent in zona pellucida.

AGGREGATION - Coming together of cells (*e.g.*, spermatozoa) without sticking, a non-reversible response comparable to chemotropism.

AGNATHUS - Absence of lower jaw.

AKINETIC - Without a kinetochore (*e.g.*, in a chromosome).

ALAR PLATE - Dorso-lateral wall of the myelencephaon, separated from the basal plate by the sulcus limitans.

ALBUGINEA OF TESTIS - The stroma of the primitive testis which forms a layer between the germinal epithelium and the seminiferous tubules.

ALECITHAL EGG - (*See* under EGG).

ALIMENTARY CASTRATION - Prolonged starvation.

ALLANTOIN - Nitrogenous portion of allantoic fluid.

ALLANTOIS - An extra embryonic sac-like extension of the hindgut of amniotes, having the dual function of excretion and respiration.

ALLELOMORPHS - Gene pairs; an allelomorph is one of two dissimilar genes which, on account of their corresponding position (locus) in corresponding (homolgous) chromosomes, are subject to alternative (Mendelian) inheritance in a diploid form. These genes may be identical (homozygous), or dissimilar (heterozygous), names similarly applied to chromosome or to individuals possessing such genes.

ALLO-HAPLOID - Androgenetic haploid.

ALLOMETRY - Study of the relative size's of parts of animals at different absolute sizes, ages, weights, or chemical compositions. Term now used in place of heterogony by Huxley and Teissier (1936).

ALLOMORPHOSIS - The physical or chemical relation of parts of an organism at some early stage to either the whole or part of a later stage, (*e.g.*, the egg size compared with the adult size or weight).

ALLO-POLYPLOID - Polyploid species hybrid.

AMBOCEPTOR - A synonym used for fertilizin in suggesting its double combination with the sperm and egg receptors in the process of fertilization. This double receptor may also receive blood inhibitors, or anti-fertilizin.

AMELOBLASTS - Cells which secrete the enamel cap of the (mammalian) tooth.

AMELUS - Failure of the extremities to develop, remaining as mere stubs.

AMNIOCARDIAC VESICLES - Paired primordia of the parietal cavity which appear in the mesoblast lateral to the head fold of the embryo, and grow beneath the foregut to give rise to the pericardial cavity. Named because of embryonic relation to both amnion and pericardium.

AMNION - Thin, double muscular membrane enclosing the embryos of some invertebrates and of reptiles, birds, and mammals. It is derived from the somatopleure in vertebrates, is filled with fluid which acts as a protective jacket.

AMNIOTIC BANDS - Fibrous bands from the amnion to the embryo due to local necrosis of foetal tissues.

AMNIOTIC RAPHÉ - Point of junction of the amniotic folds as they encircle the embryo, synonymous with sero-amniotic or chorioamniotic junction.

AMPHIMIXIS - Mixing of germinal substances accomplished during fertilization.

ANALOGOUS - Structures said to have the same function but different embryological and/or evolutionary origin. Opposed to homologous.

ANAL PLATE - A thickening and invagination of mid-ventral ectoderm just posterior to the primitive streak which meets evaginating endoderm of the hindgut, later to be perforated as the proctodeum (anus). *Syn.*, cloacal membrane.

ANAPHASE - Phase of mitosis when the paired chromosomes are separating at the equatorial plate and begin to move toward the ends of the spindle.

ANASTOMOSIS - Joining together as of blood vessels and nerves, generally forming a network.

ANDROGAMONES - The anti-fertilizins of Lillie, so named by Hartmann. An acidic protein of low molecular weight.

ANDROGENESIS - Development of the egg with paternal. (sperm) chromosomes only, accomplished by removing the egg nucleus after activation by the spermatozoon but before syngamy (Wilson, 1925). May also be accomplished by irradiation damage of egg nucleus. Non-participation of the female pronucleus in fertilization and development.

ANESTRUS - The quiescent period which follows estrus in the reproductive cycle of the female mammal.

ANEUGAMY - Abnormal ploidy of one or both pronuclei. Fertilization involving a diploid pronucleus in addition to a normal haploid pronucleus.

ANEUPLOIDY - Deviation from normal diploidy but involving partial sets of chromosomes.

ANEUPLOIDY, MULTIFORM - Complex chromosomal mosaics, possibly the result of multipolar mitoses.

ANEUROGENIC - Used in relation to organs developed without proper components of the central nervous system (*e.g.*, limb buds in embryos without spinal cords).

ANGENESIS - Regeneration of tissues.

ANGIOBLAST - Migratory mesenchyme cell associated with formation of vascular endothelium.

ANLAGÉ - A rudiment; a group of cells which indicate a prospective development into a part or organ. *Syn.*, ebauché or primordium.

ANIMALIZATION - Changing by physical or chemical means the presumptive fate of embryonic areas which normally would have become endodermal. *Syn.*, ectodermization, or animalisierung of Lindahl.

ANORMOGENESE - A course of development which deviates in a typical manner from the normal.

ANTERIOR - Toward the head; head end. *Syn.*, cephalic, cranial, rostral.

ANTERIOR INTESTINAL PORTAL - The entrance from the yolk sac into the foregut.

ANTIFERTILIZIN - A sperm substance identified as an acidic protein, found on the surface of the spermatozoon derived from the plasma membrane and combines with fertilizin of great specificity to bring about attachment of the gametes.

ANURY - Tailless mice, homozygous TT, lacking the posterior portion of the body completely, generally die by 10.5 gestation days because of failure to make vascular connection with the mother. May involve anus, urethra, and genital papilla.

ANUS - The posterior opening of the digestive tract, temporarily closed over by the proctodeal membrane.

AORTIC ARCH - Blood vessel which connects the dorsal and ventral aortae by way of the visceral arch.

AORTIC BULB - Embryonic bulge of truncus arteriosus where it swings toward the midline to give rise to the aortic arches.

APICAL REGION (or apex) - Region of head, head cap, or nucleus of the spermatozoon facing the basement membrane of the seminiferous tubule during spermatid maturation. First point of entrance into the ovum.

APROSOMUS - Featureless face due to the arrest of development, the skin covering is normal but lacking in eyes, nose, and mouth.

AQUEDUCT OF SYLVIUS - Ventricle of the mesencephalon (mesocoel) becomes the aqueduct of Sylvius, connecting with the cavities of the optic lobes. *Syn.*, iter.

AQUEOUS HUMOR - Fluid which fills the anterior and posterior chambers of the eye between the lens, probably derived from mesoderm.

ARCHENCEPHALON - Anterior portion of the brain which gives rise to the telencephalon and the diencephalon; pre-chordal brain.

ARCHIPLASM - Specific material which gives rise to the asters and spindle.

ARCUALIA - Small blocks of sclerotomal connective tissue involved in the formation of vertebrae.

AREA - A morphogenetic cell group representing one of the constituent regions of a fate-map, generally of a blastula stage or later.

AREAL - First an invisible, then a sharply differentiated region of the blastema, out of which develops a primitive organ; the organ arising from the blastema through segregation, (organogenetisches or Lehmann).

ARRHENOKARYOTIC - Refers to a blastomere of the normally fertilized egg where there has been a separation of the nuclear components; or in cases of dispermy, where the haploid chromosomes from the single sperm are isolated in the blastomere.

ARRHENOTOKY - Parthenogenetic production of males, exclusively.

ASTER - The "star-shaped structure" surrounding the centrosome, lines radiating in all directions from the centrosome during mitosis.

ASTOMUS - Complete lack or atresia of the mouth.

ASTRAL RAYS - Lines which make up the aster.

ASTROCYTES - Stellate shaped cells arising from the spongioblasts of the mantle layer, classified under the more general term of neuroglia.

ASYNTAXIA DORSALIS - Failure of the neural tube to close.

ATELIOTIC - Arrested development of the skeleton due to non-union of the epiphyses, characteristic of some dwarfs.

ATOKUS - Without offspring.

ATTACHMENT POINT - Point of chromosome to which the spindle fibre is attached and therefore the portion of the chromosome nearest the centrosome in anaphase. *Syn.*, centromere, chromocenter, kinetochore.

ATTRACTION SPHERE - (*See* CENTROSPHERE).

AURICLE - Thin walled chamber(s) of the heart which receives blood from the sinus venosus and delivers it to the ventricle on the same side.

AUTOGAMY - Self-fertilization.

AUTOSOME - Any chromosome except the so-called sex (X or Y) chromosomes.

AUXESIS - Growth by cell expansion but without cell division.

AUXOCYTE - Pre-meitoic germ cell. *Syn.*, primary cyte, meiocyte.

AXENIC - Germ free animal (*Syn.*, gnotobiotic).

AXIAL FILAMENT - The central fibre in the tail of a spermatozoon.

AXIAL MESODERM - That portion of the epimeric mesoderm nearest the notochord. *Syn.*, vertebral plate.

AXIS OF THE EMBRYO - A line representing the antero-posterior axis of the future embryo.

BÄHNUNG - Competence or labile determination.

BALFOUR'S LAW - The intervals between cleavages are longer the more yolk a cell contains in proportion to its protoplasm. "The velocity of segmentation in any part of the ovum is, roughly speaking, proportional to the concentration of the protoplasm there; and the size of the segments is inversely proportional to the concentration of the protoplasm." (Balfour - "Comparative Embryology).

BASAL PLATE - Ventro-lateral wall of myelencephalon, separated from dorso-lateral alar plate by the sulcus limitans.

BASAL ZONE - A region below the stratum spongiosum forming a strip which is contiguous with the myometrium and contains fundic cells of the compact zone, polygonal elements termed decidual cells.

BASOPHIL - Cell constituents having an affinity for basic dyes, often used as an adjective for an entire cell. (*See* acidophil.)

BIO-ELECTRIC CURRENT - An electrical potential characteristic of life, disappearing upon death, associated with activities of muscle, nerve, secretion, and early embryos.

BIOGENETIC LAW - Ontogeny is a recapitulation of the early development of ancestral phylogeny. Embryos of higher forms resemble the embryos of lower forms in certain respects but they are never like the adults of the lower (or ancestral) forms. Not to be confused with the recapitulation theory.

BIOLOGICAL INTEGRATION - Correlation of parts through neural or humoral (or both) influences, acquired during development.

BIOLOGICAL MEMORY - Ontogenetic unfolding of anlagen phyletically accumulated.

BIOLOGICAL ORDER - Fundamental basis of experimental studies, the conformity of biological processes to causal postulates.

BIORGAN - An organ in the physiological rather than the morphological sense.

BIOTONUS - The ratio between assimilation and dissimilation. A/D ratio.

BLASTEMA - An indifferent group of cells about to be organized into definite tissues, kept together by the ectoplasmic matrix of the constituent cells. Considered to be primitive, embryonic, relatively undifferentiated regenerating cell masses. Thought by some to be produced by reserve cells which were arrested during earlier embryonic development.

BLASTOCOEL - Cavity of the blastula. *Syn.*, segmentation of subgerminal cavity.

BLASTOCYST - Mammalian blastula, containing large blastocoel. *Syn.*, blastodermic vesicle.

BLASTOCYT, ACTIVE - Blastocysts on day 4 of the mouse, in normal pregnancy, ready to implant, differing from dormant blastocysts in regard to size, mitotic activity, and fine structure. Can be rendered dormant if transplanted into dormant uterus.

BLASTOCYST, DORMANT - Blastocysts during delayed implantation are slightly larger than normal size. Development and mitotic activity may cease, but such a blastocyst can become active under oestrogen treatment. Chronologically they are older than active blastocysts.

BLASTODERM - "Because the embryo chooses this as its seat and its domicile, contributing much to its configuration out of its own substance, therefore, in the future, we shall call it blastoderm." (Plander, 1817). An embryo composed of a single germ layer.

BLASTODERMIC VESICLE - (*See* Blastocyst).

BLASTOMERE - One of the cells of the early cleavage of an egg. When there is a disparity in size the smaller blastomere is a micromere; the intermediate one is a mesomere; and the larger one is a macromere, but all are blastomeres.

BLASTOTOMY - Separation of cells or groups of cells of the blastula, by any means.

BLASTULA - A stage in embryonic development between the appearance of distinct blastomeres and the end of cleavage (*i.e.*, the beginning of gastrulation); a stage generally possessing a primary embryonic cavity or vesicle known as the blastocoel; invariably monodermic, although the roof may be multi-layered.

BLOOD ISLANDS - Pre-vascular groups of mesodermal cells found in the splanchnopleure, from which arise the blood vessels and compuscles. Largely entra-embryonic.

BOWMAN'S CAPSULE - Double walled glomerular cup associated with uriniferous tubule.

BRACHYPODISM - bp in mouse linkage group 5, reduces the length of the hands and feet so that the mice are incapable of jumping. Metacarpals and metatarsals are shortened, each digit lacks a phalanx. All long bones are shortened, somewhat less than in appendages.

BRACHYURY - Short tailed condition in mice, semi-dominant, heterozygotes have shortened tails of variable length. The spinal column, ribs and sternum may be affected, the notochord is involved, as are most parts posterior to the 6th thoracic vertebra.

BRADYAUXESIS - Negative heterogony, the part grows more slowly than the whole.

BRADYGENESIS - Lengthening of certain stages in development.

BRANCHIAL - Having to do with respiration (*e.g.*, branchial vessel in gill). (*See* visceral).

BRANCHIOMERY - Type of metamerism exemplified in the visceral arches.

BRYSHTHALMIA - Eyes that are too large, may be due to oversized lenses.

BUD - An undeveloped branch, generally an anlagé of appendage (*e.g.*, limb or wing bud).

BUDDING - Reproductive process by which a small secondary part is produced from the parent organism, and which gradually grows to independence.

BULBUS ARTERIOSUS - The most anterior division of the early, tubular, embryonic heart which leads from the ventricle to the truncus arteriosus.

BURSA FABRICII - Endodermal cavity derived from the posterior portion of the embryonic cloaca and communicating with the dorsal part of the proctodeum at the level of the urodeal membrane.

CACOGENESIS - Inability to hybridize; means "bad descent" (kakogenesis).

CAENOGENETIC - Term for new stages in ontogeny which have been intercalated as an adaptation to some inevitable condition which the mode of life of the young animal imposed.

CANALS OF GÄRTNER - Remnants of mesonephric ducts in the broad ligament close to the uterus and vagina.

CAPACITATION - Physiological conditioning of sperm in the female genital tract, allowing the release of their hyaluronidase. A necessary maturing of the spermatozoon before it can penetrate the zona pellucida as well as the cumulus cells.

CAP PHASE - Period of spermiogenesis during which the head cap develops from the acrosomic granule.

CARCINOGEN - A chemical substance which is capable of causing living cells to become cancer-like in growth and behavior.

CARDINAL VEINS - Anterior, posterior, and sub-cardinal veins; anterior veins receive blood from the head, including the first three segmental veins; posterior receive blood from all pairs of trunk segmental veins and veins of the Wolffian bodies; paired sub-cardinals enlarge, fuse, left half degenerates, and the balance fuses with the developing inferior (posterior) vena cava.

CARYOLYSIS - Solution or dissolution of the nucleus.

CARYORHEXIS - Breaking up of the nucleus, or its rupture. (*Syn.*, Karyorhexis).

CAUDAL REGION - Region of sperm head near insertion of tail; side of cell on the lumen side of seminiferous epithelium in case of spermatid; in general, means region opposite the head or apex.

CAUDAL TUBE - In spermiogenesis the intracytoplasmic membrane which forms a collar around the insertion of the tail on the nucleus during the acrosome phase; is lightly stained with hematoxylin.

CAVAL FOLD - The inferior vena cava develops in the lateral of the two plicae mesogastrica which is known as the caval fold.

CELL - Protoplasmic territory under the control of a single nucleus, whether or not the territory is bounded by a discrete membrane. By this definition a syncitium is made up of many cells with physiological rather than morphological boundaries.

CELL CHAIN THEORY - Theory of neurogenesis wherein the peripheral nerve is of pluricellular origin; opposed to the outgrowth theory.

CELL-CONE - A sub-system of an ordered class of cells; a single cell (other than a zygote) and all cells derived from it in a division hierarchy.

CELL THEORY - The body of any living organism is either a structural and functional unit or is composed of a nucleus and its sphere of influence, whether or not that sphere is bounded by a morphological entity. "Omnis cellula e cellula." Virchow

CEMENTOBLASTS - Cement forming cells of tooth.

CENTRAL CANAL - (*See* NEUROCOEL)

CENTRIOLAR APPARATUS - Structure(s) derived from centrioles; ring centrioles, axial formations in neck, middle piece, and tail of spermatozoon.

CENTRIOLE - The granular core of the centrosome, the radiating area comprising the centrosphere. Appears within the centrosome during mitosis.

CENTROMERE - An element of the chromosome thread where it attaches to the spindle, free of DNA.

CENTROSOME - The dynamic center involved in mitosis, including the central granule (centriole) and the surrounding sphere of rays (centrosphere). It is the center of the aster which outlasts the astral rays. Double centrosome called diplosome.

CENTROSPHERE - The rayed portion of the centrosome; the structure in the spermatid which gives rise to the acrosome. *Syn.*, Spermatosphere, idiosome; also attraction sphere.

CEPHALIC FLEXURE - Ventral bending of the embryonic head at the level of the mid- and hindbrain.

CEPHALO-THORACOPAGUS - Fusion of the head and chest regions in conjoined twins.

CEREBRAL FLEXURE - Flexures of the head region, including the cranial, pontine, and the cervical flexures.

CEREBRAL PEDUNCLES - Longitudinal tracts in the floor of the mesencephalon.

CERVICAL CYST - Imperfect occlusion of a branchial (2nd) cleft. *Syn.*, branchial cyst.

CERVICAL FISTULA - Incomplete closure of the branchial cleft.

CERVICAL SINUS - Depression between the third and fourth oral clefts and the body.

CHALONES - Internal secretions with depressing effects, opposed to hormones.

CHEMO-NEUROTROPISM - Chemical attraction of degenerating nerve upon regenerating nerve fibers. The chemical nature of nerve orientation (growth and connections) depending upon diffusing substances which seem to attract nerve fibers.

CHIMERA - Compound embryo derived by grafting together major portions of two embryos, generally of different species, exchange of parts too great to be called a transplant. From Greek mythology: forepart a lion, middle a goat, and hindpart a dragon.

CHOANA - The opening of the internal nares into the oral cavity, primitively in the anterior part of the mouth but secondarily moved posteriorly by growth of the palatine process.

CHONDRIFICATION - The process of forming cartilage, by the secretion of an homogeneous matrix between the more primitive mesoderm cells.

CHONDRIN - Chemical substances in cartilage matrix which makes it increasingly susceptible to basic stains.

CHONDROCRANIUM - That portion of the floor of the skull which is originally cartilaginous, and is later displaced by the basal bones of the skull.

CHORDA DORSALIS - *Syn.*, notochord.

CHORION - An extra-embryonic membrane which develops from the somatopleure as a corollary of the amnion, and which encloses both the amnion and the allantois; consists of inner mesoderm and outer ectoderm. *Syn.*, serosa, false amnion.

CHORION FRONDOSUM - That portion of the mammalian chorion which forms the placenta and adheres to the decidua basalis. Its villi are long, branched, and profuse.

CHORIONIC VILLI - Primary villi are tongue-like and penetrate and erode the endometrium, secondary villi are mesodermal invasions of the primary villi. Embryonic blood vessels develop in this mesoderm and become the tertiary or definitive placental villi.

CHORION LAEVE - That portion of the mammalian chorion except the chorion frondosum.

CHOROID COAT - A mesenchymatous and sometimes pigmented coat within the sclerotic coat but surrounding the (pigmented layer of the) eye in vertebrate embryos.

CHOROID FISSURE - An inverted groove in the optic stalk whose lips later close around blood vessels and nerves that enter the eyeball.

CHOROID KNOT - A thickened region of the fused lips of the choroid fissure, near the pupil from which arise the cells of the iris.

CHOROID PLEXUS - Vascular folds of the thin roof of the telencephalon, diencephalon and myelencephalon, into their respective cavities.

CHROMAFFIN TISSUE - Tissue of the developing adrenal gland which exhibits characteristic reactions with chromic acid salts.

CHROMATID - Longitudinal half of an anaphase, interphase, or prophase chromosome at mitosis. One of four strands (in meiosis) involved in crossing over and visible after pachytene. Becomes a chromosome at metaphase of the second (usually the reduction) division.

CHROMATID BREAKS - The artificial transverse severance of one or both threads after the chromosome has divided longitudinally. When breakage of both threads occurs at the same time this is called isochromatid break.

CHROMATIN - Deeply staining substance of the nuclear network and the chromosomes, consisting of nuclein; gives Feulgen reaction and stains with basic dyes.

CHROMATOBLASTS - Potential pigment cells which, upon proper extrinsic stimulation, will exhibit pigmentation.

CHROMATOID BODY - An irregular structure found in the cytoplasm of the spermatocyte and spermatid, intensely stained with iron hematoxylin.

CHROMATOPHORE - Pigment bearing cell frequently capable of changing size, shape, and color; responsible for superficial color changes in many animals under the influence of the sympathetic nervous system and/or the neuro-humors.

CHROMIDIA - Extra-nuclear granules of chromatin.

CHROMOMERE - Unit of chromosome recognized as a chromatin granule.

CHROMONEMA - Optically single thread within the chromosome, a purely descriptive term without functional implications.

CHROMONUCLEIC ACID - One of the two types of nucleic acid detected in chromatin only. *Syn.*, desoxyribose nucleoprotein, thymonucleic acid. (See plasmonucleic acid.)

CHROMOSOME - The chromatic or deeply staining bodies derived from nuclear network, which are conspicuous during mitotic cell division and which are represented in all of the somatic cells of an organism in a number characteristic for
the species; bearers of the genes, distinct for high content of DNA.

CHROMOSOME ABERRATION - An irregularity in the constitution or the number of
chromosomes which may produce modifications in the normal course of development.

CHROMOSOME, ACENTRIC - Chromosome without a centromere.

CHROMOSOME BREAKS - Artificial transverse severance of a chromosome before its
natural (longitudinal) division into the chromatid threads.

CHROMOSOME DELETION - Loss of an acentric segment of a chromosome.

CHROMOSOME, DICENTRIC - Chromosome with two centromeres.

CHROMOSOME DUPLICATION - The occurrence of a segment twice in the same chromosome.

CHROMOSOME EXCHANGE, INTERCHANGE, OR INTRACHANGE - If two different
chromosomes are broken, illegitimate union between the four broken ends leads
to interchange. If each new chromosome contains a centromere (symmetrical
interchange) there is no mechanical disadvantage attaching to the new formation.
If new chromosome has both centromeres (asymmetrical interchange) the acentric fragment is likely to be lost while the dicentric chromosome may have mechanical difficulties in the later stages of division. Interchanges may occur when
one chromosome is broken in two places and the fragments rejoin in a manner
differing from the original alignments.

CHROMOSOME, HOMOLOGOUS - Similar to allelomorphic chromosomes.

CHROMOSOME INVERSION - The presence in a chromosome of a segment whose parts
are in reverse linear order to that which is normal for that chromosome.

CHROMOSOMES, SEX - In many organisms sex is determined by a special chromosomal mechanism. In man and the fruit fly Drosophila, an embryo destined to develop into a male will have two chromosomes of the "sex" pair as distinguishable,
known as an "X" and a "Y" chromosome. The "Y" is contributed by the spermatozoon, but half of the sperm carry "X" chromosomes. When an "X" bearing
sperm fertilizes an ovum, it then becomes an embryo with "XX" which will become
a female. This seems to be true also for the mouse. In birds, moths, and butter-
flies, the female has the dissimilar pair of sex chromosomes.

CHROMOSOME STRUCTURAL CHANGE - Any alteration in the order or arrangement
of the component particles of one or more chromosomes. May be gross or minute. May include inversions, duplications, deficiencies involving small portions of any chromosome.

CHROMOSOME TRANSLOCATION - The change in position of a segment of a chromosome either to a different chromosome or to another part of the same chromosome.

CHROMOPHOBE - Cells whose constituents are non-stainable; no affinity for dyes.

CILIARY PROCESS - Supporting and contractile elements of the iris which originate
from the lenticular zone.

CIRCLE OF WILLIS - An arterial circle formed by anastomoses between the internal
carotids and the basilar artery; surrounds the pituitary gland.

CIRCULATORY ARCS - Intra-embryonic; vitelline; allantoic circulatory channels, each
involving afferent and efferent tracts and interpolated capillary bed.

CLEAVAGE - The mitotic division of an egg resulting in blastomeres. *Syn.*, segmentation.

CLEAVAGE, IMMEDIATE - Sometimes the second polar body is as large as the maturing egg and may be fertilizable, giving rise to a mosaic individual if fertilized
by another spermatozoon.

CLEAVAGE NUCLEUS - The nucleus which controls cleavage. This may be the syngamic nucleus of normal fertilization; the egg nucleus of parthenogenetic or gynogenetic eggs; or the sperm nucleus of androgenetic development.

CLEAVAGE PATH - Path taken by the syngamic nucleus to the position awaiting the first division.

CLITORIS - The small conical structure of the female mammal which is comparable both in position and in embryological origin to the penis of the male.

CLOACAL MEMBRANE - The endoderm of the large terminal cavity of the hindgut which fuses with the superficial ectoderm at the base of the tail to form a membrane that closes the cloaca to the outside, later to rupture as the anus. (See ANAL PLATE).

COCHLEA - Portion of the original otic vesicle associated with sense of hearing; supplied by vestibular ganglion of eighth cranial nerve, having to do with equilibration.

COELOM - Mesodermal body cavity of chordates, from the walls of which develop the gonads. It is subdivided in higher forms into pericardial, pleural, and peritoneal cavities. Extended as the exocoel or extra embryonic body cavity.

COENOBLAST - The layer which will give rise to the endoderm and mesoderm (obsolete).

COITUS - The sexual act of insertion of the male penis into the vagina of the female and the ejaculation of semen from the penis, sometimes resulting in fertilization.

COLLECTING TUBULE - Portion of nephric tubule system leading to the nephric duct (Wolffian, etc.); term also used to refer to tubules which conduct spermatozoa from the seminiferous tubule to the vasa efferentia, within the testis.

COLLICULI, INFERIOR - Posterior pair of thickenings of dorso-lateral walls of mesencephalon (two or four; corpora bigemina or quadrigemina) containing synaptic centers for auditory reflexes (mammal).

COLLICULI, SUPERIOR - Anterior pair of thickening of dorso-lateral walls of mesencephalon (two or four; corpora bigemina or quadrigemina) containing synaptic centers for visual reflexes in mammals. Syn., optic lobes.

COLLOID - Dispersed substance whose particles are not smaller than 1μ nor larger than 100μ approximately. Physical state of protoplasm.

COMMISSURE, ANTERIOR - Axis cylinders, of spinal cord neuroblasts, which originate in the mantle layer, grow ventrally and cross over to the opposite side of the cord. Refers also to the fibers connecting the cerebral hemispheres, developing in the torus transversus of the lamina terminalis.

COMMISSURE, INFERIOR - Floor of the diencephalon between the mammillary tubercles.

COMMISSURE, POSTERIOR - Roof of the brain between the anterior limit of the mesencephalon and the synencephalon of the forebrain.

COMMISSURE, TROCHLEARIS - Region of the dorsal isthmus, the roof between the metencephalon and the mesencephalon.

COMPETENCE - State of reactivity, of disequilibrium in a complex system of reactants. Possessing labile determination or having reaction possibility. Competencies may appear simultaneously or in sequence within a given area, some to disappear later even without function. Embryonic competence seems to be lost in all adult tissues but may be reclaimed in a blastema. It is a name for the state of the cell area at or before the time when irritability is resolved and a developmental path is chosen. The word supercedes the older words of "potence", "potency", or "potentiality".

CONES OF GROWTH - The enlarged outgrowth of the neuroblast forms the axis cylinder or axone of the nerve fiber and is termed the cone of growth because the growth processes by which the axone increases in length are supposed to be located there.

CONGENITAL - An adjective which refers to (abnormal) conditions acquired during embryonic or fetal life, to be contrasted with genetic and post-natal acquired characters.

COPULATION PATH - The second portion of sperm migration path through the egg toward the egg nucleus, when there is any deviation from the entrance or penetration path; the path of the spermatozoon which results in syngamy.

CORDS, MEDULLARY - Structures which give rise to the urogenital connections and take part in the formation of the seminiferous tubules, and are derived from the blastema of the mesonephric cords.

CORDS, SEX - Strands of somatic cells and primordial germ cells growing from the cortex toward the medulla of the gonad primordium. Best seen in early phases of testes development.

CORNEA - Transparent head ectoderm plus underlying mesenchyme forms a layer directly over the eye.

CORONA RADIATA - Layer of follicle cells which immediately surround the egg of the mammal, and which cells are elongated and acquire intercellular canals radiating outwardly from the egg to the surrounding theca. *Syn.*, follicular epithelium.

CORPORA QUADRIGEMINA - Two pairs of rounded elevations arising from the dorso-lateral thickened walls of the mesencephalon associated with centers of hearing and vision.

CORPUS ALBICANS - Scar tissue over the point of rupture of an ovarian follicle following absorption of the corpus luteum.

CORPUS HAEMORRHAGICUM - Blood clot found in recently ruptured ovarian follicle.

CORPUS LUTEUM - A yellow granular substance derived from the zona granulosa found in the empty follicular vesicle after ovulation in mammals; endocrine function relative to gestation and ovulation. Substance called lutein.

CORPUS STRIATUM - Center of coordination of certain complex muscular activities, derived from the mantle layer of the ventro-lateral walls of the mammalian telencephalon.

CORPUS VITREUM - (*See* VITREOUS HUMOR).

COWPER'S GLANDS - Derivatives of the urethral epithelium, adjacent to the prostate, providing fluid for spermatozoa. *Syn.*, bulbo-urethral glands.

CRANIAL - Relative to the head; "craniad" means toward the head. *Syn.*, rostral, cephalad.

CRANIAL FLEXURE - Forward bending of the forebrain with the angle of the bend occuring transversely at the level of the midbrain. (*See* CEPHALIC FLEXURES).

CRANIOPAGUS - Cranial union in conjoined twins.

CRANIOSCHISIS - Open-roofed skull associated with undeveloped brain.

CREST, NEURAL - Paired cell masses derived from ectoderm cells along the edge of the former neural plate, and wedged into the space between the dorso-lateral wall of the closed neural tube and the integument. Gives rise to spinal ganglia after segmentation.

CREST SEGMENT - The original neural crests which become divided into segments, with the aid of the somites, from which develop the spinal and possibly also some cranial ganglia.

CROSS-FERTILIZATION - Union of gametes produced by different individuals which, if they are of different species, may produce hybrids with variable viability.

CROSSING OVER - Mutual exchange of portions of allelomorphic pairs of chromosomes during the process of synapsis in maturation.

CRYPORCHISM - An exceptional condition in which the testes of the male mammal fail to descend into the scrotum, usually causing sterility because of the body heat.

CRYPTS - Depressions found in the uterine wall for the reception of chorionic cotyledons.

CUMULUS OÖPHROUS - The aggregation of cells which immediately surrounded the mammalian egg within its Graafian follicle.

CUSHION SEPTUM - Two endothelial thickenings arise within the auricular canal and grow together to form a partition known as the cushion septum of the heart.

CUTIS PLATE - (*See* DERMATOME).

CYANOSIS - Mixing of arterial and venous blood in the newborn due to the failure of the ductus Botalli and the incompletely formed interauricular canal, resulting in "blue babies" of high mortality. Bluing of the skin as a result of a circulatory defect.

CYCLE OF SEMINIFEROUS EPITHELIUM - Successive cellular changes which appear in any area of a seminiferous tubule, consisting of 14 stages in the rat, each consisting of typical cell associations. A cycle comprises the period between which two appearances of the same cell association occurs in the same area.

CYCLE, SEXUAL - Cyclic breeding activity, most evident in the female, associated with definite changes in the genital and endocrine systems.

CYCLOPIA - Failure of the eyes to separate; median fusion of the eyes which may be due to suppression of the rostral block of tissue which ordinarily separates the eyes.

CYSTIC DUCT - Narrow proximal portion of embryonic bile duct leading from the gall bladder to the common bile duct.

CYTASTERS - Asters arising independently of the nucleus in the cytoplasm. May contain centrosomes, and achromatic figure with attraction sphere and astral rays and may divide and even cause the cytoplasm around them to divide. Activity and structure unrelated to chromosomal material.

CYTE - A suffix meaning "cell" as oö-cyte (egg forming cell), spermato-cyte (sperm forming cell), or osteo-cyte (bone forming cell).

CYTOCHROME - An oxidizable pigment found in nearly all cells exhibiting definite spectral bands in reduced form, discovered by Keilin (1925). Insoluble in water, poisoned by HCN, CO_2, and H_2S.

CYTOLEOSIS - Process by which a cell, already irreversibly differentiated, proceeds to its final specialization.

CYTOLISTHESIS - Tendency of embryonic cells to aggregate and to fill up disruptions of their union even in the absence of a common surface membrane, due to surface tension and selective adhesiveness (Roux, 1894). Moving of cells over one another by sliding, rotation, or both processes.

CYTOLOGY - The study of cells.

CYTOLYSIS - Breakdown of the cell, indicated by dispersal of formed components.

CYTOPLASM - The material of the cell exclusive of the nucleus; protoplasm apart from nucleoplasm.

CYTOSOME - Cytoplasmic mass exclusive of the nucleus.

CYTOTROPISM - Inevitable movement of a cell in response to external forces.

DECIDUA - The portion of the uterine wall cast off at the time of parturition.

DECIDUA BASALIS - That portion of the uterine wall to which the placenta is attached. *Syn.*, decidua serotina.

DECIDUA CAPSULARIS - The portion of the uterine mucosa and epithelium which covers the mammalian blastocyst opposite the placenta. *Syn.*, decidua reflexa.

DECIDUA VERA - The portion of the uterine wall aside from that associated with either the decidua basalis or the decidua capsularis, but which will eventually be in contact with the decidua capsularis with the expansion of the embryo.

DECIDUAL REACTION - The tissue changes which occur following the burrowing of the fertilized egg into the endometrium of the uterus.

DECIDUATE - The type of placentation which is characterized by the destruction of material (uterine) tissue, and hemorrhage at parturition. True placenta.

DELAMINATION - A separation (of cell layers) by splitting, a process in mesoderm formation.

DENTAL PAPILLA - The mesenchymal portion of the tooth primoridium.

DENTAL RIDGE - A plate of cells which grows into the future gums from the lining of the mouth of the embryo, giving rise to the enamel-forming cells of the tooth,

DENTINE - The main portion of the tooth derived from mesoderm.

DERMAL BONES - Bony plates which originate in the dermis and cover the cartilaginous skull.

DERMATOME - The outer unthickened wall of the somite which gives rise to the dermis. *Syn.*, cutis plate.

DERMIS - The deeper layers of the skin entirely derived from mesoderm (dermatome).

DERMOCRANIUM - The portion of the skul which does not go through an intermediate cartilaginous stage in development. Syn., membrano-cranium.

DETERMINATION - Process of development indicated when a tissue, whether treated as an isolate or a transplant, still develops in the originally predicted manner.

DETERMINATION OF SEX - Method by which the sex of the unborn is revealed. Not to be confused with sex determination which is generally achieved at the time of fertilization.

DEUTOPLASM - Yolk or secondary food substance of the egg cytoplasm, non-living.

DEVELOPMENT - Gradual transformalion of dependent differentiation into self-differentiation; transformation of invisible multiplicity into a visible mosaic, elaboration of components in successive spatial hierarchies.

DIAKENISIS - A stage in maturation when the double chromosomal threads fuse to form the haploid number, generally in curious shapes, and then line up for the first two maturational divisions. The nuclear membrane is still present.

DIAPHRAGM - (See SEPTUM TRANSVERSUM).

DICHIRUS - Partial duplication of digits in hand or foot, possibly inherited. A type of polydactyly.

DIENCEPHALON - The portion of the forebrain posterior to the telencephalon, including the second and third neuromeres.

DIESTRUS - Short period of quiescence immediately following estrus in the mammalian sexual cycle.

DIFFERENTIATING CENTER - Area responsible for the localization and determination of various regions of the embryo, resulting in harmonious proportioning of parts.

DIFFERENTIATION - Acquisition of specialized features which distinguish areas from each other; progressive increase in complexity and organization, visible and invisible; elaboration of diversity through determination leading to histogenesis; production of morphogenetic heterogeneity. Syn., differenzierung.

DIFFERENTIATION, CELLULAR - The process which results in specialization of a cell as measured by its distinctive, actual, and potential functions.

DIFFERENTIATION, CORPORATIVE - Differentiation resulting from the physiological functioning of parts.

DIFFERENTIATION, DEPENDENT - All differentiation that is not self-differentiation; the development of parts of the organism under mutual influences, such influences being activating, limiting, or inhibiting. Inability of parts of the organism to develop independently of other parts. Such a period in ontogeny always precedes that of irreversible determination. "Experimental embryology is a study of the differentiations which are dependent, causally effected." (Roux, 1912). Syn., correlative differentiation, Abhangige Differenzierung, Différentiation provoqueé.

DIFFERENTIATION, FUNCTIONAL - Differentiation of tissues resulting from forces associated with functions (stresses and strains) which they are performing.

DIFFERENTIATION, POTENCY - The total repertoire of differentiations, cytological and histological, available to a given cell. Wider significance than prospective fate.

DIFFERENTIATION, SELF - The perseverance in a definite course of development of a part of an embryo, regardless of its altered surroundings. Syn., différentiation spontanée.

DIFFUSE PLACENTA - Placenta in which the individual villi are scattered over the intercotyledonary areas (e.g., Giraffe).

DIKINETIC - Dicentric, having two kinetochores.

DIMEGALY - Possessing spermatozoa or ova of two sizes.

DIOCOEL - The cavity of the diencephalon, the ultimate third ventricle.

DIPHYGENIC - Having two types of development.

DIPLICHROMOSOME - Two identical chromosomes, held together at the kinetochore and originated by doubling of chromosomes without separation of daughter chromosomes.

DIPLOID - Normal number of chromosomes double the gametic or haploid; complete set of paired chromosomes as in the fertilized egg or somatic cell.

DIPLOTENE - The stage in maturation following the pachytene when the chromosomes again appear double and do not converge toward the centrosome. Sometimes refers to split individual chromosomes.

DIPYGUS - (*See* duplicatus inferior).

DISCOIDAL PLACENTA - Placenta developing on only one side of the blastocyst (*e.g.*, Rodents) in the general shape of a button or disc. Opposed to zonary placenta.

DISCUS PROLIGERUS - The oöcyte of the mammal is attached to the inner wall of the follicle by its neck which, along with the cells surrounding the ovum, are together known as the discus proligerus.

DISTAL - Farther from any point of reference, away from the main body mass.

DIVERTICULUM - The blind outpocketing of a tubular structure (*e.g.*, liver or thyroid anlagé).

DIVISION HIERARCHY - "Four dimensional array of cells of which one and only one member (the zygote) is before all other members in time, and is the only one to which every other term stands in a relation which is some power of D (*i.e.*, the relation is Dpa)." (Bertelanffy & Woodger, 1933).

DNA - A double helix consisting of two polynucleotide chains linked by pairs of purine and pyrimidine bases. Genetic information is coded in the order in which these bases are dispoded. There are two functions: self-replicating and heterocatalytic.

DOMINANCE - A relationship between two allelomorphic genes in a heterozygote. It may be complete or incomplete, in the latter case a heterozygote could be distinguished from a homozygote dominant.

DOMINANT - When (A) of a gene pair (A + a) in a heterozygote produces a phenotype resembling that of the homozygote (AA), "A" is regarded as dominant over "a". The alternative is recessive ("a" recessive to "A").

DOPA - 3:4:dioxyphenylalanin, an intermediate oxidation product of tyrosine and one that appears as a precursor of melanin pigment.

DORSAL ROOT GANGLION - The aggregation of neuroblasts which are derived from the neural crests and which send their processes into the dorsal horns of the spinal cord.

DORSAL THICKENING - The roof of the mesencephalon which gives rise to the optic lobes.

DUCT - (*See* ducts under specific names).

DUCTUS ARTERIOSUS - (*See* DUCTUS BOTALLI).

DUCTUS BOTALLI - The dorsal portion of the sixth pair of aortic arches which normally becomes occluded after birth, the remainder of the arch giving rise to the pulmonary arteries. *Syn.*, ductus arteriosus.

DUCTUS CHOLEDOCHUS - Common chamber associated with the duodenum into which the three pancreatic ducts, the hepatoenteric duct and the cystic duct (from the gall bladder) all empty, prior to their developing separate openings into the gut.

DUCTUS COCHLEARIS - The connection between the lagena and the utricle.

DUCTUS CUVIERI - Union of all somatic veins which empty directly into the heart, specifically the vein which unites the common cardinals and the sinus venosus. Sometimes regarded as synonymous with common cardinal.

DUCTUS CYSTO ENTERICUS - The original caudal duct which is derived principally from the right lobe of the liver.

DUCTUS ENDOLYMPHATICUS - The dorsal portion of the original otic vessels which has lost all connection with the epidermis, and which is partially constricted from the region which will form the semi-circular canals.

DUCTUS HEPATO-ENTERICUS - The original cephalic duct which is derived principally from the left lobe of the liver.

DUCTUS VENOSUS - The anastomising sinusoids of the umbilical blood stream form a major channel through the substance of the mammalian liver, to receive blood from the left and right hepatic (omphalomesenteric) veins as it leaves the liver to join the posterior vena cava and then enter the right auricle.

DUODENUM - Portion of the embryonic gut associated with the outgrowths of the pancreas and the liver (bile) ducts.

DUPLICITAS INFERIOR - Conjoined twins fused anteriorly, having two rumps. *Syn.*, dipygus.

DYADS - Aggregations of chromosomes consisting of two rather than four (tetrad) parts, term used to describe condition during maturation process.

DYSTELEOLOGY - Apparent lack of purpose in organic processes or structures although they may ultimately be shown to be teleological.

ECTODERM - Primary germ layer from which are derived the skin and all of its derivatives, the nervous system and sense organs and parts of the mouth and anus.

ECTRODACTYLY - Absence of hallux and sometimes one or two adjacent digits (*Syn.*, oligodactylism).

EDEMA - General puffiness or swelling of parts due to the accumulation of fluid in the tissues.

EFFERENT DUCTULES - Some mesonephric tubules adjacent to the testes which, together with a portion of the mesonephric duct, become the epididymis.

EGG MEMBRANES - Includes all egg coverings such as vitelline membrane, chorion, and the teritary coverings.

EGG RECEPTOR - Part of Lillie's scheme picturing parts that go into the fertilization reaction involving fertilizin. Egg receptor plus amboceptor plus sperm receptor gives fertilization.

EJACULATION - The forcible emission of mature spermatozoa from the body of the male.

EJACULATORY DUCT - The short portion of the mesonephric duct between the seminal vesicles and the urethra.

ELECTRODYNAMIC THEORY OF DEVELOPMENT - Theory that cell mitoses establish a definite differential potential capable of orienting growing nerve roots (axis cylinders) and thereby directing them (*e.g.*, toward the brain).

EMANCIPATION - Dynamic segregation from "autonomisation" establishment of local autonomy within embryonic areas.

EMBRYO - Any stage in the ontogeny of the fertilized egg, generally limited to the period prior to independent food-getting. The stage between the second week and the second month of the human embryo, or between about 4.5 and 15 days for the mouse. (*See* Fetus).

EMBRYOMA - (*See* teratoma).

EMBRYONIC DISC - The portion of the early mammalian embryo where the ectoderm and endoderm are in close contact with each other.

EMBRYONIC FIELD - Region of formative processes within the embryo, larger than the area of ultimate realization of structures concerned.

EMBRYONIC KNOB - The inner, trophoblastic, cell mass of the mammalian embryo (ectodermal).

EMBRYOTHROPH - The materials obtained when the tissues are broken down by the mammalian embryo, prior to the establishment of the placental circulation.

EMBRYOTROPHY - The means or the actual nourishment of the embryo.

ENCAPSIS - Superordinate system within the embryo. Processes may be purposeful for a subordinate system and yet destroy another system to which it itself is subordinate. These relations are called encapsis.

ENAMEL ORGAN - The ectodermal portion of the tooth anlagé.

ENDOCARDIAL CUSHION OF ATRIO-VENTRICULAR CANAL - A median partition dividing the atrium into right and left channels.

ENDOCARDIUM - Delicate endothelial tissue forming the lining of the heart.

ENDOCHRONDRAL BONE - Bone pre-formed in cartilage. *Syn.*, cartilage bone.

ENDODERM - The outermost layer of the didermic mouse gastrula or trophoblast and the germ layer from which are derived the lining of the digestive tract and all of its derivatives.

ENCOLYMPHATIC DUCT - (*See* DUCTUS ENDOLYMPHATICUS).

ENDOLYMPHATIC SAC - (*See* SACCUSS ENDOLYMPHATICUS).

ENDOMETRIUM - (*See* UTERINE MUCOSA).

ENTERON - The definitive gut of the embryo, always lined with endoderm.

ENTYPY - A method of amnion formation in the mammal in which the trophoblast above the embryonic knob is never interrupted; a method of gastrulation (rodents) wherein the endoderm comes to lie externally to the amniotic ectoderm.

EPENDYMAL CELLS - Narrow zone of non-nervous and ciliated cells which surround the central canal (neurocoel), from the outer ends of which branching processes extend to the periphery, such processes forming a framework for other cellular elements in the spinal cord and brain.

EPICARDIUM - Outer thin layer covering the myo-cardium; originally part of the epi-myocardium.

EPIDERMIS - The ectodermal portion of the skin including the cutaneous glands, hair, feathers, nails, hoofs, and some types of horns and scales in various vertebrates.

EPIDIDYMIS - A tubular portion of the male genital system, derived from the anterior part of the Wolffian body, which conducts sperm from the testis to penis.

EPIGAMIC - Tending to attract the opposite sex.

EPIGENESIS - Developing of systems starting with primitive, homogeneous, lowly organized condition and achieving great diversification. Term coined by Harvey, the antithesis of preformation.

EPIGENETIC CRISIS - Change in speed (rate of development) or in direction of a part, rendering it more vulnerable to environmental variables. May be confused with "critical period", a developmental process with interactions. May be correlated with increase in nitrogen synthesis in cells of a specific area.

EPIMERE - The most dorsal mesoderm, that lying on either side of the nerve and notochord, which gives rise to the somites. *Syn.*, axial mesoderm.

EPIMORPHOSIS - Proliferation of material precedes the development of new parts.

EPIPHYSIS - An evagination of the anterior diencephalon of vertebrates which becomes separated from the brain as the pineal (endocrine) gland of the adult.

EPIPLOIC FORAMEN - Opening from the peritoneal cavity into the omental bursa, formed with change in position of the stomach. *Syn.*, foramen of Winslow.

EPITHELIUM - A thin covering layer of cells, may be ectodermal, endodermal, or mesodermal.

EPOOPHORON - The anterior portion of the mesonephros rudimentary in the female but becoming the epididymis in the male bird and mammal.

EQUATIONAL MATURATION DIVISION - The maturational divisions in which there is no (qualitative) reduction in the chromosomal complex, similar in results to mitosis.

EQUATORIAL PLATE - The lateral view of the chromosomes, lined up on the mitotic spindle, prior to any anaphase movement.

ERGASTOPLASM - Basophilic parts of the cytoplasm, mitochondria of cytologists.

ESTRIN - An hormone found in the mammalian ovarian follicle.

ESTROGEN - Secretion product of the ovary which controls estrus and endometrial growth.

ESTROUS CYCLE - The periodic series of changes which occur in the mammalian uterus, related to the preparation of the uterus for implantation of the ovum, and to repair.

ESTRUS - Period of the reproductive cycle of the mammal when the uterus is prepared for implantation of the ovum.

EUCHROMATIN - The part of the regular chromatic structure of the nucleus which is rich in thymonucleic acid, and presumably the genes, alternating (in the chromosomes) with achromatic regions. It is in the form of discs, and takes methyl green stain.

EUPLOIDY - Deviation from the normal diploid condition but involving complete sets of chromosomes.

EUSTACHIAN TUBE - Vestige of the endodermal portion of the hyomandibular pouch connecting middle ear and pharyngeal cavity, lined with endoderm.

EVAGINATION - The growth from any surface outward.

EXOCOEL - The cavity within mesoderm beyond the limits of the embryo, continuous with the coelom. *Syn.*, extra embryonic body cavity; coelom; seroamniotic cavity.

EXOGENOUS - Originating from without the organism.

"EX OVO OMNIA" - All life comes from the egg (Harvey, 1657).

EXTRA EMBRYONIC - Refers to structures apart from the embryonic body, such as the membranes.

FACTOR - The hypothetical determinant of a character in respect of which an organism may show alternative (Mendelian) inheritance.

F_1, F_2 - First and second filial generation.

FALCIFORM LIGAMENT - Portion of the original ventral mesentery between the liver and the ventral body wall, supporting the liver.

FALLOPIAN TUBES - The oviducts of the mammal.

FALSE AMNIOTIC CAVITY - A temporary cavity arising in the dorsal trophoblast of the mammalian embryo, having no connection with the true amniotic cavity.

FERTILIZATION - Activation of the egg by a spermatozoön and syngamy of the two pronuclei; union of male and female gamete nuclei; amphimixis.

FERTILIZATION, PARTIAL - Male pronucleus fails to meet the female pronucleus, and eventually unites with a blastomere nucleus of the 2 or 4 cell stage.

FERTILIZATION, SOMATIC - Sperm penetration of cells of the female genital tract, even without syngamy. May be confused with phagocytosis.

FERTILIZATION MEMBRANE - A non-living membrane seen to be distinct from the egg shortly after fertilization, very probably the vitelline membrane elevated off of the egg (or from which the egg has shrunken away by exosmosis).

FERTILIZIN - A substance in the jelly coat of the egg, a glyco-protein with a molecular weight of 82,000 or more. In solution it is multivalent, the molecules being capable of linking together several spermatozoa. Functions to attract homologous spermatozoa to the jelly coat surface, a necessary preliminary to fertilization. May also be a component of the plasma membrane of the egg.

FETUS - Embryo after it has attained definite characteristics of the species; the embryo of the mammal when it has attained sufficient development to survive independently of the mother, about 17 days for the mouse.

FETUS PAPYRACEUS - Compressed fetus, abnormal: "paper-doll fetus".

FEULGEN REACTION - Schiff's aldehyde test accomplished by hydrolysis of thymonucleic acid to yield the aldehyde which reacts with fuchsin given a brilliant violet or pink color, a specific test for the thymonucleic acid of chromosomes.

FIBRILLATION - Process of formation of (collagenous) fibers by the aggregation of ultramicrons whose axes are nearly parallel. May be the method of axis formation in limb rudiments.

FIELD - Mosaic of spatio-temporal activities within the developing organism constitute fields; areas of instability with positional relations to the whole organism, with which specific differentiations are about to take place (*e.g.*, heart or limb fields). Dynamic system of interrelated parts in perfect equilibrium in the undifferentiated organism. Not a definite circumscribed area (like a stone in a mosaic) but a center of differentiation with intensity diminishing with the distance from the center, and with different fields overlapping. A system of patterned conditions in a self-sustaining configuration. Has a material substratum which may be reduced without fundamentally altering the original field pattern. Field is both heteroaxial and heteropolar. "Morphe concept" of Gurwitsch (1914).

FIELD, MORPHOGENETIC - Embryonic area out of which specific structures will develop; fields which determine the development of form in a unitary structure.

FIELD ORGAN - Area in which a specific organ of the embryo will develop (*e.g.*, eye field.

FLEXURE - Refers to a bending such as the cranial, cervical and pontine flexures. Also dorsal and lumbo-sacral flexures of the pig.

FOLLICLE - A cellular sac within which the egg generally goes through the maturation stages from oögonium to ovum; made up of follicle cells, theca interna and externa.

FOLLICLE, GRAAFIAN - The follicle of the mammalian ovary, including a double-layered capsule; the membrana granulosa, discus proligerus, corona radiata, and follicular fluid.

FONTANELLES - The area on top of the fetus's skull that is not completely covered by bones, the so-called "soft spot" on the newborn child's head.

FORAMEN - (*See* under specific names as INTERATRIAL, EPIPLOIC, MUNRO, etc.)

FORAMEN OVALE - An opening between the embryonic auricular chambers of the heart.

FOREBRAIN - The most anterior of the first three primary brain vesicles, associated with the lateral opticoels. *Syn.*, prosencephalon.

FOREGUT - The more anterior portion of the enteric canal, the first to appear, aided by the development of the head fold. Its margin is the anterior intestinal portal.

FREEMARTIN - Mammalian intersex due to masculinization of a female by its male partner when the foetal circulations are continuous and the sex hormones are intermingled, as in parabiosis.

FRONTAL - A plane at right angles to both the transverse and sagittal, dividing the dorsal from the ventral. *Syn.*, coronal.

FURCHUNG - Division of the egg cell into blastomeres by mitosis.

GAMETE - A differentiated (mature) germ cell, capable of functioning in fertilization. (*e.g.*, spermatozoön, ovum) *Syn.*, germ cell.

GAMETOGENESIS - The process of developing and maturing germ cells.

GANGLION, ACOUSTIC - Eighth (VIII) cranial ganglion from which the fibres of the eighth cranial nerve arise, purely sensory. Ganglion later divides into vestibular and spiral ganglia.

GANGLION, ACUSTICO-FACIALIS - Early undifferentiated association of the 7th and 8th cranial ganglia.

GANGLION, GASSERIAN - The fifth (V) cranial ganglion, carrying both sensory and motor fibres. *Syn.*, trigeminal ganglion, semilunar ganglion. (*See* Trigeminal ganglion.)

GANGLION, GENICULATE - The ganglion at the root of the facial (VII) cranial nerve, carrying both sensory and motor fibres. Later divides into vestibular and spiral ganglia.

GANGLION, NODOSAL - Ganglion associated with the vagus (X) cranial nerve which carries afferent fibres to pharynx, larynx, trachea, oesophagus, thoracic and abdominal viscera.

GANGLION, PETROSAL - Ganglion associated with the glossopharyngeal (IX) cranial nerve, more peripheral than the superior ganglion carrying sensory fibres from pharynx and root of tongue. From myelencephalon.

GARTNER'S CANAL - Remains of the mesonephric duct in female mammals.

GASSERIAN GANGLION - The fifth cranial or trigeminal ganglion, derived from the hindbrain.

GASTRO-HEPATIC OMENTUM - Portion of the ventral mesentery between the liver and the stomach which persists. *Syn.*, ventral mesogastrium.

GASTROCOEL - Major cavity formed during the process of gastrulation. *Syn.*, archenteron.

GASTROSCHISIS - Improper closure of the body wall along the mid-ventral line, persistent externalized viscera.

GASTRULA - The didermic or double-layered embryo. The two layers are ectoderm and endoderm with only positional significance when first formed. In the mouse the endoderm is at first external to ectoderm.

GASTRULATION - Dynamic processes involving cell movements which change the embryo from a monodermic to a di- or tri-dermic form, generally involving inward movement of cells to form the enteric endoderm. Process varies in detail in different forms, but may include epiboly, concrescence, confluence, involution, invagination, extension, convergence - all of which are descriptive terms for morphogenetic movements.

GEL - A system in which there is a reduction in the amount of solvent relative to the amount of solid substance, thereby causing the whole to become viscous (*e.g.*, asters).

GENE - Self-producing molecule transmitted by the chromosome which determines the development of the characters of the individual, some of which may be solely embryonic.

GENES, LINKED - Genes showing the tendency to assort in parental combinations in a heterozygote instead of assorting independently (*See* linkage).

GENETIC LIMITATION - Each cell must react exclusively in accordance with the standards of the species which it represents.

GENITAL - Refers to the reproductive organs or processes, or both.

GENITAL DUCTS - Any ducts which convey gametes from their point of origin to the region of insemination. *e.g.*, collecting tubules, vas deferens, vas efferens, epididymis, seminal vesicle, oviduct (Fallopian tub) uterus, etc. *Syn.*, gonoduct.

GENITAL RIDGE - Initial elevation for the development of the external genitalia; paired mesodermal thickenings between the meso-nephros and the dorsal mesentery of all vertebrates, which are the gonad primordia.

GENITAL TUBERCLE - The ridge at the base of the phallus, also the primordium of the mammalian labioscrotal swellings, primordium of either penis or clitoris.

GENITALIA - Refers to the sexual organs, external or internal.

GENOME - Haploid gene complex; minimum (haploid) number of chromosomes with their genes derived from a gamete.

GENOTYPE - The actual genetic make-up of an individual, regardless of its appearance (opposed to phenotype).

GERM - The egg throughout its development, or at any stage.

GERM CELL - A cell capable of sharing in the reproductive process, in contrast with the somatic cell. (*e.g.*, spermatozoön or ovum.) *Syn.*, gamete.

GERM LAYER - A more-or-less artificial spatial and histogenic distinction of cell groups beginning in the gastrula stage, consisting of ectoderm, endoderm and mesodermal layers. No permanent or clear cut distinctions, as shown by transplantation experiments.

GERM PLASM - The hereditary material, generally referring specifically to the genotype. Opposed to somatoplasm.

GERMINAL EPITHELIUM - The peritoneal epithelium out of which the reproductive cells of both the male and female develop. *Syn.*, germinal ridges, gonadal ridges.

GERMINAL SPOT - *Syn.*, nucleolus of ovum.

GERMINAL VESICLE - The pre-maturation nucleus of the egg.

GESTALTEN - A system of configurations consisting of a ladder of levels; electron, atom, molecule, cell, tissue, organ, and organism, each one of which exhibits specifically new modes of action that cannot be understood as mere additive phenomena of the previous levels. With each higher level new concepts become necessary. The parts of a cell cannot exist independently, hence the cell is more than a mere aggregation of its parts, it is a patterned whole. Coherent unit reaching a final configuration in space. Gestaltung means formation.

GESTATION - Period of carrying the young within the uterus, 19 to 21 days in the various strains of mice.

GIANT CELLS (in placenta) - Contiguous with the spongiotrophoblasts; nuclei invaginated in many places and the recesses contain cytoplasm. Cytoplasm complex, vacuoles and membranes have general appearance of phagocytes. Contains mucopolysaccharides, are fibrous, and sometimes seem to be continuous with Reichert's membrane. Surface is often in direct contact with maternal blood.

GLIA CELLS - Small rounded supporting cells of the spinal cord, derived from the germinal cells of the neural ectoderm.

GLOMERULUS - An aggregation of capillaries associated with the branches of dorsal aorta but lying within the substance of the functional kidney; function is excretory.

GLOMUS - The vascular aggregations within the head kidney or pronephros, never to become a glomerulus.

GLOTTIS - The opening between the pharynx and the larynx.

GNOTOBIOTIC - Animal that is germ free (*Syn.*, axemic).

GOLGI APPARATUS - May be a dense reticular mass of argentophilic or osmiophilic bodies, enclosing argentophobic or osmiophogic inclusions, disposed near the nucleus and often around the centrosome. May also be scattered in small bodies, associated with ribosomes, vesicles, and vacuoles in electron micrographs.

GONAD - The organ within which germ cells are produced and generally matured, (*e.g.* ovary or testis.) *Syn.*, sex or germ gland.

GONIUM - Suffix referring to a stage in the maturation of a germ cell prior to any maturation divisions. *e.g.*, spermatogonium, or oögonium.

GONODUCT - (*See* GENITAL DUCTS).

GONOSOMIC MOSAIC - Individual that may be phenotypically normal, karyotypically normal in tissues examined, but imbalanced in the reproductive cells and meiosis.

GRAAFIAN FOLLICLE - (*See* FOLLICLE, GRAAFIAN).

GRADIENT - Gradual variation of developmental forces along an axis; scaled regions of preference. (*See* AXIS).

GRANULES, CORTICAL - Granules 0.1 to 0.5 microns in diameter, in the egg surface, which disappear upon fertilization. May release a membrane-modifying agent, with progressive changes in the zona pellucida.

GRANULOSA - The layer of follicle cells which surround the mammalian ovum, so called because of their granular appearance when crowded.

GROWTH - Cell proliferation; a developmental (synthetic) increase in total mass of protoplasm at the expense of raw materials; an embryonic process generally following differentiation (*See* heterogony).

GROWTH, ACCRETIONARY - Growth involving increase in non-living structural matter.

GROWTH, AUXETIC - Growth involving increase in cell size alone.

GROWTH, ISOGONIC - Similar rates of growth in different regions of the embryo.

GROWTH, MULTIPLICATIVE - Growth involving increase in the number of nuclei and of cells. *Syn.*, meristic growth.

GROWTH, PARTITION COEFFICIENTS OF - Inherent growth rates (*e.g.*, in limb rudiments) involving changes in proportions.

GROWTH COEFFICIENT - Growth rate of a part relative to the growth rate of the whole (organism) depending on factors inherent in the tissues concerned.

GROWTH POTENTIALS - Capabilities or predispositions for growth.

GROWTH REGULATION - A substance (R) postulated by Harrison, distinct from nutritional factors, present in the circulating medium of the organism, which controls growth.

GUBERNACULUM - The fibrous cord which draws the testis down into the scrotum of the male mammal just before birth.

GYNOGAMONES - Highly acidic, polysaccharide containing protein of low nitrogen content, and elongate, gel-forming molecular structure. Possibly the fertilizins of Lillie, but so named by Hartmann.

GYNOGENESIS - Development of an egg with the egg nucleus alone. This may be brought about by rendering the sperm nucleus functionless for syngamy by irradiation or other means, or by surgical removal. Opposed to androgenesis.

HAEMOTROPHE - The nutritive substances supplied to the embryo from the maternal blood stream.

HAPLOID - Having a single complete set of chromosomes, none of which appear in pairs, the condition in the gametic nucleus. Opposed to diploid, or twice the haploid, where the chromosomes appear as pairs (*e.g.*, as in somatic cells).

HARDERIAN GLAND - A solid ingrowth of ectodermal cells of the conjunctival sac appearing at the innermost angle of the nictitating membrane.

HEAD CAP - Thin membrane arising from the acrosomic granule and extending over a large portion of the nuclear surface of the mature spermatozoon; staining with PA-PSA.

HEAD OF SPERMATOZOÖN - Nucleus, head, cap, and acrosome in spermatids at the acrosome and maturation phases, and in spermatozoa.

HEAT, PERIOD OF - Period of strong mating impulse in some female mammals. *Syn.*, estrus.

HEMIKARYOTIC - Haploid. In merogony, hemikaryotic, arrhenokaryotic, androgenetic or in artificial parthenogenesis, hemikaryotic, thelykaryotic, gynogenetic.

HEMIMELIA - Reduction or absence of one bone of a normal pair such as radius-ulna, tibia-fibula; the entire limb and girdle may be involved. Some evidence as genetic dominant, may be lethal.

HEMIMELUS - Failure of distal portion of appendages to develop.

HEMIPLACENTA - The chorion, the yolk sac, and generally the allantois which together serve as an organ of nutritional supply to the uterine young of marsupials.

HEMIZYGOTE - A diploid organism in which one chromosome is present only once, as in the case of the sex ("X") chromosome in the male of many species.

HEMOTROPHE - Nutritive materials absorbed by the placenta or fetal membranes directly from the circulation of maternal flood. Type of nutrition found in hemochorial placentas.

HEPATIC PORTAL VEINS- Remnants of the posterior portions of the left vitelline or omphalomesenteric vein, supplied with blood mainly from the placenta but also from veins of the alimentary canal and connected with the hepatic veins only through sinuses within the liver. Function eventually assumed by the mesenteric vein.

HEPATIC SINUSOIDS - Maze of dilated and irregular capillaries between the loosely packed framework of hepatic tubules.

HEPATIC VEINS - Veins from the liver to the heart, originating as the anterior portions of the vitelline or the omphalomesenteric veins.

HERMAPHRODITE - An individual capable of producing both spermatozoa and ova.

HERTWIG'S LAW - The nucleus tends to place itself in the center of its sphere of activity; the longitudinal axis of the mitotic spindle tends to lie in the longitudinal axis of the yolk-free cytoplasm of the cell.

HETEROCHROMATIN - Part of the chromatic structure which seems to be related to the formation of the nucleolus. Takes a violet stain after methyl green but is digested away by ribonuclease. Probably represents both thymo- and ribo-nucleic acids.

HETEROPLASIA - Development of a tissue from one of a different kind.

HETEROPLOIDY - Any deviation from the normal diploid number of chromosomes.

HETEROPYCNOSIS - Character of X and Y chromosomes before the first meiotic division in spermatogenesis; X and Y have end-to-end association in synapsis; heteropycnotic X may be genetically inactive.

HETEROZYGOUS - Condition where the zygote is composed of gametes bearing allelomorphic genes. Opposed to homozygous.

HINDBRAIN - The most posterior of the three original brain divisions, the first neuromere of which is larger than the succeeding neuromeres, there being a total of five. *Syn.*, rhombencephalon.

HINDGUT - Portion of embryonic gut just posterior to the posterior intestinal portal. Level of origin of the rectum, cloaca, post-anal gut, allantois, and caudal portions of the uro-genital systems.

HISTOGENESIS - The appearance, during embryonic development, of histological differentiation; the development of tissue differentiation.

HISTOLYSIS - The destruction of tissues.

HISTOTELEOSIS - Process by which a cell-line, already irreversibly differentiated, proceeds to its final histological specialization.

HISTOTROPHE - The nutritive substances supplied to the embryos of viviparous forms from sources other than the maternal blood stream (*e.g.*, from uterine glands). Secretion and degradation product of the endometrium as well as extravasated maternal blood, which undergo absorption. Found in epitheliochorial placental types.

HOMOIOTHERMAL - Refers to condition where the temperature of the body of the organism is under the control of an internal mechanism; the body temperature is regulated under any environmental conditions. Opposed to poikilothermal. *Syn.*, warm blooded (animals).

HOMOLOGOUS - Organs having the same embryonic development and/or evolutionary origin, but not necessarily the same function.

HOMOLOGY - Similarity in structure based upon similar embryonic origin.

HOMOZYGOUS - Condition where the zygote is composed of gametes bearing identical rather than allelomorphic genes.

HORIZONTAL - An unsatisfactory term sometimes used synonymously with frontal, longitudinal, and even sagittal plane or section. Actually means across the lines of gravitational force.

HORMONE - A secretion of a ductless gland which can stimulate or inhibit the activity of a distant part of the biological system already formed.

HUMORAL SYSTEM - Body fluids carrying specific chemical substances which may circulate in formed channels (blood vessels or lymphatics) or diffuse freely in the body cavities or tissue spaces, (*e.g.*, neurohumors of Parker which act on the pigmentary system).

HYALOPLASM - Ground substance of the cell apart from the contained bodies.

HYALURONIDASE - Enzyme from plasma membrane of the sperm, derived from spermatogenic epithelium, and aid to penetrating cumulus oöphorus. When sperm die they release their quanta of this enzyme, but living sperm cannot release it until they have undergone physiological conditioning known as capacitation.

HYBRID - A successful cross between different species, although organism may be sterile (*e.g.*, mule).

HYBRIDIZATION - Fertilization of an egg by sperm of a different species.

HYDROCEPHALUS, CONGENITAL - Generalized disturbance of the membranous skeleton (recessive and lethal at birth), due to a shortening of the chondrocranium. Affects the entire skeleton because the cartilage is abnormal, or delayed in development.

HYDRODYNAMICS - Process by which the detailed architecture of the blood vessels is derived, such details as size, angles or branching, courses to be followed, etc. The internal water pressure may be the cause of specific developmental procedure.

HYMEN - A membrane of varying thickness in different individuals which closes the lower portion of the vagina, and is generally ruptured during the initial coitus. The mouse vagina is generally covered through weaning period.

HYOID ARCH - The mesodermal mass between the hyomandibular and the first branchial cleft, or between the first and second visceral pouches or clefts which gives rise to parts of the hyoid apparatus. *Syn.*, second visceral arch.

HYOMANDIBULAR - Refers to the pouch, cleft, or slit between the mandibular and the hyoid arches.

HYPERINNERVATION - Supplying an organ with more than a single (normal) nerve fiber.

HYPERPLASIA - Overgrowth; abnormal or unusual increase in elements composing a part.

HYPERTROPHY - Increase in size due to increase in demands upon the part concerned.

HYPERTROPHY, COMPENSATORY - Increase in size of part or a whole organ due to loss or removal of part or the whole of an organ (generally hypertrophy in one member of the pair of organs).

HYPOMERE - The most ventral segment of mesoderm out of which develop the somatopleure, splanchnopleure, and coelom. *Syn.*, lateral plate mesoderm.

HYPOMORPHIC - Cells or tissues which are subordinate to formative processes.

HYPOMORPHOSIS - Harmonious underdevelopment.

HYPOPHYSIS - An ectodermally derived structure arising anterior to the stomodeum and growing inwardly toward the infundibulum to give rise to the anterior and intermediate parts of the pituitary gland. *Syn.*, Rathke's pocket.

HYPOPLASIA - Undergrowth or deficiency in the elements composing a part.

HYPOTHALAMUS - Ventral portion of lateral thickening of diencephalon, not clearly distinguishable from the mesothalamic portion of the optic thalamus.

HYPOTHESIS - A complemental supposition; a presumption based on fragmentary but suggestive data offered to bridge a gap in incomplete knowledge of the facts. May even be offered as an explanation of facts unproven, to be used as a basis of expectations to be subject to verification or disproof.

HYSTEROTELY - Formation of a structure is relatively delayed.

IDIOZOME - The material out of which the acrosome is formed during the metamorphosis of spermatid to spermatozoön.

IMPLANTATION - The process of adding, superimposing, or placing a graft within a host without removal of any part of the host; the attachment of the mammalian blastocyst.

IMPLANTATION, DELAYED - Implantation of blastocyst may be delayed in females that are suckling a previous litter. Can also be induced in pregnant, non-suckling mice if they have been ovariectomized after the current ovulation, and are given daily injections of progesterone. Estrogens terminate the delay and implantation follows.

INDECIDUATE PLACENTA - The type of placenta in which each villus simply fits into a crypt, as a plug fits into a socket, from which it is withdrawn at birth without serious hemorrhage or destruction of maternal tissues. *Syn.*, nondeciduate.

INDUCTION - Causing cells to form an embryonic structure which neither the inductor nor the reacting cells would form if not combined; the calling forth of a morphogenetic functional state in a competent blastema as a result of contact. In contrast with evocation, induction is successive, and purposeful in the sense that one structure leads to another. Sometimes loosely used to include evocator influences from non-living materials. Originally meant diversion of development from epidermis toward medullary plate.

INFUNDIBULUM OF THE BRAIN - Funnel-like evagination of the floor of the diencephalon which, along with the hypophysis, will give rise to the pituitary gland of the adult.

INFUNDIBULUM OF THE OVIDUCT - (*See* OSTIUM ABDOMINALE).

INGUINAL CANAL - The canal connecting the body or abdominal cavity with the scrotum in the male, through which the testes may descend just before birth.

INHIBITION, CONTACT - When two normal tissues meet, as in a culture, their forward movement ceases and they form a stable region of attachment at their surfaces of contact.

INNER CELL MASS - The spherical cells on the inner side of the mammalian blastocyst.

INSEMINATION - The process of impregnation; to fertilize.

INSTINCT - "The overt behavior of the organism as a whole" ..... which is in physiological condition to act according to its genetically determined neuromuscular structure when adequate internal and external stimuli act upon it." (Hartmann, 1942, Psychosomatic Med. 4:206.)

INTERATRIAL FORAMEN - (*See* FORAMEN OVALE).

INTERAURICULAR SEPTUM - A longitudinal sheet of (mesodermal) tissue which grows ventrally from the roof of the auricular chamber to divide it into right and left halves.

INTERKINESIS - Resting stage between mitotic divisions.

INTERMEDIATE CELL MASS - The narrow strip of mesoderm which, for a time, joins the dorsal epimere with the ventral hypomere, being made up of a dorsal portion continuous with the dorsal wall of the somite and the somatic mesoderm and a ventral portion continuous with the ventral wall of the somite and the splanchnic mesoderm. Source of origin of the excretory system. *Syn.*, nephrotome or middle plate.

INTERNAL LIMITING MEMBRANE - A membrane which develops on the innermost surface of the inner wall of the optic cup.

INTERSEX - An individual without sufficient sexual differentiation to diagnose as male or female.

INTERSTITIAL CELLS - Specialized cells between the seminiferous tubules of the testes, produce hormones.

INTERSTITIAL TISSUE OF TESTIS - Cell aggregates between the seminiferous tubules of the testis which elaborate a male sex hormone.

INTERVENTRICULAR SEPTUM - A partition growing anteriorly from the apex of the ventricle, which extends from the auricle to the bulbus arteriosus and divides the ventricle.

INTERVERTEBRAL FISSURE - A cleft between the caudal and cephalic divisions of the sclerotome.

INTERZONAL FIBRE - Portion of the spindle fibres located between chromosome groups in the anaphase and telophase stages.

INTESTINAL PORTAL - An opening from the midgut into either the anterior or posterior levels of the formed gut.

INVAGINATION - Movement by in-sinking of the egg surface and forward migration involving displacement of inner materials. The folding or impushing of a layer of cells into a preformed cavity as one of the methods of gastrulation. Not to be confused with involution.

INVOLUTION - Rotation of a sheet of cells upon itself; movement directed toward the interior of an egg; the rolling inward or turning in of cells over a rim. One of the movements of gastrulation. *Syn.*, embolic invagination; einrollung, or umschlag.

IRIS - The narrow zone bounding the pupil of the eye in which two layers of the optic cup become blended so that the pigment from the outer layer invades the material of the inner layer, giving the eye a specific color by variable reflection.

ISAUXESIS - Relative growth comparisons in which the rate of the part is the same as that of the whole. (*Syn.*, isogony.).

ISO-AGGLUTININ - (*Syn.*, for fertilizin.)

ISOGONY - Proportionate growth of parts so that growth coefficient is unity and there are constant relative size differences. Equivalent relative growth rate.

ISO-HISTOGENIC - Uniform with regard to the histologically compatible factors.

ISOMETRY - Study of relative sizes of parts of animals of the same age.

ISOPYCNOSIS - Condition characteristic of all chromosomes of the male, including the X. In females only one X is isopycnotic with the autosomes, the others are heteropycnotic hence the females often react as heterozygotes with regard to the X chromosome.

ISOTROPIC - Synonym for pluripotent.

ISOTROPY - Originally used to mean absence of predetermined axes within the egg; now means condition of egg where any part can give rise to any part of the embryo (*e.g.*, equivalence of all parts of the egg protoplasm).

ITER - (*See* AQUEDUCT OF SYLVIUS).

JACOBSON'S ORGAN - Ventro-medial evaginations from the olfactory pits which later become the glandular and sensitive olfactory epithelia.

JANICEPS - Janus monster, face to face union of conjoined twins.

JANUS EMBRYO - Double monster with faces turned in opposite directions. *Syn.*, duplicatas cruciata typica.

JUGULAR VEINS - Veins which bring blood from the head; the superior or internal jugular being the anterior cardinal veins and the inferior jugular veins growing toward the lower jaw and mouth from the base of each ductus Curieri.

KARYOPLASM - Protoplasm within the confines of the nucleus.

KERN-PLASMA RELATION - Ratio of the amount of nuclear and of cytoplasmic materials present in the cell. It seems to be a function of cleavage to restore the kern-plasma relation from the unbalanced condition of the ovum (with its excessive yolk and cytoplasm) to that of the somatic cell.

KINETOCHORE - Spindle fiber attachment region. *Syn.*, centromere.

LABIA - Latin for "lips", referring to the lateral folds of skin around the orifice of the vagina, possibly homologous to the scrotum.

LABIO-SCROTAL SWELLINGS - The primordia of the labia majora (female) or of part of the scrotum (male) in mammals. Mound-like swellings on either side of the genital tubercle that give rise to the labia of the female and scrotum of the male.

LACHRYMAL GROOVE - A shallow groove between the lateral nasal and the maxillary processes.

LACUNA - Literally a "little lake", referring to the numerous gaps in the mammalian trophoderm into which maternal blood is emptied when the vessels are destroyed.

LAMINA TERMINALIS - The point of suture of the anterior neural folds (*i.e.*, the anterior neuropore) where they are finally separated from the head ectoderm, consisting of a median ventral thickening at the anterior limit of the telencephalon from the anterior side of the optic recess to the beginning of the velum transversum and including the anterior commissure of the torus transversus.

LANGHANS CELLS - Found in the secondary and tertiary chorionic villi, closely opposed on their outer surfaces to the syncitium and their basal surfaces to the basement membrane. Chromophobic or slightly basophilic, meager endoplasmic reticulum (ergastoplasm), contain glycogen early in gestation, no lipids and few mitochondira.

LANUGO - The fine hairy covering of the fetal mammal.

LARYNGO-TRACHEAL GROOVE - A transverse narrowing of the post-branchial region of the embryonic pharynx with consequent formation of a groove which lead posteriorly to the lung primordia.

LARYNX - The anterior part of the original laryngo-tracheal groove which becomes a tube opening into the pharynx by way of the glottis.

LATERAL - Either the right (dextral) or left (sinistral) side; Lateral means toward the side.

LATERAL AMNIOTIC FOLDS - Folds of the amnion extending up over the sides of the embryo, developing as corollaries to, or in consequence of, the head and tail amniotic folds.

LATERAL LIMITING SULCUS - (*See* LIMITING SULCUS).

LATERAL MESOCARDIUM - Septum posterior to the heart extending from the base of each vitelline vein obliquely upward to the dorso-lateral body wall, representing one of the three parts of the septum transversum.

LATERAL MESODERM - (*See* LATERAL PLATE MESODERM).

LATERAL NEURAL FOLDS - (*See* MEDULLARY FOLDS).

LATERAL PLATES OR LATERAL PLATE MESODERM - The lateral mesoblast within which the body cavity (coelom and exocoel) arises. *Syn.*, lateral mesoderm.

LATERAL VENTRICLES OF THE BRAIN - The thick-walled and laterally compressed cavity of the prosencephalon which opens into the third ventricle by way of the foramin of Monro; the walls will become the cerebral hemispheres.

LECITHIN - Organismic fat which is phosphorized in the form of phosphatides.

LENS - A thickening in the head ectoderm opposite the optic cup which becomes a placode; invaginates to acquire a vesicle; and then pinches off into the space of the optic cup as a lens. Inner surface convex; substance fibrous.

LENS PLACODE - The early thickened ectodermal primordium of the lens.

LENTICULAR ZONE - The portion around the rim of the optic cup adjacent to the pupil, separated from the retinal zone in later development by the ora serrata.

LEPTOTENE - A stage in maturation which follows the last gonial division and is prior to the synaptene stage, structurally similar to the resting cell stage. The chromatin material is in the form of a spireme. The term means thin, diffuse.

LESSER PERITONEAL CAVITY - The growth of the liver to the right mesonephros and finally to the portal vein cuts off a portion of the peritoneal cavity giving rise later to the greater and lesser omental spaces on either side of the coeliac fold. *Syn.*, bursa omenti (major and minor).

LIMB BUDS - Swellings on the sides of vertebrate embryos which will eventually give rise to the appendages.

LINKAGE - Exists between two factors when they do not assort independently of one another in a double heterozygote; measured among gametes by frequency of recombination by crossing over.

LIPIDS - Fats and fatty substances such as oil and yolk (lecithin) found in eggs. E.g. cholesterol, ergosterol.

LIPIN - Fats and fatty substances such as oil and yolk (*e.g.*, lecithin) in eggs, important as water holding device in cells as well as insuring cell immiscibility with surrounding media. (*e.g.*, cholesterol, ergosterol).

LIPOGENESIS - Omission of certain stages in ontogeny.

LIPOPHORES - Pigmented cells in the dermis and epidermis, derived from neural crests and characterized by having diffuse yellow (lipochrome) pigment in solution.

LIPOSOMES - Droplets of yellow oil which may be formed by the coalescence of droplets of broken down lipochondria.

LIQUOR FOLLICULI - The fluid of the mammalian Graäfian follicle into which the matured ovum is freed when the discus proligerus is broken, finally to be liberated from the ovary at the time of ovulation. Contains hormones.

LITHOPEDION - Mummified or calcified fetuses; "stone-child".

LOBSTER CLAW - Missing digits in hands or feet, or split hand or foot; probably inherited.

LOCULUS - A local enlargement of the uterus which contains an embryo and its associated membranes.

LUMBO-SACRAL FLEXURE - The most posterior of the four flexures of the embryo.

LUTEIN - The yellow colored material contained in the cells which fill the empty mammalian Graäfian follicles. (*See* CORPUS LUTEUM).

MACROMERE - Larger of blastomeres where there is a conspicuous size difference.

MACROSOMIA - Gigantism, enlarged skeleton due to disturbed function of the pituitary and possibly also the thyroid glands.

MACROSTOMUS - Failure of the primitive mouth slit to reduce normally.

MALFORMATION, CONGENITAL - Abnormality at birth attributed to faulty development; may have a genetic basis but more frequently extrinsic trauma.

MALPIGHIAN BODY - A unit of the functional kidney including Bowman's capsule and the glomerulus. *Syn.*, renal corpuscle, Malphighian corpuscle.

MAMMARY GLANDS - Multiple milk glands of the typical mammal, derived from the milk ridges.

MANDIBULAR ARCH - The rudiment of the lower jaw or mandible, mesodermal, and anterior to the first or hyomandibular pouch. Gives rise to the palato-quadrate and to Meckel's cartilage.

MANDIBULAR GLANDS - Series of solid ingrowths of the oral mucosa extending on both sides of the base of the tongue to near the mandibular symphysis.

MANTLE FIBRES - Those fibres of the mitotic spindle which attach the chromosome to the centrosomes.

MANTLE LAYER OF THE CORD - Layer of the developing spinal cord with densely packed nuclei, slightly peripheral to the germinal cells from which they are derived. Includes the elongated cells of the ependyma.

MARGINAL LAYER OF THE CORD - Layer of the spinal cord peripheral to the mantle layer, practically devoid of nuclei.

MASSA INTERMEDIA - Fusion of thick lateral walls of diencephalon across the third ventricle.

MATERNAL PLACENTA - Uterine mucosal portions of the typical mammalian placenta.

MATRIX - Ground substance surrounding the chromonemata, usually less chromatic and making up the body of the chromosome. *Syn.*, kalymma or hyalonema.

MATRIX, INTERCELLULAR - The cytoplasmic wall substance of cells in a whole blastema which forms an integrated foam structure and because of its continuity, shows a very definite syncitial character.

MATROCLINUS - Tendency to resemble the mother, may occur in either male or female offspring hence is neither sex linked nor sex limited.

MATURATION - The process of transforming a primordial germ cell (spermatogonium or öogonium) into a functionally mature germ cell, the process involving two special divisions, one of which is always meiotic or reductional.

MATURATION PHASE - Sometimes used to designate the last period of spermiogenesis, leading to the release of free spermatozoa. This period begins when a new crop of spermatids arises from secondary spermatocytes in the same region of the seminiferous tubule.

MEATUS VENOSUS - The junction of the primitive omphalomesenteric veins posterior to the sinus venosus, around which develops the substance of the liver. Later it will also receive the left umbilical vein.

MECKEL'S CARTILAGE - The core of the lower jaw derived from the ventral part of the cartilaginous mandibular arch.

MECOMIUM - A green, pasty-like mass of necrotic cells, mucous and bile which accumulate in the digestive tract of the fetus.

MEDIAN PLANE - "Middle" plane (of the embryo). May be median sagittal or median frontal.

MEDULLA OBLONGATA - That portion of the adult brain derived from myelencephalon.

MEDULLARY - (*See* terms under NEURAL, such as canal, groove, plate, substance, tissue, tube.)

MEDULLARY CORDS - That portion of the suprarenal glands which is derived from the sympathetic nervous system; central cords. Also that portion of the embryonic gonad presumably derived from pre-migratory germ cells upon reaching the genital ridge.

MEIOSIS - A process of nuclear division found in the maturation of germ cells, involving a separation of members of pairs of chromosomes. *Syn.*, reductional division.

MELANOKINS - Stimuli which act upon melanophores, such as temperature, humidity, light, hormones, and certain pharmacological agents.

MELANOPHORE, ADEPIDERMAL - Dermal melanophore.

MELANOPHORES - Cell with brown or black (melanin) pigment granules or rods, found in every class of vertebrates. Derived from the neural crests and migrating throughout the body.

MEMBRANA GRANULOSA - The layers of follicle cells which bound the mammalian follicular cavity.

MEMBRANA REUNIENS - A membrance which extends dorsally from the neural arches around the upper part of the neural tube; the line of later chondrification.

MEMBRANE, FERTILIZATION - A membrane representing either the elevated vitelline membrane or a newly formed membrane found at the surface of an egg immediately upon fertilization or following artificial parthenogenetic stimulation (activation); generally considered an adequate criterion of successful activation of the egg. First seen by Fol (1876) on the starfish egg.

MEMBRANE BONE - Bone developed in regions occupied by connective tissue, not cartilage.

MEMBRANE VITELLINE - (*See* VITELLINE MEMBRANE).

MEMBRANES - (*See* EGG MEMBRANES).

MEMBRANOUS LABYRINTH - The parts of the internal ear, lined with ectodermal epithelium and filled with endolymphatic fluid; including the ductus endolymphaticus, the pars superior labyrinthii, and the part inferiro labyrinthii.

MENSTRUATION - Process in Primates caused by decrease in secretion of estrogen and progestone, involving loss of endometrium and some hemorrhage; indication that implantation did not take place.

MEROMORPHOSIS - The new part regenerated is less than the part removed.

MESENCEPHALON - The section of the primary brain between the posterior level of the prosencephalon and an imaginary line drawn from the tuberculum posterius to a point just posterior to the dorsal thickening. Gives rise to the optic lobes, crura cerebri, and the aqueduct of Sylvius. *Syn.*, midbrain.

MESENCHYME - The form of embryonic mesoderm or mesoblast in which migrating cells unite secondarily to form a syncitium or network having nuclei in thickened nodes between intercellular space filled with fluid.

MESENTERY - Sheet of (mesoderm) tissue generally supporting organ systems. (*e.g.*, mesorchium, mesocardium).

MESIAL - *Syn.*, median, medial, middle.

MESOCARDIUM - The mesentery of the heart; may be dorsal, ventral, or lateral. *See* under LATERAL MESOCARDIA.

MESOCOLON - That portion of the embryonic dorsal mesentery which supports the colon.

MESODERM - The third primary germ layer developed in point of time, may be derived from endoderm in some forms and from ectoderm in others. *See* other terms such as MESOBLAST, MESENCHYME, LATERAL PLATE MESODERM, EPIMERE, MESOMERE, HYPOMERE, GASTRAL, PERISTOMIAL, AXIAL, etc.

MESOMERE - Cells of intermediate size when there are cells of various sizes (macromeres being the largest and micromeres the smallest, respectively). Also used as synonym for intermediate cell mass which gives rise to the nephric system.

MESOMETRIUM - Attachment of the uterus to be the coelomic wall.

MESONEPHRIC DUCT - The duct which grows posteriorly from the mesonephros to the cloaca and later becomes vas deferens. (*Syn.*, Wolffian duct).

MESONEPHRIC TUBULES - Primary, secondary, and sometimes tertiary tubules developing in the Wolffian body, functioning in the mammal.

MESONEPHROS - The Wolffian body, or intermediate kidney, functional as kidney in the embryonic mammal. (*See* EPIDIDYMIS. VASA EFFERENTIA).

MESORCHIUM - Mesentery (mesodermal) whiich surrounds and supports the testis to the body wall.

MESOTHELIUM - Epithelial layers or membranes of mesodermal origin.

MESOVARIUM - Mesentery (mesodermal) which suspends the ovary from the dorsal body wall.

METAMERISM - Serial segmentation, as seen in the nervous, muscular and circulatory systems.

METAMORPHOSIS - Change in structure without retention of original form, as in the change from spermatid to spermatozoön.

METANEPHRIC DIVERTICULUM - Bud-like outgrowth of the mesonephric duct just anterior to its juncture with the cloaca.

METANEPHROS - The permanent kidney of mammals, derived from the nephrogenous tissue of the most posterior somite level (renal corpuscles and secreting tubules) and from a diverticulum of the posterior somite level (renal corpuscles and secreting tubules) and from a diverticulum of the posterior portion of the Wolffian duct (collecting tubules and definitive ureter).

METAPHASE - Stage in mitosis when the paired chromosomes are lined up on the equatorial plate midway between the amphiasters, supported by the mitotic spindle, prior to any anaphase movement.

METAPLASIA - Permanent and irreversible change in both type and character of cells; transformation of potencies of an embryonic tissue into several directions, generally an indication of a pathological condition (*e.g.*, bone formation in the lung). It is thought that some differentiated tissue may become undifferentiated and then undergo a new differentiation in a different direction.

METATELA - Thin roof of 4th ventricle of the brain.

METENCEPHALON - The anterior part of the hindbrain (rhombencephalon) which gives rise to the cerebellum and the pons of the adult brain, separated from the mesencephalon by the isthmus, and including neuromere #6.

METESTRUS - Short period of regressive changes in the uterine mucosa in which the evidence of fruitless preparation for pregnancy disappear. *Syn.*, postestrus.

MICROCEPHALUS - Small or pin-headed; a condition due to the arrested development of the cranium and the brain, accompanied by reduced mentality.

MICROGNATHUS - Retarding of lower jaw in the new born.

MICROMERE - Smaller of the cells when there is variation in the size of blastomeres.

MICROMETRY - Measurement of a microscopic object, using an ocular micrometer.

MICROPHTHALMIA - Eyes that are too small, often due to undersized lenses.

MICROSOMIA - Dwarfism, reduced skeleton, due possibly to disturbed function of the pituitary and thyroid glands.

MICROSTOMUS - Small mouth; excessive closure of the mouth.

MICROSURGERY - Procedures described by Spemann, Chambers, Harrison, and others, where steel and glass instruments of microscopic dimensions are used to operate on small embryos.

MIDBRAIN - (*See* MESENCEPHALON).

MIDDLE PIECE - The region of the tail in immature or free spermatozoa which is located between the base of the neck and the ring centriole and in which a spiral-like structure of mitochondrial origin may be found.

MIDDLE PIECE BODY - A cytoplasmic remnant attached to the neck region in the maturing spermatids immediately before their release into the lumin of the seminiferous tubule as spermatozoa. This remnant is no longer visible in the spermatozoa found in the epididymis.

MIDGUT - That portion of the archenteron which will give rise to the intestines and to the yolk stalk; bounded in the early embryo by the anterior and the posterior intestinal portals.

MILIEU - Term used to include all of the physico-chemical and biological factors surrounding a living system (*e.g.*, external or internal milieu).

MILK-RIDGE - A band of tissue between the somites (dorsally) and the level of the heart, liver, and mesonephros (ventrally) in the embryo, which gives rise to the mammary glands.

MITOCHONDRIA - Small, permanent cytoplasmic granules which stain with Janus Green B, Janus Red, Janus Blue, Janus Black 1, Rhodamin B, Diethylsafranin, dilute methylene blue, and which have powers of growth and division and are probably lipoid in nature, and may contain proteins, nucleic acids, and even enzymes. *Syn.*, plastens.

MITOSIS - The process of cell division in somatic cells as distinct from germ cells, in which each of the daughter cells is provided with a set of chromosomes similar to one another and to that possessed by the parent cell; consists of prophase, metaphase, anaphase, and telophase.

MITOTIC INDEX - The number of cells, in each thousand, which are in active mitosis at any one time and place in an organism; the percentage of actively dividing cells.

MONOSPERMY - Fertilization accomplished by only one sperm. Opposed to polyspermy.

MONRO, FORAMINA OF - Tubular connections between the single third and the paired lateral ventricles of the forebrain.

MONSTER, AUTOSITE-PARASITE - Double embryos with great size discrepancy so that the smaller one bears a parasitic relationship to the larger; variously produced.

MONSTER, DICEPHALUS - Double-headed abnormality, produced by any means.

MONSTER, ISCHIOPAGUS - Double embryos, widely separated except at the tail; produced by any means.

MORPHOGENESIS - All of the topogenetic processes which result in structure formation; the origin of characteristic structure (form) in an organ or in an organism compounded of organs.

MORPHOGENETIC MOVEMENTS - Cell or cell area movements concerned with the formation of germ layer (*e.g.*, during gastrulation) or of organ primordia. *Syn.*, Gestaltungsbewegungen.

MORULA - A spherical mass of cells, as yet without segmentation cavity.

MOVEMENT, FORMATIVE - Localized changes in cell areas resulting in the formation of specifically recognizable embryonic regions.

MOVEMENT, HOMOLOGOUS - Movement of homologous muscles in transplanted limbs, the synchronous contraction of muscles.

MÜLLERIAN DUCT - (*See* OVIDUCT).

MUTATION - The inception of a heritable change, of rare occurrence; usually deterimental, but a few may be beneficial.

MYELENCEPHALIC TELA - The thin roof of the myelencephalon. *Syn.*, metatela.

MYELENCEPHALON - The posterior portion of the hindbrain (rhombencephalon) which has a thin roof that becomes the choroid plexus of the fourth ventricle and thick ventral and ventro-lateral walls which give rise to the medulla oblongata. The cranial ganglia 5 to 12 inclusive are associated with this portion of the brain.

MYELOBLASTS - Muscle forming (embryonic) cells.

MYELOCOEL - Cavity of the myelencephalon; fourth ventricle.

MYOBLASTS - Formative cells within the myotome or muscle plate which will give rise to the true striated muscles of the adult.

MYOCARDIUM - The muscular part of the heart arising from the splanchnic mesoblast.

MYOCOEL - Temporary cavities within the myotomes which may have been connected with the coelom.

MYOMATA - Benign tumor, derived from muscle fibers.

MYOTOME - The thickened primordium of the muscle found in each somite. *Syn.*, muscle plate.

NACHBARSCHAFT - Morphogenetic effects produced by contact with other tissues or structures of a developing organ; contiguity effects.

NARES, EXTERNAL - The external openings of the tubes which are connected with the olfactory vesicles.

NARES, INTERNAL - The openings of the tubular organ from the olfactory placodes into the anterior part of the pharynx.

NASAL CHOANAE - Openings of the olfactory chambers into the mouth.

NASAL PIT - (*See* OLFACTORY PIT).

NASO-FRONTAL PROCESS - A median projection overhanging the mouth and separating the olfactory pits.

NASO-LACRYMAL GROOVE - Groove between junction of naso-lateral and maxillary processes, extending to the mesial angle of the eye. Portion becomes tear (naso-lacrymal) duct which drains fluid from conjunctival sac into the nose.

NASO-LATERAL PROCESS - Lateral elevation dorsal to nasal (olfactory) pit.

NASO-OPTIC FURROW - *Syn.*, naso-lacrymal groove.

NEBENKERN - Cytological structure near the nucleus of the early spermatid.

NECK - Small space between the base of the nucleus and the middle piece in maturing spermatid and spermatozoön. Only the axis filament of the tail is visible there in most species.

NECROHORMONES - The chemical substances produced by degenerating nuclei which cause the premature and incomplete divisions of oöcytes in sexually mature mammals.

NECROSIS - Local death of a cell or group of cells, not the whole body.

NEIGHBORWISE - The reaction of a transplant appropriate to its new environment, indicating its plasticity, pluripotency, or lack of determination. *Syn.*, Artsgemäss.

NEMAMERE - One of the physical units composing a gene-string or genonema, which carries the genes. May be composed of several genes, or a single gene may extend over several nemameres. Governs biophysical reactions of the gene-string.

NEOPLASM - A new growth, generally a tumor. Histologically and structurally an atypical new formation.

NEPHROCOEL - The cavity, found in the nephrotome or intermediate cell mass, which temporarily joins the myocoel and the coelom.

NEPHROGENIC CORD - Continuous band of intermediate mesoderm (mesomere) without apparent segmentation, prior to budding off of mesonephric tubules.

NEPHROGENIC TISSUE - The intermediate cell mass, mesomere, or nephrotome which will give rise to the excretory system.

NEPHROSTOME - The funnel-shaped opening of kidney tubules into the coelom; the outer tubules of the mesonephric kidney acquire ciliated nephrostomal openings from the coelom and shift their connections to the renal portal sinus.

NEPHROTOME - The intermediate cell mass.

NEPHROTOMIC PLATE - *Syn.*, intermediate mesoderm, mesomere.

NERVE, ABDUCENS - Sixth (VI) cranial nerve arising from the basal plate of the myelencephalon which controls the external rectus muscles of the eye.

NERVE, ACCESSORY - Eleventh (XI) cranial nerve, motor, its fibers arising from posterior myelencephalon and first six segments of the spinal cord.

NERVE, AUDITORY - Eighth (VIII) cranial nerve, purely sensory, arising from acoustic ganglion and associated with the geniculate ganglion of the seventh nerve.

NERVE, FACIAL - Seventh (VII) cranial nerve, both sensory and motor related to taste buds and facial muscles. Arises from plate of myelencephalon.

NERVE, GLOSSOPHARYNGEAL - Ninth (IX) cranial nerve, somatic motor, arising superior and petrossal ganglia.

NERVE, HYPOGLOSSAL - Twelfth (XII) cranial nerve, somatic motor, arising from posterior myelencephalon and extending to the muscles of the tongue.

NERVE, OCULOMOTOR - The third cranial nerve which arises from neuroblasts in the ventral zone of the midbrain near the median line.

NERVE, OLFACTORY - First (I) cranial nerve, sensory, without ganglion and with non-medullated fibers, which arise from the epithelial linings of the olfactory pits and have synaptic connections in olfactory bulbs with nerves to brain.

NERVE, OPTIC - Second (II) cranial nerve, sensory, nerves arise from neuroblasts of sensory layer of retina, pass through choroid fissure to enter brain at diencephalic floor. In contrast with other cranial nerves these intersect so that each eye has connections with both sides of the brain (optic chiasma).

NERVE, VAGUS - A tenth (X) cranial nerve, mixed, arising from the myelencephalon and associated with jugular ganglion.

NERVES, CRANIAL, PERIPHERAL, SPINAL - Designated purely with respect to morphological position.

NEURAL ARCH - The ossified cartilages which extend dorsally from ten centrum around the nerve cord, involving both the caudal and the cephalic sclerotomes. The cephalic arch of one sclerotome fuses with the caudal of the next to form a single arch which corresponds to a vertebra. *Syn.*, vertebral arch.

NEURAL CANAL - (*See* NEUROCOEL and NEURAL TUBE).

NEURAL CREST - A continuous cord of ectodermally derived cells lying on each side in the angle between the neural tube and the body ectoderm separated from the ectoderm at the time of closure of the neural tube and extending from the extreme anterior to the posterior end of the embryo; material out of which the spinal and possibly some of the cranial ganglia develop, and related to the development of the sympathetic ganglia and parts of the adrenal gland by cell migration.

NEURAL FOLD - Elevation of ectoderm on either side of the thickened and depressing medullary plate; folds which close dorsally to form the neural tube. *Syn.*, medullary folds.

NEURAL GROOVE - The sinking in of the center of the medullary plate to form a longitudinal groove, later to be incorporated within the neural tube (spinal cord) as the central canal. *Syn.*, medullary groove.

NEURAL PLATE - Thickened broad strip of ectoderm along the future dorsal side of all vertebrate embryos, later to give rise to central nervous system. *Syn.*, medullary plate.

NEURAL TUBE - The tube formed by the dorsal fusion of the neural folds, the rudiment of the nerve or spinal cord.

NEUROBIOTAXIS - Concentration of nervous tissue takes place in the region of greatest stimulation.

NEUROBLASTS - Primitive or formative nerve cells, probably derived (along with epithelial and glia cells) from the germinal cells of the neural tube.

NEUROCOEL - The cavity of the neural tube, formed simultaneously with the closure of the neural folds. *Syn.*, central canal, neural canal.

NEUROCRANIUM - The dorsal portion of the skull associated with the brain and sense organs.

NEUROGEN - An evocator which causes neural induction in vertebrates. May include the organizer, chemical substances, carcinogens, estrogens, etc.

NEUROGENESIS, MECHANICAL HYPOTHESIS OF - Mechanical tension of plasm medium in any definite direction is said to orient and aggregate the fibrin micellae in a corresponding direction.

NEUROGLIA - (*See* GLIA CELLS).

NEUROHUMORS - Hormone-like chemical substance produced by nervous tissue, particularly the ends of developing nerves which consequently act as stimulating agents.

NEUROMERE - Apparent metamerism of the embryonic brain, the divisions being prosencephalon-3, mesencephalon-2, and rhombencephalon-6.

NEUROPORE - A temporary opening into the neural canal due to a lag in the fusion of the neural folds at the anterior extremity; in the vicinity of the epiphysis.

NEURULA - Stage in embryonic development which follows gastrulation and during which the neural axis is formed and histogenesis proceeds rapidly. The notochord and neural plate are already differentiated, and the basic vertebrate pattern is indicated.

NON-DECIDUOUS PLACENTA - *See* INDECIDUATE PLACENTA.

NORMALIZING - Formative action anchored in the organization associated with the determination of development, not super-material entelechy but an integral part of the organism itself. Integrating and balancing tendencies.

NOTOCHORD - Rod of vacuolated cells representing the axis of all vertebrates, found beneath the neural tube and dorsal to the archenteron. Origin variable or doubtful, in most cases thought to be derived from or simultaneously with the endoderm.

NOTOCHORDAL CANAL - An exaggerated primitive pit in some mammals which extends into the head process. May be homologous to neurenteric canal.

NOTOCHORDAL SHEATH - Double mesodermal sheath around the notochord consisting of an outer elastic sheath developed from superficial chorda cells and an inner secondary or fibrous sheath from chorda epithelium.

NUCLEOFUGAL - Refers to outgrowth in two or more directions from the nuclear region as a center, such as in the formation of myelin around a nerve fiber, starting at the sheath cell nucleus as a center and growing in two directions.

NUCLEOLONEME - Filamentous parts of the nucleolus, the other being amorphous.

NUCLEOLUS - An oval constituent of the nucleus which fades and disappears just before mitosis as the chromosomes condense and reappear after mitosis when the chromosomes resume their extended state.

NUCLEOPLASM - The gel-like and fluid plasm of the nucleus which is largely protein and some ribonucleic acid (RNA).

NUCLEUS, DEITER'S - Synaptic center at the boundary between the myelencephalon and the metencephalon, where sensory neurones of semicircular canals lead.

NUCLEUS, RED - Synaptic center of the mid-brain which acts as a coordinating pathway for synergic type of muscular control.

ODONTOBLASTS - Dentine forming embryonic cells; columnar shaped outer cells of the mammalian dental papilla.

OEDEMA - Excessive accumulation of water (lymph) in the tissues and cavities of the body; may be subcutaneous and/or intracellular. Due to a block in drainage channels and generally associated with cardiac inefficiency. (EDEMA).

OESOPHAGUS - Elongated portion of the foregut between the future glottis and the opening of the bile duct.

OLFACTORY CAPSULE - The extreme anterior ends of the skull trabeculae which form the cartilaginous capsules around the olfactory organs.

OLFACTORY LOBES - The anterior extremities of the telencephalic cerebral lobes, partially constricted, associated with the first pair of cranial nerves.

OLFACTORY PIT - Depressions within the olfactory placodes which will become the olfactory organs (external nares).

OLFACTORY PLACODE - The thickened ectoderm lateral to the stomodeal region primordia of the olfactory pits.

OLIGODACTYLY - Condition in which the fingers or toes are congenitally fewer than normal.

OLIGOSYNDACTYLISM - Numerical reduction of digits by fusion (syndactylism) or elimination or loss of a digit (oligodactylism). A semi-dominant condition in the mouse with manifestations on all four appendages. If homozygous, it dies in utero.

OMENTAL BURSA - Pouch formed in the dorsal mesogastrium as the embryonic stomach changes its position.

OMENTUM - *Syn.*, mesogastrium; gastro-hepatic omentum. Dorsal membrane which supports the gut.

"OMNE VIVUM E VIVO" - All life is derived from pre-existing life (Pasteur).

OMPHALOMESENTERIC VEINS - The vitelline veins as they enter the body at the level of the anterior intestinal portal; united posterior to the sinus venosus as the meatus venosus. Venous rings join these vessels dorsal to the gut to become a single curved vessel emptying through the liver into the meatus venosus. The vitelline vein is continuous with the omphalomesenterics, and is therefore extra-embryonic.

ONTOGENY - Developmental history of an organism; the sequence of stages in the early development of an organism.

OÖCYTE - The presumptive egg cell after the initiation of the growth phase of maturation. *Syn.*, ovocyte.

OÖGENESIS - The process of maturation of the ovum; transformation of the oögonium to the mature ovum. *Syn.*, ovogenesis.

OÖGONIA - The multiplication (mitotic) stage prior to maturation of the presumptive egg cell (ovum), found most frequently in the peripheral germinal epithelium.

OÖPLASM - Cytoplasmic substances connected with building rather than reserve materials utilized in the developmental process.

OPTIC CHIASMA - Thickening in the forebrain ventral to the infundibulum, found as a bunch of optic nerve fibres in the future diencephalon.

OPTIC CUP - Invagination of the outer wall of the primary optic vesicle to form a secondary optic vesicle made up of two layers; a thick internal or retinal layer continuous at the pupil and the choroid fissure, and a thin external layer which is pigmented. The cavity of the cup becomes the future posterior chamber of the eye.

OPTIC LOBES - The thickened, evaginated, dorso-lateral walls of the mesencephalon. (*See* COLLICULI, SUPERIOR).

OPTIC RECESS - A depression in the forebrain anterior to the optic chiasma which leads to the optic stalks.

OPTIC STALK - The attachement of the optic vesicle to the forebrain, at first a tubular connection between the optic vesicle and the diencephalon. The lumen is later obliterated by the development of optic nerve fibres.

OPTIC THALAMI - The thickened lateral walls of the diencephalon.

OPTIC VESICLE - Evaginations of forebrain ectoderm to form the primary optic vesicles which in turn invaginate to form the secondary optic vesicles or optic cups of the eyes.

OPTICOEL - The cavity of the primary optic cup.

OPTICO-OCULAR APPARATUS - Includes all the structures related to the eye: optic vesicles, optic stalks, and primary optic chiasma, which develop from the simple median anlagé precociously found in the medullary plate.

ORA SERRATA - The line of separation between the retinal and lenticular zones of the eye cup.

ORAL PLATE - Fused stomodeal ectoderm and pharyngeal endoderm to form the oral membrane. Breaks through to form the mouth. Syn., pharyngeal membrane, oral membrane, stomodeal plate.

ORGANIZATION - Indicated by the inter-dependence of parts and the whole. "When elements of a certain degree of complexity become organized into an entity belonging to higher level of organization," says Waddington, "we must suppose that the coherence of the higher level depends on properties which the isolated elements indeed possessed but which could not be exhibited until the elements entered into certain relations with one another." Relations beyond mere chemical equations; bordering on the philosophical idea. Process of differentiation or specialization which takes place according to a definite pattern in space and time, not chaotically in the direction of haphazard distribution (See Gestalten).

ORGANOGENESIS - Emancipation of parts from the whole; appearance or origin of morphological differentiation.

OSSEIN FIBRES - Organic fibres in bone which give it strength and resilience.

OSSIFICATION - The process of bone formation, occurring extensively in the embryo but persisting throughout life, balanced by simultaneous degeneration of bone; occurs in or around cartilage or in membranous form in looser connective tissue.

OSTEOBLASTS - Mesenchymal cells which actively secrete a calcareous material in the formation of bone; bone-forming cells.

OSTEOCLASTS - Bone destroying cells; cells which appear in and tend to destroy formed bone; constantly active, even in the embryo.

OSTIUM ABDOMINALE TUBAE - The most anterior, fimbriated end of the oviduct in female vertebrates; the point of entrance of the ovulated egg into the oviduct. Syn., infundibulum of the oviduct. (See TUBAL RIDGE).

OTIC VESICLE - Syn., auditory vesicle; otocyst.

OTOCEPHALY - Tendency to fusion or approximation of ears, accompanying cyclopia.

OTOCYST - The original auditory vesicle appearing at the level of the rhombencephalon, forming first as a placode. Syn., auditory vesicle.

OTOLITH - Granular concretion found within the (embryonic) ear.

OUTGROWTH NEURONE THEORY - The cells found along the course of a nerve fiber, the fiber developing as a protoplasmic outgrowth (extension) from a single ganglion cell.

OVARY - Sex gland in which ova are produced, characteristic of the female.

OVIDUCTS - The paired Müllerian ducts in both males and females, which generally degenerate in the males.

OVIGEROUS CORDS - Columns or strands of tissue which divide the germinal epithelium of the primordium of the ovary, carrying primordial germ cells with them and later breaking up into nexts of cells each of which contains an oögonium. Syn., egg tubes or cords of Pflüger (mammal).

OVIPOSITION - The process of laying eggs.

OVOGENESIS - (See OÖGENESIS).

OVOGONIA - (See OÖGONIA).

OVOPHILE - Presumed receptor portion of amboceptor suitable to receive the egg receptor, anti-fertilizin, or blood inhibors, in the fertilizin reaction (Lillie).

OVULATION - The release of eggs from the ovary, not necessarily from the body.

OVUM - Latin for egg.

PACHYTENE - Stage in maturation when the allelomorphic pairs of chromosomes are fused (telosynapsis or parasynapsis) so as to appear haploid, during which process crossing over may occur; stage just prior to diplotene. Syn., diplonema. The term means "thick" or "condensed".

PALATINE GLANDS - Oral glands anterior, lateral, and posterior to the choanae.

PALATO-QUADRATE - True ossified bone developing from the proximal parts of the first three visceral arches, a portion of which gives rise to the annulus tympanicus.

PALLIUM - Outer, thickened walls of the telencephalic vesicles which will give rise to the cerebral hemisphere; dorsal and posterior to the sulcus rhinalis.

PANCREAS - Digestive and endocrine glands arising as single dorsal and paired ventral primordia in the vicinity of the liver.

PAPILLARY MUSCLES - Muscles which arise from trabeculae carneae in the heart and later, in conjunction with the tendenous cords, control the heart valves.

PARABIOSIS - Lateral fusion of embryos by injuring their mirror surfaces and approximating them so that they grow together (*See* telobiosis).

PARAPHYSIS - A pouch-like evagination of the telencephalon median.

PARASYNAPSIS - Lateral fusion, term applied to chromosomes of maturation stages in pachytene or to experimental fusion of embryos. (*See* TELOSYNAPSIS).

PARATHYROIDS - Endocrine glands derived from endoderm of the third and fourth pairs of visceral pouches and which control calcium metabolism.

PARENCEPHALON - The anterior (ventral) portion of the diencephalon, separated from the synencephalon by the epiphysis.

PARIETAL CAVITY - (*See* PERICARDIAL CAVITY).

PARIETAL RECESS - Passage between the pericardial and periotoneal cavities of embryos.

PARTHENOGENESIS - Development of the egg without benefit of spermatozoa; development stimulated by artificial means.

PARTHENOGENESIS, ARTIFICIAL - Activation of an egg by chemical or physical means (*e.g.*, butyric acid, hypertonic solutions, irradiation, needle prick, etc.)

PARTITION-COEFFICIENT - The factor which determines the size of any part at any time by parcelling out materials; relative capacity for various parts of the embryo to absorb food from a common supply at different times. Such coefficients are expressions of intrinsic growth potentials, so balanced in normal development that no single structure can monopolize the nutriment to the detriment of other structures.

PATROCLINUS - Paternal influence where the F_1 tend to resemble their fathers more than their mothers.

PENETRANCE - A genetic term referring to the degree or percentage of genotypically abnormal.

PENIS - Elongated genital tubercle, enclosed in genital folds (prepuce) and associated with genital swellings (scrotal pouches) in male; organ of transfer of genital products from the testes of the male to the vaginal cavity of the female. *Syn.*, Phallus.

PERFORATORIUM - A structure lying between the acrosome and the nucleus of the sperm head, presumably derived from the nucleus.

PERIAXIAL CORDS - The primordia of the trigeminus and acustico-facialis ganglia and later mark the paths of the trigeminal and facial nerves. Distinguished as more deeply stained and concentrated masses than mesenchyme just posterior to the optic vesicles.

PERICARDIAL CAVITY - The cavity or membranous sac which encloses the heart, representing a cephalic portion of the coelom of the original amnio-cardiac vesicles within the embryonic body, bounded by the proamnion and posteriorly by the omphalomesenteric veins. *Syn.*, parietal cavity.

PERICARDIUM - The thin mesodermal membrane which encloses the pericardial cavity and heart

PERICHONDRIUM - Mesenchymal layer immediately around forming cartilage.

PERIOSTEUM - Mesenchymal layer, often originally perichondrium, which will be found immediately around forming bone.

PERITONEAL CAVITY - The body cavity (coelom) separated from the pleural cavity by the pleuro-peritoneal septum, including the septum transversum.

PERITONEUM - Coelomic mesothelium of the abdominal region reinforced by connective tissue.

PERIVEITELLINE SPACE - The space between the vitelline (fertilization) membrane and the contained egg, generally filled with a fluid. It is the space between the zona radiata and the egg.

PERIVITELLINE MEMBRANE - (*See* VITELLINE MEMBRANE).

PFLUGER, CORDS OF - The ovigerous layer which grows into the stroma of the ovary as ovigerous cords, carrying primitive ova with them.

PFLUGER'S LAW - The dividing nucleus elongates in the direction of the least resistance.

pH - Method of stating the measure of the hydrogen ion concentration, expressed as the log of the reciprocal of the hydrogen ion concentration in gram-mols per liter. The negative value of the power of 10 equivalent to the concentration of hydrogen ions in gram-molecules per liter. The neutral solution (neither acidic nor basic) has a pH value of 7: pH values less than 7 are acid and those more than 7 are alkaline.

PHAGOCYTOSIS - Process of engulfment of solid particles, generally by specific types of (white) cells known as phagocytes.

PHALLUS - *Syn.*, penis.

PHENOTYPE - The outward appearance of an organism regardless of its genetic make-up, opposed to genotype.

PHEROMONES - Substances produced by animals which elicit responses in other members of the same species. Female mice produce a pheromone which inhibits estrous cycles in other females. Male mice produce a pheromone which induces early puberty in females and more frequent estrus. In sequence they synchronize estrus in mice groups.

PHOCOMELIA - Failure of proximal portion of appendages to develop, distal parts may be normal. Disproportionate shortening of the limbs (micromelia) which may be associated with an otherwise normal axial skeleton, sometimes due to a single recessive gene but may also be caused by drugs (thalidomide). Homozygotes die soon after birth because of extensive median cleft palate which interferes with respiration and sucking.

PHYLOGENY - Series of stages in the history of the race; the origin of phyla.

PIGMENT LAYER OF OPTIC CUP - Thin outer wall of the primary optic cup, posterior to the retina, which never fuses with the rods of the retina.

PINEAL - (*See* EPIPHYSIS).

PINOCYTOSIS - Process of engulfment of water and solutes.

PITUITARY - (*See* HYPOPHYSIS).

PLACENTA - An extra embryonic vascular structure of placental mammals, which serves as an organ of nutritive and respiratory exchange between the fetus and the mother. Basically consists of uterine vascular endothelium, uterine stroma, uterine epithelium, fetal trophoblast, fetal stroma, and fetal capillary endothelium. When all of these structures are present it is known as an epitheliochorial placenta, found in pig and mare.

PLACENTA, ENDOTHELIOCHORIAL - There is loss of maternal connective tissue in addition to the maternal uterine epithelium. Found in carnivores, sloths, insectivores, and bats.

PLACENTA, EPITHELIOCHORIAL - Simple apposition of chorion to the endometrial epithelium without erosion, all membranes present (Cetacea, lemurs, mare, and pig).

PLACENTA, HEMOCHORIAL - Type in which there is ultimate loss of maternal endothelium with erosion of the uterine wall; the trophoblast is bathed directly by circulating maternal blood. Found in rodents, anthropoids, and man.

PLACENTA, HEMOENDOTHELIAL - The placental barrier becomes reduced to a layer of endothelium and a very thin plasmodium, the latter absent in some places.

PLACENTA, SYNDESMOCHORIAL - The uterine epithelium disappears through invasive activity of the trophoblast in cow and sheep.

PLACENTA, YOLK-SAC - Chorio-vitelline, found in mouse, in addition to hemochorial placenta, which engages actively throughout gestation in the absorption of transudate and secretion representing histiotrophe derived from the endometrium. A combination of advanced (hemochorial) and primitive (yolk sac) placentas. Main mediator of antibodies and glucose (mouse).

PLACODE - Plate or button-like thickening of ectoderm from which will arise sensory or nervous structures (*e.g.*, olfactory placode).

PLANE - (*See* "section".)

PLASMODESMATA - Protoplasmic bridges claimed to be the means of nerve fiber growth; plasmodesmata supposedly incorporated into the substance of the axone during its origin.

PLASMONUCLEIC ACID - One of the two types of nucleic acid, this one occuring in the cytoplasm, in the plasmosome (nucleolus), and possibly in minute quantities in the chromatin.

PLASMOSOME - A true nucleolus (*See* NUCLEOLUS).

PLASMOTROPHODERM - Outer layer of syncytial cells of the trophoderm following implantation. *Syn.*, syncytiotrophoderm.

PLASTENS - (*See* mitochondria.)

PLEIOTROPISM - Multiple effects of a single gene due to effects upon metabolism.

PLEURA - Membrane enclosing the cavity surrounding the lungs, consisting of splanchnic mesoderm.

PLEURAL CAVITY - The portion of the coelomic cavity separated ventrally by the septum transversum (pleuro-peritoneal septum) from the peritoneal cavity, and one into which the primary lung buds grow.

PLEXUS, CHOROID - Vascular folds in the roof of the prosencephalon, diencephalon, and rhombencephalon.

POLAR BODY - Relatively minute, discarded nucleus of the maturing oöcyte (generally three). *Syn.*, polocytes.

POLYANDROUS SYNGAMY (polyandry) - Development of two or more small pronuclei leading to the union of these with the female pronucleus.

POLYANDRY - Presence of one female and two male pronuclei in fertilization, arising from polyspermy. May be more common when coitus occurs late in estrus. Opposed to polygyny.

POLYDACTYLY - Extra digits in hands or feet; probably inherited.

POLYEMBRYONY - Natural isolation of blastomeres leading to the production of multiple embryos; development of several embryos from a single zygote.

POLYESTROUS CYCLE - Reproductive cycles which occur at least several times a year.

POLYESTRUS - Mammals which have more than a single estrus cycle in one year.

POLYGYNY - Failure of emission of first or second polar body with the result that the pronucleus unites with two female pronuclei. Opposed to polyandry.

POLYHYDRAMNIOS - Condition where the amniotic fluid exceeds the normal

POLYMEGALY - Several sizes, as for ova or sperm from one strain.

POLYPLOID - Possessing a multiple number of chromosomes, such as triploid (3 times the haploid number) tetraploid (4 times the haploid), etc. Always more than the normal diploid number of the type zygote.

POLYPLOIDOGEN - A chemical substance which brings about the polyploid condition, usually by inhibiting certain phases of nuclear division.

POLYSPERMY - Insemination of an egg with more than a single sperm, although but a single sperm nucleus is functional, in syngamy.

PONS VAROLII - The thickened floor and ventro-lateral zones of the metencephalon.

PONTINE FLEXURE - Cephalic flexure indicated by a ventral bulge in the floor of the myelencephalon.

POST-ANAL GUT - A posteriorly projecting blind pocket of the hindgut; that portion of the hindgut posterior to the anal plate or proctodeal plate. *Syn.*, post-cloacal gut.

POSTERIOR INTESTINAL PORTAL - Opening from the unformed midgut into the formed hindgut.

POSTERIOR TUBERCLE - (*See* TUBERCULUM POSTERIUS).

POSTESTRUS - Short period of regressive changes in the uterine mucosa following fruitless preparation for pregnancy.

POST-REDUCTION - Maturation in which the equational and reductional divisions occur in that order.

POTENCY - Ability to develop embryologically; capacity for completing destiny; ability to perform an action: "future development verbally transformed to an earlier stage" (Waddington). The test of potency is actual realization in development. It is an explanatory rather than a descriptive term for developmental possibility. A piece of an embryo has the possibility of a certain fate before determination, and the power to pursue it afterwards.

POTENCY, PROSPECTIVE - The sum total of developmental possibilities, the full range of developmental performance of which a given area (or germ) is capable. Somehow more than, and inclusive of, prospective fate and prospective value. Connotes possibility, not power. Not to be confused with competence.

PREFORMATION - Theory that the adult is represented in miniature within the egg or sperm, and that development is simply enlargement.

PREGNANCY - Condition of actually bearing an embryo of fetus within the uterus or Fallopian tubes.

PRE-MIGRATORY GERM CELL - Yolk laden cells of splanchnopleuric origin which migrate by way of blood vessels to the gonad primordia. Believed by some to be the precursors of gonad stroma and/or functional germ cells.

PRENATAL - Term used to refer to any stage in the development of the mammalian prior to delivery at birth, to distinguish from neo-natal or post-natal.

PRE-ORAL GUT - The extension of the pharynx anterior to the oral plate and behind the hypophysis, which flattens out and finally disappears. *Syn.*, Seessel's pocket.

PRE-REDUCTION - Maturation in which the reductional and equational divisions occur in that order.

PRIMARY OÖCYTE - The termination of the growth phase in the maturation of the ovum from the oögonial stage, prior to any maturational divisions.

PRIMARY SPERMATOCYTE - Stage in spermatogenesis whose division results in secondary spermatocytes; stage beginning with growth of the spermatogonia.

PRIMORDIAL GERM CELLS - The diploid cells which are destined to become germ cells. *Syn.*, primitive germ cells. *e.g.*, oögonia and spermatogonia.

PRIMORDIUM - The beginning or earliest discernible indication of an organ. *Syn.*, rudiment, anlagé.

PROACROSOMIC GRANULES - Small corpuscles appearing in the idiosome of the young spermatids soon after the beginning of spermiogenesis; stain purple with PA-PSA.

PROCESSUS VAGINALIS - Peritoneal recess into scrotal pouch.

PROCTODEUM - An ectodermal pit in the region of the future cloaca which invaginates to fuse with hindgut endoderm to form the anal or proctodeal plate, later to rupture and form the anus.

PROESTRUS - Period of active preparation of the uterine mucosa (endometrium) leading to estrus.

PROGESTERONE - Endocrine secretion from the corpus luteum which causes the thickening of the endometrium.

PROGESTIN - A hormone from the corpus luteum found in all Placentalia.

PRONEPHRIC CAPSULE - Mesodermal connective tissue covering of the pronephric masses derived from adjacent myotomes and somatic mesoderm.

PRONEPHRIC CHAMBER - A portion of the coelomic cavity open anteriorly and posteriorly but closed ventrally by the development of the lungs.

PRONEPHRIC DUCT - The outer portion of the pronephric nephrotomes which develops a lumen connected posteriorly with the mesonephric or Wolffian duct. *Syn.*, segmental duct.

PRONEPHRIC TUBULES - The lateral outgrowths of the most anterior nephrotomal masses which acquire cavities connected with the pronephric duct. Possibly become infundibulum of oviduct.

PRONEPHROS - The embryonic kidney of all vertebrates and consisting of as many primitive tubules as somites concerned; completely lost in all adult vertebrates except a few bony fish. *Syn.*, head kidney.

PRONUCLEI OF MOUSE - Nucleoli of pronucleus of mouse are devoid of RNA; may arise from the metamorphic head of the sperm, the nuclear membrane, or within the nuclear membrane. The nucleo-cytoplasmic ratio (N/C at this stage is relatively 1/30 and cleavage restores the somatic ratio of 1/7).

PRONUCLEUS - Either of the gametic nuclei in the egg after fertilization and before syngamy; female pronucleus is the mature egg nucleus after the elimination of the polar bodies, distinct from the germinal vesicle which is the pre-maturation nucleus.

PROPHASE - The first stage in the mitotic cycle when the spireme is broken up into definite chromosomes, prior to lining up on the metaphase (equatorial) plate.

PROSENCEPHALON - (*See* FOREBRAIN).

PROSOCOEL - Cavity of the prosencephalon.

PROSPECTIVE SIGNIFICANCE - The normal fate of any part of an embryo at the beginning of development. *Syn.*, prospective Bedeutung, Potentialite reele.

PROSTATE GLAND - Sex gland derived from the urethral epithelium surrounding the urethra near the neck of the bladder; secretes fluid for transport and activation of spermatozoa in mammal.

PROSIMAL - Nearer the point of reference, toward the main body mass.

PUPIL - The opening into the secondary optic vesicle, occluded in part by the lens, and regulated in diameter by the ciliary muscles of the iris.

PYCNOSIS - Increase in density of the nucleus (or the cytoplasm) which may be hyperchromatic. Pycnotic cells in the central nervous system are called chromophile cells. Such cells have an increased affinity for haematoxylin and methylene blue.

PYGOPAGUS - Rump union in conjoined twins.

RACHISCHISIS - Cleft spine, due to failure to close completely.

RAMUS-COMMUNICANS - The connection between the sympathetic ganglion and the spinal nerve, as numerous as the ganglia in any vertebrate; probably originating from the crest cells. Ramus means branch.

RATHKE'S POCKET - The tubular ectodermal hypophysis of the chick (*See* HYPOPHYSIS).

RAUBER'S CELLS - The remnant of the trophoblast at the point of junction with the embryonic knob.

RECAPITULATION THEORY - Theory that embryonic development reviews the major steps in evolutionary history (*See* qualifications under BIOGENETIC LAW).

RECESSIVE - Opposite of dominant, refers to gene which can be masked in the phenotype by the presence of its allelomorph.

RECTUM - Narrowed posterior portion of the hindgut, lined with thickened endodermal epithelium, which opens directly into the cloaca.

REDUCTIONAL MATURATION DIVISION - One of the two important divisions in the maturation of gametes which results in the separation of allelomorphic (homologous) pairs of chromosomes so that the resulting cells are invariably haploid. *Syn.*, meiotic division, disjunctional division. Opposed to equational division.

REFERTILIZATION - By injecting versene solutions into the Fallopian tube of rodents after the normal time of sperm penetration, a secondary penetration may occur but generally fails to form a pronucleus and degenerates.

REGENERATION - Repair or replacement of lost part or parts by growth and differentiation past the phase of primordial development. The vast organizing potencies of the different regions of the early embryo are lost after the completion of development and there remain only certain regions of the body which are said to be capable of regeneration. Regenerative powers are more extensive among embryos and adults of phyletically low forms.

REICHERT'S MEMBRANE - A thin, pink-staining non-cellular membrane between the distal endoderm and the trophectoderm, appearing just at 6.5 days gestation. It is continuous over the entire inner surface of the traphectoderm, but is probably derived from the distal endoderm. Its function is to maintain the separation of maternal and fetal tissues.

REGIONS, PRESUMPTIVE - Regions of the blastula which, by previous experimentation, have been demonstrated to develop in certain specific directions under normal entogenetic influences.

REGULATION - A reorganization toward the whole; the power of pre-gastrula embryos to utilize materials remaining, after partial excision, to bring about normal conditions; more flexible power than regeneration.

RENAL CORPUSCLES - Derivatives of the intermediate cell-mass, located adjacent to the median face of Wolffian body. *Syn.*, Malpighian body.

RENAL PORTAL SYSTEM - The venous system which carries blood to the kidneys, involving the lateral portions of the caval veins, (really parts of the posterior cardinals) the iliacs, and the dorso-lumbars.

RESIDUAL BODY - A large and irregular portion of the cytoplasm which is shed off just before the release of the maturing spermatids as spermatozoa. Usually it is poorly staining cytoplasm surrounding a mass stained with iron hematoxylin which may represent an accumulation of granules previously identified in the cytoplasmic body.

RESTITUTION - The rejoining of broken chromosome threads to produce a structure superficially indistinguishable from a normal chromosome.

RETE CORDS - Strands of epithelial cells, containing many primordial germ cells which connect with the seminiferous tubules and later become the vasa efferentia, in the bird. *Syn.*, rete testis.

RETINAL ZONE - Ectodermal derivatives of the optic cup consisting of the internal limiting membrane, retinal and lenticular zones, and outer pigmented layer. The retina proper includes portions from internal limiting membrane to the rods and cones inclusive. (No cones in the mouse).

RHINENCEPHALON - Most primitive part of the telencephalon, concerned with the olfactory sense and including the olfactory bulb, olfactory tract, and pyriform lobe.

RHOMBENCEPHALON - (*See* HINDBRAIN).

RHOMBOIDAL SINUS - (*See* SINUS RHOMBOIDALIS).

RIBOSOME - Cytoplasmic bodies consisting largely of RNA and protein and responsible for the basophilia of the cytoplasm.

RNA, MESSENGER - Template RNA, unstable, M.W. 500,000 including 1500 nucleotides, carries the transcription of that part of DNA which is concerned with the structure of proteins, each nolecule of which directs the synthesis of 10 to 20 molecules of protein, and can be associated with as many as 5 ribosomes.

RNA, RIBOSOMAL - Stable, M.W. 550,000 or 1,100,000 equalling 1650 or 3300 nucleotides, the major portion of RNA found in most cells, composed of 2 basic units, functions in protein synthesis but are non-specific with regard to type of protein synthesized.

RNA, TRANSFER - Soluble RNA M.W. 24,000 equalling 74 nucleotides, stable (double helical structure). For each amino acid there is a specific enzyme and one or more specific transfer RNA's. Function is to align amino acids in correct sequence in polypeptide chains.

RUNT DISEASE - Results from cellular homografts containing relatively large proportions of immunilogically competent cells injected into genetically foreign hosts that are unable to reject them promptly. Injected cells persist, proliferate, colonize, and often kill the host. Development is retarded, there is diarrhea, hyperplasia of lymphoid tissues and organs, especially the spleen, with hepatomegaly, effects on blood, marrow, and skin.

SACCULE - The outer and ventral portion of the inner ear from which is derived the cochlea. Associated with the VIII auditory nerve. *Syn.*, sacculus.

SACCUS ENDOLYMPHATICUS - The original endolymphatic duct, closed off from the exterior, which grows up over the myelencephalon to join the other sac and form a vascular covering of the brain.

SACCUS VAGINALIS - Scrotal pouch, peritoneal pocket in scrotal pouch.

SACH'S LAW - All cells tend to divide into equal parts and each new plane of division tends to intersect the preceding one at right angles.

SAGITTAL - A mesial plane, or any plane parallel to it, dividing the right parts of the body from the left. Right angles to both the frontal and transverse planes.

SCLEROTIC COAT - A tough mesenchymatous and partially cartilaginous coat outside of the choroid coat of the vertebrate eye. *Syn.*, sclera.

SCLEROTOME - Loose mesenchymal cells proliferated off from the inner and ventral edges of the myotomes which contributes to the formation of the axial skeleton. Three parts are distinguished; the narrow undivided perichordal part; the dense aggregations of caudolateral cells; and the cephalic portion, all of which contribute to the axial skeleton.

SCROTUM or SCROTAL SAC - A single or subdivided chamber, external to the body proper, within which the testes are retained, the internal body temperature being too high for survival of the spermatozoa.

SECONDARY OÖCYTE - The stage in oögenesis between primary oöcyte and ovum, may be either haploid or diploid, depending upon species considered and which maturation division occurs first.

SECONDARY SPERMATOCYTE - The stage in spermatogenesis whose next division results in haploid spermatids, these spermatocytes being either haploid or diploid depending upon species considered. (*See* POST and PRE-REDUCTION DIVISIONS).

SECRETORY TUBULE - The portion of the kidney tubule actually involved in excretory process.

SECTION - Generally a slice of an embryo, often of microscopic dimensions, taken in any one of the various planes such as frontal, transverse, or sagittal. (*See* SERIAL SECTIONS.)

SECTION, CROSS - Cut made at right angles to the long axis of the embryo. *Syn.*, transverse section.

SECTION, FRONTAL - Cut made parallel to the longitudinal axis of the embryo and separating the more dorsal from the more ventral.

SECTION, SAGITTAL - Cut made parallel to the longitudinal axis of the embryo but separating the right from the left portions. Term often confused with "median" or "longitudinal" which really mean no more than "axial, " hence could also be "frontal".

SECTIONS, SERIAL - Thin (microscopic) slices of an embryo laid on the slide in sequence (generally from left to right, as one reads) so that the beginning of the embryo is at one side (left) and the end of the embryo at the opposite side (right) of the slide.

SEGMENTATION - Term used synonymously with cleavage. Also means serial repetition of embryonic rudiments (structural patterns) in successive levels of regular spacing, as in the case of somites, and spinal nerves. *Syn.*, cleavage.

SEGMENTATION CAVITY - The cavity of the blastula; *Syn.*, subgerminal cavity, blastocoel.

SEGREGATION, EMBRYONIC - Progressive restriction of original potencies in the embryo; the process of step by step repartitioning of the originally homogeneous zygote into the separate parts of the presumptive embryo.

SEMEN - Mixture of secretions from the bulbo-urethral glands (Cowper's glands), prostate gland, seminal vesicles and the suspended spermatozoa. Composite ejaculate of male during coitus which contains spermatozoa and hyaluronidase which liquefies the matrix of the cumulus oöphorus surrounding the ovulated eggs in the ampulla of the oviduct. The enzyme is derived from the spermatogenetic epithelium, not from the accessory organs, but semen does include secretions from male accessory organs.

SEMICIRCULAR CANALS - Tubular derivatives of the utricle lined with ectoderm from the otocyst, which constitute the accessory balancing mechanisms.

SEMI-LUNAR VALVES - Cup-like pockets within the aortic and pulmonary divisions of the bulbus, which prevent the back flow of blood.

SEMINAL VESICLE - Glandular dilatation of the distal end of the ductus deferens where spermatozoa are temporarily collected prior to ejaculation.

SEMINATION - The act of fertilizing, by the discharge of spermatozoa.

SEMINIFEROUS TUBULE - Tubular divisions of the testis derived from sexual (rete) cords, covered by a connective tissue theca and containing supporting (Sertoli) cells and (all) stages of spermatogenesis.

SEMIPLACENTA - Type of placenta in which the uterine mucosa and the chorion do not actually grow together so that there is no tearing at birth. *Syn.*, contact placenta.

SENESCENCE - The progressive loss of growth power; old age.

SENSORY LOAD - Determined by the number of receptor organs associated with a specific nerve.

SEPTUM - A partition.

SEPTUM SPURIUM - An embryonic partition or ridge which soon undergoes retrogression, but which is a prolongation of the dorsal wall of the atrium from one of the valvulae venosae, that effectively guards the sinus orifice against the backflow of blood to the heart. *Syn.*, false septum.

SEPTUM TRANSVERSUM - A partition which separates the peritoneal and pericardial cavities composed of three parts: a median mass made up of the liver, sinus and ductus venosus, and the dorsal and ventral ligaments; the lateral mesocardia; and also the lateral closing fold which extends from the mesocardia to the ventrolateral body wall. Beginning of diaphragm.

SERIAL SECTIONS - Thin (often of microscopic dimensions) sections of embryos which are mounted on slides in the order of their removal from the embryo, so that a study in sequence will provide an understanding of all organ systems from one region of the embryo to the other.

SEROSA - *Syn.*, chorion.

SERTOLI CELL - Derivative of the sexual cords of the testis, found within the seminiferous tubule and functionally similar to the follicle cell in the ovary in that it is the nutritive, supporting, or nurse cell of the maturing spermatozoa. The heads of adult spermatozoa may be seen embedded in the cytoplasm of Sertoli cells.

SEX, HETERODYNAMIC - The sex in which the gametes are of two kinds with respect to the possession of specific sex influencing chromosomes, such as the X-chromosome in Drosophila. The frog, mouse, and human male are presumably heterogametic.

SEX CELL CORD - Division of the sex-cell ridge or gonad primordium, not to be confused with sexual (rete) cords.

SEX CHROMATIN - Generally refers to the X chromosome; the maximum number of sex chromatin masses is 1 less than the number of X chromosomes. One chromosome is consistently isopycnotic and the others are all heteropycnotic. It replicates later and faster than the other chromosomes, probably because its constituent chromonemata must uncoil before DNA synthesis can take place. These are never seen in embryonic cells before the blastocyst stage, at the time of implantation.

SEX DETERMINATION - Generally means either the conditions of fertilization which determine the ultimate sex of the embryo, or the predetermination of sex by experimental means.

SEX LINKED - Genes borne on a sex chromosome and therefore conditioned in their heredity by the distribution of the sex chromosomes.

SEXUAL CORDS - Derivatives of the germinal epithelium from which they become separated and give rise to the bulk of the gonads of both sexes.

SEXUAL CORDS OF THE OVARY - Sex cords of the originally indifferent gonad primordium which form only the cords of the medulla of the ovary, the functional follicles coming from the germinal epithelium.

SEXUAL CORDS OF THE TESTIS - Sex cords of the originally indifferent gonad primordium which give rise to the seminiferous tubules of the testis, forming a rather solid mesenchymatous reticulum when cavities begin to appear lined with spermatogonia (from primordial germ cells) and Sertoli (from peritoneal cells), the whole constituting the seminiferous tubules.

SEXUAL CYCLE - Periodic sequence of changes in the uterine mucosa of the female placenta. Mammals, apart from pregnancy, include uterine changes as follows:
Diestrus - period of quiescence, if short.
Anestrus - period of quiescence, if long.
Proestrus - period of construction, destruction, and some repair.
Estrus - period of repair.

SHEATH, MYELIN - Myelin covering of axones in the so-called white matter of the spinal cord.

SINUS RHOMBOIDALIS - The region of the receding primitive streak around which the posterior ends of the neural folds are diverted in the chick embryo.

SINUS VENOSUS - The point of fusion of vitelline veins of the embryo or the most posterior of the original four chambers of the heart; bilaterally symmetrical and related to the ducts of Cuvieri and the ductus venosus to 45 hours; right horn elongated and left reduced with shifting of sino-auricular aperture and has the shape of horseshoe between atrium and the septum transversum. The most anterior part involved in formation of right auricle.

SITUS INVERSUS - An inversion of the bilateral symmetry; reversal of right and left symmetry.

SITUS INVERSUS VISCERUM - Twisting of the digestive tract and sometimes the heart, occurring naturally (rarely) or as a result of shifting of embryonic parts.

SKELETOGENOUS SHEATH - Sclerotomal cells which first form a continuous layer around both the notochord and nerve cord.

SKIN - (See DERMIS and EPIDERMIS). *Syn.*, integument.

SOMATIC - Relating to body in contrast with germinal cells; or relating to the outer body in contrast to inner splanchnic mesoderm.

SOMATIC DOUBLING - Doubling of the initial number of chromosomes with which the egg begins development, occurring (probably in most cases) at the first or early mitotic divisions (cleavages) of the egg, after fertilization.

SOMATIC UMBILICUS - A short, thick hollow stalk which connects the chick embryo with the underlying yolk-sac and the extra-embryonic membranes, composed of ectoderm and somatic mesoderm continuous with the amnion.

SOMATOBLAST - Blastomeres with specific germ layer predisposition, *i.e.*, ectodermal somatoblasts.

SOMATOPLEURE - The layer of somatic mesoderm and closely associated ectoderm, the extension of which (from the body wall) gives rise to both the amnion and chorion.

SOMITE - Blocks of paraxial mesoblast metamerically separated by transverse clefts, derived from enterocoelic or gastral mesoderm and giving rise to the dermatome, myotome, and sclerotome.

SPERM - The germ cell characteristically produced by the male. *Syn.*, spermatozoön, sperm cell, male gamete, spermatosome.

SPERM RECEPTOR - Chemical associated with the spermatozoa, reacting with fertilizin (amboceptor) in Lilli's side chain hyopthesis of the fertilizin reaction.

SPERMATELEOSIS - Metamorphic changes from spermatid to spermatozoon.

SPERMATID - The products of the second maturation division in spermatogenesis, the spermatids having certain cytological characteristics and being invariably haploid; cells which go through a metamorphosis into functionally mature spermatozoa.

SPERMATOCYTE - Stages in spermatogenesis between the time the primordial germ cell (spermatogonium) begins to grow, without division, until after the divisions which results in spermatids. (*See* PRIMARY and SECONDARY SPERMATOCYTES).

SPERMATOGENESIS - The entire process which results in the maturation of the spermatozoön.

SPERMATOGONIA: TYPE A - Cells with ovoid and pale nucleus, with thin nuclear membrane and fine chromatin granules; one large chromatin granule may also be found present at all stages of spermatogenetic cycle.

SPERMATOGONIA: INTERMEDIATE TYPE - Cells appearing near the end of initial stage (I) at which time they can still be confused with Type A cells, but they gradually take on the appearance of Type B cells as follows: Nuclear membrane thickens, nuclear contents darken; a few chromatin flakes appear close to the nuclear membrane and the nucleus progressively changes in shape from ovoid to round.

SPERMATOGONIA: TYPE B - Cells with a spherical and dark nucleus, coarse chromatin granules attached to the nuclear membrane; cells resulting from the division of the intermediate cell type spermatogonia and which ultimately produce primary spermatocytes.

SPERMATOGONIUM - The primordial germ cell of the male gonad, indistinguishable from somatic cells, both of which are diploid; stage prior to maturation when the presumptive spermatozoön undergoes rapid multiplication by mitosis.

SPERMATOSPHERE - (*See* IDIOZOME).

SPERMOPHILE GROUP - Portion of the amboceptor in Lillie's fertilizin hypothesis into which sperm receptors fit in the fertilization reaction.

SPINA BIFIDA - Split tail caused by a variety of abnormal environmental conditions any of which may prevent the proper gastrulation and neurulation which lead to this split-tailed condition.

SPINAL CORD - That portion of the central nervous system, excluding the brain, which is derived from the epithelial and neural ectoderm of the original blastula, consisting of ependyma, glia, neuroblasts and their derivatives and connecting cells.

SPINDLE - A group of fibres between the centrosomes during mitosis, to which the chromosomes are attached and by means of which (mantle fibre portion) the chromosomes are drawn to their respective poles.

SPINOUS PROCESS - Prolongation of neural processes fused dorsally to the neural canal; becomes dorsal spine of vertebra.

SPIREME - A continous chromatin thread characteristic of the so-called resting cell nucleus. Existence questioned by current cytoplogists.

SPLANCHNIC - Refers to the viscera, opposed to somatic or body.

SPLANCHNIC MESODERM - The visceral mesoderm, or that nearest the embryonic axis in the lateral plate.

SPLANCHNOCOEL - That portion of the enterocoel which lies between the somatic and splanchnic mesoderm within the body. *Syn.*, coelom.

SPLANCHNOCRANIUM - That portion of the skull which is preformed in cartilage and which arises from the first three pair of visceral arches. Opposed to neurocranium.

SPLANCHOPLEURE - The layer of endoderm and inner mesoderm (splanchnic) within which develop the numerous blood vessels of the area vasculosa and later the yolk-sac septa; the layers within the body which give rise to the lining and to the musculature of the alimentary canal.

SPLEEN - This organ arises as a proliferation from the peritoneum covering the left side of the dorsal mesentery just anterior to the pancreas.

SPONGIOBLASTS - Cells of the mantle layer of the developing spinal cord destined to form merely supporting tissue.

SPONGIOSA - The glandular layer of the uterus adjacent to the muscularis to which the trophoderm is attached the other portion of the uterine mucosa being the compacta.

STEM CELL RENEWAL - Phenomenon suggesting that a small number of spermatagonia undergo only one or two mitoses and then return to their original stem cell state; possibly later to undergo spermatogenetic mitosis.

STERILITY - Inability to breed. Condition may be temporary or permanent.

STOMODEUM - Ectodermal invagination (pit) which fuses with the pharyngeal endoderm to form the oral plate, which later ruptures to form the margins of the mouth cavity. The stomodeal portion of the mouth lining is therefore ectodermal.

STRATUM COMPACTUM - Outer cellular portion of the endometrium surrounding the blastocyst, subjacent to the junctional zone.

STRATUM GRANULOSUM - Layer of follicle cells surrounding the mammalian ovum.

STRATUM SPONGIOSUM - Deeper glandular part of the endometrium surrounding the blastocyst, with many secreting glands. Extends to the basal zone.

STROMA - The mesodermally derived, medullary, supporting tissues of an organ.

SUB-CARDINAL VEINS - Embryonic veins ventral to the nephric tissue the posterior portions of which fuse to contribute to the formation of the inferior vena cava.

SUB-CLAVIAN VEINS - These arise primitively as branches of the posterior cardinals.

SUBSTANTIA PROPRIA - Mesenchyme of the cornea continuous with the sclera.

SUBSTRATE - The substance which is acted upon by an enzyme.

SUB-ZONAL LAYER - The morula lies within a sub-zonal layer of cells later called the trophoblast.

SULCUS LIMITANS - Longitudinal groove between the dorsal alar plate and the ventral basal plate best seen at the level of the myelencephalon; and mark for locating the nuclei and fibre tracts.

SUPERNUMERARY NUCLEI - Nuclei of excess spermatozoa which invade the egg at fertilization, many of which cause accessory cleavages and then degenerate.

SUSTENTACULAR CELL - A cell which provides nourishment for another, such as the Sertoli or follicle cells of the gonads.

SYLVIUS, AQUEDUCT OF - (*See* AQUEDUCT OF SYLVIUS).

SYMPATHETIC SYSTEM - Originating either from mesenchymal element arising in situ or, more probably, from ectodermal elements emanating from the neural crests; to organize as a chain of ganglia near the dorsal aorata and controlling the involuntary (visceral) musculature.

SYMPODIA - Fusion, to varying degrees, of the legs (*e.g.*, mermaid or siren condition).

SYNAPSIS - Union, such as the lateral (parasynapsis) or terminal (telosynapsis) union of embryos; or pairing of homologous chromosomes.

SYNAPTENE STAGE - The stage in maturation between the leptotone and the synezesis (contraction) stage wherein the chromatin is in the form of long threads, intertwined in homologous pairs. *Syn.*, zygotene of amphitene.

SYNCYTIAL TROPHODERM - Layer of trophodermal cells of the embryo outside of but probably derived from the cell layer of Langhans.

SYNCTIUM - Propagation of nuclei with cytoplasmic growth but without cytoplasmic division so that there results a mass of protoplasm with many and scattered nuclei but with inadequate cell boundaries.

SYNDACTYLY - Either bony fusion or fleshy webbing of the digits, generally the second and their digits being involved. Probably inherited.

SYNENCEPHALON - The third or most posterior of the three primary divisions of the forebrain which gives rise to the posterior portion of the diencephalon. (The other divisions are parencephalon and telencephalon.)

SYNERESIS - A segregation of the colloidal phases, a corollary of ageing.

SYNEZESIS - The stage in maturation between synaptene and pachytene when the chrommatin threads are short and thick and the ends away from the centrosome are tangled.

SYNGAMY - Specifically the fusion of the gamete pronuclei, but also the union of gametes at fertilization. *Syn.*, zygotogenesis, fertilization.

SYNOPHTHALMIA - Fusion of the eyes as in cyclopia.

SYNTONIC FACTOR - Some regulating force which enables a particular cell to live harmoniously with other cells of the same type so that an organ will develop, not found in tissue cultures of cells isolated prior to differentiation, present during organogenesis.

SYNTONY - Indwelling integration of parts; a natural force within and between cells developing from the specific organization of living matter.

TAIL FOLD - A sulcus begins to develop beneath the posterior end of the embryo, giving rise to a tail fold similar to the head fold except that from the beginning it is made up of ectoderm and somatic mesoderm. The tail fold often appears prior to the head fold and is longer. *Syn.*, amniotic tail fold.

TELA CHORIOIDEA - *Syn.*, Thin roof of third and fourth brain ventricles.

TELENCEPHALON - The portion of the forebrain (ventricle) anterior to a plane which includes the posterior side of the choroid plexus and the anterior side of the optic recess or the portion of the forebrain ventral to a plane passing from the posterior wall ventral to the optic recess to the anterior wall in the center of the velum transversum. Gives rise to torus transversus (anterior commissure), cerebral hemispheres, corpora striata, paraphysis, anterior choroid plexus, olfactory lobes, lateral ventricles and part of the foramina of Monro.

TELOBIOSIS - Fusion of embryos end-to-end. (*See* Parabiosis).

TELOCOEL - Cavity of the telencephalon.

TELOPHASE - Last phase in mitosis when the respective chromosome groups have reached their respective astral centers and are beginning to re-form a resting cell nucleus, the stage often accompanied by the beginning of cytoplasmic division.

TELOSYNAPSIS - End-to-end fusion of chromosomes. (*See* PARASYNAPSIS).

TERATOGENETIC - Abnormality producing.

TERATOLOGY - Study of the causes of monster and abnormality formation.

TERATOMA - Structure which results from random differentiations; malignant assembly of tissues, often well differentiated histologically, generally embedded in an otherwise healthy organ. Some use term embryoma to refer to embryological differentiation and teratoma to mean both histological and morphological differentiation of the abnormal growth.

TESTES - Sex organs of the male in which spermatagonia are produced and matured, a distinguishing primary sex character of the male.

TETRADS - Paired (homologous) chromosomes which have become duplicated longitudinally in anticipation of the meiotic (reductional) division. When viewed from one end will appear as a group of four chromosomes, hence a tetrad.

THALAMUS - Dorso-lateral wall of the diencephalon which becomes thickened by the development of fibers passing from the cord to the more posterior parts of the cerebral hemispheres.

THECA - Connective tissue covering, generally refers to covering of ovarian follicle.

THECA EXTERNA - The outermost of the coverings of the ovarian follicle, rather loose connective tissue with abundant blood supply. Continuous with ovarian stroma.

THECA FOLLICULI - Refers to membraneous and cellular coverings of the ovum within the ovary.

THECA INTERNA - The layer of connective tissue consisting of closely packed fibres, possibly some of smooth muscle, immediately external to the ovarian follicle of birds and mammals. Less vascular and more compact than theca externa.

THORACO-GASTROSCHISIS - Failure of the body wall to close along the med-ventral line, including the thoracic region.

THORACOPAGUS - Thoracic union of conjoined twins.

THYMUS - Derivatives of first pair of branchial pouches of the embryo which separate from the pouches and migrate to a position directly anterior to the heart. Known as gland of adolescence because it recedes upon attaining sexual maturity. Endocrine function.

THYROGLOSSAL DUCT - A temporary tubular connection between the thyroid anlage and the pharynx near the base of the tongue.

THYROID BODY or GLAND - Originates as an endodermal thickening in the floor of the pharynx between the second pair of visceral arches; evaginates to form a vesicle temporarily connected with the gut by the thyroglossal duct; separates from gut becomes divided and migrates to junction of subclavian and common carotid arteries. Somewhat similar history in all vertebrate embryos. Endocrine function.

TOPOGENESIS - All of the processes of movement which result in structure formation.

TOLERANCE, ACTIVELY ACQUIRED - If mammals, birds and amphibia are exposed to living homologous tissue cells at an early stage in development (before their immunological response faculty has become functionally mature), a partial or complete specific central failure of the mechanism of immunological response is indiced.

TONGUE - Solid mesodermal mass, covered with endoderm, derived by cell proliferation from the floor of the pharynx.

TONSILS - Lymphatic structures derived from the endoderm and mesoderm of the second pair of visceral pouches.

TOOTH GERMS - Conical bands of tissue which develop in the oral cavity of mammals, generally a single one for each of the temporary and permanent teeth.

TORSION - The twisting of the embryo so that it lies on its side.

TORUS TRANSVERSUS - Thickening in the median ventro-anterior wall of the lamina terminalis of the telencephalon, just exterior to the optic recess, representing the rudiment of the anterior commissure.

TOTIPOTENCY - Related to theory that the isolated blastomere is capable of producing a complete organism. Roux (1912) included several faculties such as (1) for self-differentiation; (2) for influencing differentiation or induction of other parts; (3) for specific reaction to differentiating influences as in dependent differentiation.

TRACHEA - That portion of the respiratory tract between the larynx and the lung buds, lined with endoderm, probably derived from the posterior portion of the original laryngo-traxheal groove.

TRACHEAL GROOVE - (See LARYNGO-TRACHEAL GROOVE).

TRANSLOCATION, RECIPROCAL - A balanced change in the chromosomes, segmental rearrangements of chromosomes.

TRANSVERSE - A plane (or section) which divides the anterior-posterior axis at right angles, separating the more anterior from the more posterior. Syn., cross section, but this synonym is not generally satisfactory.

TRANSVERSE NEURAL FOLD - The continuation of the lateral neural folds (ridge) of the early frog embryo around the anterior and (i.e., region of face), the region of the temporary anterior neuropore. Syn., transverse medullary fold or ridge.

TRIASTER - Abnormal mitotic figure possessing three asters generally causing irregular distribution of chromosomes and abnormal cleavages. Other multiple aster conditions noted, (e.g. tetraster, etc.)

TRIGEMINAL GANGLION - Cranial (V) ganglia which consist of motor and sensory portions and arise from the most anterior crest segments in conjunction with cells from the inner (ganglionic) portion of the corresponding placode. Give rise to ophthalmic, mandibular, and maxillary branches; associated with the myelencephalon at the level of the greatest width of the 4th ventricle.

TRITOGENY - One-third of a fragment. (See merogony).

TROCHLEARIS NERVE - Cranial (IV) motor nerves arise from the dorsal surface of the brain near the isthmus, coming from medullary neuroblasts and innervating the superior oblique muscles of the eye.

TROPHECTODERM - Region of continuity of ectoderm and outer layer of trophoblast; extra-embryonic ectoderm following germ layer differentiation.

TROPHIC - The action of the nervous system in the absence of which the muscle tonus fails and in consequence, regeneration is impossible.

TROPHOBLAST - Thin layer of cells which constitute the wall of the mammalian blastocyste; outer layer of blastocyst prior to differentiation of the primary germ layers. The cellular parenchyma of the chorion and placenta, cells generally with large nuclei. Chords of the tropho-blast grow out from the initially smooth surface of chorion to penetrate and erode the endometrium to give rise to the primary chorionic villi. May produce steroid hormones, chorionic gonadotrophins, and other hormones. Mediates the metabolic exchange between mother and fetus.

TROPHOBLASTIC CELL COLUMNS - Distal ends of chorionic and placental villi; continue to grow in the form of columns of cytotrophoblast preceding the differentiation of the mesoderm in them.

TROPHOBLASTIC SHELL - (Peripheral cytotrophoblast). During the first phases of gestation the cytotrophoblast and trophoblastic shell produce proteolytic and cytolytic substances which are capable of attacking the endometrium.

TROPHOBLASTIC VILLI - Finger-like projections of chorionic mesoderm comprising the trophoderm of mammalian embryos.

TROPHOCHROMATIN - Nutritive chromatin of the nucleus.

TROPHODERM - Trophectoderm reinforced by a layer of somatic mesoderm. Syn., extra-embryonic somatopleure, serosa.

TRUE KNOT - Slipping of the fetus through a looped umbilical cord to produce a true knot, distinguished from looped blood-vessels which cause external bulgings called false knots.

TRUE PLACENTA - A placenta in which the chorion and uterine mucusa are intimately associated, in contrast with the primitive contact type. *Syn.*, burrowing placenta.

TRUNCUS ARTERIOSUS - Anterior continuation of the bulbus arteriosus beneath the foregut, divided in antero-posterior direction by a septum which is continuous through the bulbus to the ventricle; gives off the external carotids to the mandibular arches and the second, third, and fourth aortic arches which join the dorsal aorata. *Syn.*, ventral aorta.

TUBAL FISSURE - Longitudinal slit in the roof of the pharynx which connects the median chamber of the tubo-tympanic cavity with the oral cavity.

TUBAL RIDGE - The primordia of the Müllerian ducts or oviducts arising in the embryo lateral to each mesonephros and adjacent to the respective Wolffian duct, the anterior ends of which (in the female) become the ostia abdominalia tubae.

TUBERCULUM POSTERIUS - A thickening in the floor of the brain at the region of the anterior end of the notochord, representing the posterior margin of the diencephalon.

TUBO-TYMPANIC CAVITY - Remnants of the dorsal parts of the first pair of visceral (hyomandibular) pouches and the lateral walls of the pharynx, connecting the pharynx and the middle ear, represented by the Eustachian tube of the adult bird or mammal.

TUBULES - (*See* under specific names such as COLLECTING, MESONEPHRIC, PRONEPHRIC, SEMINIFEROUS.)

TUNICA ALBUGINEA - (*See* ALBUGINEA OF TESTIS).

TWINS, IDENTICAL - True twins, from a single egg and having common membranes and umbilicus.

TWINS, ORDINARY - Pleural pregnancy resulting from the fertilization of separate ova simultaneously liberated from individual follicles. Separate development, implantation, decidua capsularis, and fetal membranes.

TYMPANIC CAVITY - Cavity of the middle ear, a vestige of the hyomandibular pouch. (*See* TUBO-TYMPANIC CAVITY).

TYMPANIC MEMBRANE - Membrane made up of ectoderm, mesenchyme, and endoderm which separates the tympanic cavity from the exterior. *Syn.*, ear drum.

ULTIMOBRANCHIAL BODY - (*See* POSTBRANCHIAL BODY).

UMBILICAL ARTERIES - Branches of the sciatic arteries of the embryo, the right member being the smaller; carry blood to the allantois.

UMBILICAL VEINS - At first paired veins in the lateral body wall of the embryo which bring blood from the allantois and join the ducts of Cuvier, right vein disappearing and the left changing its connection to join the anterior half of the ductus venosus. Only the proximal portion persists as a vein in the ventral body wall.

UMBILICUS - The stalk-like connection between the embryo and all extra-embryonic structures, including the somatic stalk, allantoic stalk plus its arteries and veins, and the yolk stalk with its arteries and veins.

UMBILICUS, SOMATIC - (*See* SOMATIC UMBILICUS.)

UMBILICUS, YOLK SAC - (*See* YOLK-SAC UMBILICUS.)

URACHUS - The canal which connects the allantois and the urinary bladder in embryos.

URETER - Diverticulum from the posterior end of the Wolffian (mesonephric) duct appearing in the embryo; functioning as an excretory duct of the adult.

URETHRA - Single duct of the male mammal which discharges urine and also through which semen is liberated from the male genital tract into the genital tract of the female during coitus; mesodermal.

URINARY BLADDER - An endodermally lined vesicle derived from the hindgut, homologous to the allantois of the chick, connected with ureters.

URINIFEROUS TUBULE - Functional kidney tubule of both mesonephros and metanephros.

UROGENITAL SYSTEM - The entire excretory and reproductive systems, some embryonic parts of which degenerate before birth of mammals. Shows various degrees of common origin and ultimate function. (*See* specific excretory and reproductive components.)

UTERINE GLANDS - Glands within the uterine mucosa which secrete fat, proteids, and glycogen, a source of embryonic nutrition.

UTERINE MILK - A viscid fluid secreted by the uterine mucosa consisting of fats, proteids, and glycogen, a source of embryonic nutrition.

UTERINE MUCOSA - Mucosal lining of the uterus which shows cyclic changes associated with reproductive activity. *Syn.*, endometrium.

UTERUS - Thick walled muscular structure of the female sexual system which is lined with highly secretory endometrium, and serves as an implantation site for the developing mammal (traditionally called the womb).

UTERUS, ACTIVE (OR SENSITIVE) - Uteri on days 4 to 5 of normal pregnancy when implantation is the normal event. Endocrinologically attuned to implantation, competent to produce a decidua as a result of blastocyst stimulation.

UTERUS, DORMANT - The uterus of delayed implantation, can be made active by oestrogen stimulation. Will inhibit maturation of transplanted trophoblast, but does not interfere with shedding of the zona pellucida. Zona shedding is age dependent, following trophoblast maturation.

UTRICLE - A vesicle, generally referring to the superior portion of the otocyst which gives rise to the various semicircular canals of the ear, and into which these canals open. Lined with ectoderm.

VAGINA - The cavity of the female mammal possessing, at its external boundaries, a homologue of the male penis called the clitoris.

VALVES, SEMI-LUNAR - (*See* SEMI-LUNAR VALVES).

VASA DEFERENTIA - Mesonephric or Wolffian ducts which persist as the male gonoducts of the mammal, connecting with the testes through the vasa efferentia and epididymis and functioning as sperm ducts after the degeneration of the embryonic mesonephros and the development of the gonads. *Syn.*, vas deferens.

VASA EFFERENTIA - Derivatives of the anterior half of the mesonephric or Wolffian body which become the epididymis. *Syn.*, vas efferens.

VEIN - (*See* under specific names).

VELUM TRANSVERSUM - Depressed roof of the telencephalon just anterior to the lamina terminalis, which later becomes much folded and vascular as the anterior roof of the third ventricle. The division point between the tel-and diencephalon.

VENA CAVA ANTERIOR - Junction of inferior jugular (anterior cardinal) and (in the chick) the subclavian and vertebral veins which empty into the ductus Cuvieri, and later the right auricle. *Syn.*, superior vena cava, or superior caval veins.

VENA CAVA POSTERIOR - The single median ventral vein which represents the remnant of the anterior right cardinal and which later receives the hepatic vein prior to joining the ductus Cuvieri, and later joins the right auricle directly.

VENTRAL - Belly surface. Ventrad means toward the belly surface.

VENTRAL LIGAMENT OF LIVER - *Syn.*, falciform ligament.

VENTRAL MESENTERY - Double layer of splanchnic mesoblasts which connects the alimentary canal with the extra-embryonic splanchnopleure in the embryo; in the region of the hindgut includes both somatic and splanchnic mesoderm as a thick mass of mesoblast which binds the hindgut to the somatopleure. In the region of the fore- and midgut of the later embryo this includes the meatus venosus and the liver material.

VENTRICLE III - Main cavity (diocoel) of the forebrain, related to paired lateral ventricles or telocoesls, by way of the foramina of Monro.

VENTRICLE IV - Main cavity of the hindbrain (rhombencephalon) connected anteriorly with the aqueduct of Sylvius and posteriorly with the neural canal and extending through both the metencephalon and myelencephalon, having as a roof the vascular posterior choroid plexus.

VENTRICLE, LATERAL - (*See* LATERAL VENTRICLES OF THE BRAIN).

VENTRICLE OF THE HEART - Double and very muscular chamber of the heart developing from the anterior myocardium, subdivided by septa and provided with valves; connected with bulbus arteriosus anteriorly.

VER - Ventral ectodermal ridge in the (mouse embryo) tail tip which is thickened or columnar rather than of pavement variety which covers the tail elsewhere. It extends about 300 microns from the tail tip craniad and resembles the apical ectodermal ridge of limb buds; gives positive alkaline phosphatase reaction which is a sign of functional hypertrophy of the cells. Has something to do with the outgrowth of the tail and is not peculiar to the mouse. Seen from 9 to 11 days, gone by 12 days.

VERNIX CAAEORA - A cheese-like material which sometimes covers the skin of the fetus, and is derived from dead cells and fat.

VERTEBRA - Derivatives of the sclerotome which surround the nerve cord and notochord, and finally incorporate the notochord by chondrification and ossification (centrum).

VERTEBRAL ARCH - (*See* NEURAL ARCH).

VERTEBRAL PLATE - (*See* AXIAL MESODERM).   *Syn.*, segmental plate.

VESICLE - (*See* under specific names).

VESICLE, GERMINAL - Nucleus of the egg while it is a distinct entity and before the elimination of either of the polar bodies.

VESTIBULE - A shallow basin into which the vagina and the urethra of the female mammal open.

VILLUS - A finger-like projection, such as the chorionic villus which is a projection of the mammalian chorion into folds of the uterine mucosa.

VISCERAL - Pertaining to the viscera.

VISCERAL ARCHES - Generally six pairs of mesodermal masses between the visceral pouches and lateral to the pharynx of all vertebrate embryos, including the mandibular, hyoid, and four branchial arches. Each arch is bounded by the endoderm on the pharyngeal side and ectoderm on the outside.

VISCERAL FURROW - Ectodermal invaginations which may meet endodermal pharyngeal evaginations to form visceral clefts. Syn., visceral groove.

VISCERAL GROOVE - (*See* VISCERAL FURROW).

VISCERAL MESODERM - (*See* SPLANCHNIC MESODERM and SPLANCHNOPLEURE).

VISCERAL PLEXUS - An aggregation of sympathetic neurones which control the viscera, having migrated posteriorly from the 10th (vagus) cranial ganglia.

VISCERAL POUCH - Endodermal evagination of the pharynx which, if they meet the corresponding visceral furrow, often breaks through to form the visceral cleft. *Syn.*, pharyngeal pouch.

VITALISM - A philosophical approach to biological phenomena which bases its proof on the present inability of scientists to explain all the phenomena of development. Idea that biological activities are directed by forces neither physical nor chemical, but which must be supra-scientific or super-natural. Effective guidance in development by some non-material agency (*See* mechanism).

VITELLIN - Egg-yolk phospho-protein.

VITELLINE - Adj., pertains to yolk, vein, or membrane.

VITELLINE ARTERY - Paired omphalomesenteric vessels which later fuse (as the dorsal mesentery forms) to go out through the umbilical stalk to the yolk as the vitelline (yolk) arteries, originating from the dorsal aorta.

VITELLINE MEMBRANE - The cytoplasm and yolk of the mouse egg is limited by a plasma or permeability membrane which is generally called the vitelline membrane, but is not to be confused with the membrane of the same name among invertebrates. During maturation of the oocyte this membrane becomes thrown into microvilli, some of which interdigitate with the surrounding follicle cells. As the zona pellucida is formed these microvilli seem to extend through the pellucida while follicle cell processes extend also through the pellucid area toward the vitellus. Its function, aside from being the plasma membrane of the ovum, is to aid in the engulfment of the invading spermatozoon.

VITELLINE SUBSTANCE - Yolk.

VITELLINE VEIN - (*See* OMPHALOMESENTERIC VEIN).

VITREOUS HUMOR - The rather viscous fluid of the eye chamber posterior to the lens, formed by cells budded from the retinal wall and from the inner side of the lens, hence ectodermal and probably also mesenchymal in origin. (See AQUEOUS HUMOR.)

VIVIPAROUS - Animals which bring forth young in advanced state of development, more advanced than eggs.

WAVE OF SEMINIFEROUS EPITHELIUM - Complete series of the successive cell associations found along a seminiferous tubule; the length of the wave being the distance between two successive, identical cell associations. The sequence of pictures along a wave is similar in sequence of events taking place in one given area during a cycle of the seminiferous epithelium.

WEBER'S LAW - The degree of sensitivity to a stimulus in any reacting system is not constant but depends, not alone on the nature of the stimulus, but upon the period of life and the strength of an already existing stimulus. A stimulus therefore represents a change, but a reacting system takes into account any pre-existing stimulus upon which this change is built. Theory that equal relative differences between stimuli of the same kind are equally perceptible.

WOLFFIAN BODY - (See MESONEPHROS).

WOLFFIAN DUCT - (See MESONEPHRIC DUCT, URINOGENITAL DUCT AND VAS DEFERENS).

WOLF SNOUT - Projecting of the premaxilla beyond the surface of the face, accompanying double (hare) lip and sometimes a cleft palate.

XANTHOLEUCOPHORE - Crystals and soluble yellow pigment; cells bearing such.

XANTHOPHORES - Yellow pigment in solution; cells bearing this yellow pigment.

X-CHROMOSOME - Female identifying chromosome when diploid (paired) but contains many genes unrelated to sex determination. In the female the two X chromosomes are different, one being isopycnotic and the other heteropycnotic.

XIPHOPAGUS - Xiphoid fusion of conjoined twins; sometimes the skin alone.

Y-CHROMOSOME - Short chromosome easily distinguished from autosomes and acrocentric. Presence with X chromosome typical for the male.

YOLK - Highly nutritious food (metaplasm) consisting of non-nucleated spheres and globules of fatty material found in all except alecithal eggs. *Syn.*, lecithin.

YOLK SAC - Extra-embryonic splanchnopleure.

ZONA PELLUCIDA - Transparent, non-cellular, secreted layer immediately surrounding the ovum, corresponding to the vitelline membranes of lower forms.

ZONA RADIATA - Zona pellucida which exhibits radial striations, not be confused with the corona radiata. *Syn.*, zona pellucida of the mammalian egg.

ZONA REACTION - A change occurs in the zona pellucida which renders it impermeable to further penetrations by active spermatozoa. The other mechanism of the prevention of polyspermy being a change involving the dissolution of the cortical granules of the ovum.

ZYGOTE - The diploid cell formed by the union of two gametes. *Syn.*, fertilized egg.

References

Adams, D. H. 1953. "Some studies on liver catalase in embryonic and immature chickens and mice." Brit. Jour. Cancer. 7:501-508.

Adams, F. W., and H. H. Hillemann. 1950. "Morphogenesis of the vitelline and allantoic placentae of the golden hamster." Anat. Rec. 108:363-383.

Ader, R., and P. M. Conklin. "Handling of pregnant rats: Effect on emotionality of their offspring." Science. 142:411-412.

Agduhr, E. 1927. "Studies on the structure and development of the bursa ovariea and the tuba uterina in the mouse." Acta Zool. Stockholm. 8:1-133.

Alden, R. H. 1948. "Implantation of the rat egg. III. Origin and development of primary trophoblast giant cells. Am. J. Anat. 83:143-181.

Alfert, M. 1950. "A cytochemical study of oögenesis and cleavage in the mouse." J. Cell. & Comp. Physiol. 36:381-409.

Allen, E. 1922. "The estrous cycle of the mouse." Am. J. Anat. 30:297-371.

––––––. 1923. "Ovogenesis during sexual maturity." Am. J. Anat. 31:439-470.

––––––, and E. A. Doisy. 1923. An ovarian hormone: preliminary report on its localization, extraction and partial purification and action in test animals. J. Am. Med. Assn. 81:819-821.

Allen, Ezra, and E. C. McDowell. 1940. "Variations in mouse embryos of 8 days gestation." Anat. Rec. 77:165-171.

Allen, J. M. 1958. "A chemical and histochemical study of aliesterase in the adrenal gland of the developing mouse." Anat. Rec. 132:195-207.

Amoroso, E. C., and A. S. Parkes. 1947. "Effects on embryonic development of x-irradiation of rabbit spermatozoa in vitro." Proc. Royal Soc. London B. 134:57-78.

Angevine, L. B., Jr., and R. L. Sidman. 1961. "Autoradiographic study of cell migration during histogenesis of cerebral cortex in the mouse." Nature (Lond.). 192:766-768.

––––––. 1962. "Autoradiographic study of histogenesis in the cerebral cortex of the mouse." (Abstr.) Anat. Rec. 142:210.

Arey, L. B. 1965. "Developmental Anatomy." W. B. Saunders Co., 695 pps.

Asayama, S., and M. Furusawa. 1960. "Culture *in vitro* of prospective gonads and gonad primordia of mouse embryo." Dolutsugoho Zosshi (Zool.-Soc.-Japan) 69:280.

Asscher, A. W., and C. J. Turner. 1955. Vaginal sulphhydryl and disulphide groups during the oestrous cycle of the mouse. Nature (Lond.). 175:900-901.

Atkinson, W. B., and C. W. Hooker. 1945. The day to day level of estrogen throughout pregnancy and pseudopregnancy in the mouse. Anat. Rec. 93:75-89.

Atlas, M., and V. P. Bond. 1965. "The cell generation cycle of the eleven-day mouse embryo." J. Cell. Biol. 26:19-24.

Auerbach, R. 1954. "Analysis of the developmental effects of a lethal mutation in the house mouse." J. Exp. Zool. 127:305-330.

––––––. 1960. "Morphogenetic interactions in the development of the mouse thymus gland." Developmental Biol. 2:271-284.

––––––. 1961. "Genetic control of thymus lymphoid differentiation." Proc. Nat. Acad. Sci. 47:1175-1181.

––––––. 1961. "Experimental analysis of the origin cell types in the development of the mouse thymus." Developmental Biol. 3:336-354.

* This reference list cannot be complete, but is nearly so. The author invites readers to apprise him of any appropriate and unlisted references. No apology is offered for the inclusion of some few references on the rat or related forms where comparable references on the mouse are not available. For late references, see special listing starting on page 407.

Auerbach, R. 1963. "Developmental studies of mouse thymus and spleen." N. C. I. Monograph. 11:23.

_____. 1964. "On the function of the embryonic thymus." Wistar Inst. Sympos. Monograph. 2:1-8.

_____. 1964. "Experimental analysis of mouse thymus and spleen morphogenesis" (R. A. Good & A. E. Gabrielsen ed.) "The thymus in immunobiology." Hoeber, N.Y. p. 95-111.

Austin, C. R. 1948. "Number of sperms required for fertilization." Nature (Lond.). 162:534-535.

_____. 1948. "Functions of hyaluronidase." Nature (Lond.). 162:63-64.

_____. 1949. "The fragmentation of eggs following induced ovulation in immature rats." J. Endocrinology. 6:104-111.

_____. 1951. "Observations on the penetration of the sperm into the mammalian egg." Austr. J. Sci. Res. B. 4:581-596.

_____. 1951. "Activation and the correlation between male and female elements in fertilization." Nature (Lond.). 168:558-559.

_____. 1952. "The capacitation of the mammalian sperm." Nature (Lond.). 170:326.

_____. 1955. "Polyspermy after induced hypothermia in rats." Nature (Lond.). 175: 1038.

_____. 1957. "Fate of spermatozoa in the uterus of the mouse and rat." J. Endocrin. 14:335-342.

_____. 1959. "Entry of spermatozoa into the Fallopian tube mucosa." Nature (Lond.). 183:908.

_____. 1959. "The role of fertilization." Perspectives Biol. Med. 3:44.

_____. 1960. "Fate of spermatozoa in the female genital tract." J. Reprod. Fert. 1:151.

_____. 1960. "Capacitation and the release of hyaluronidase from spermatozoon." J. Reprod. Fert. 3:310-311.

_____. 1960. "Anomalies of fertilization leading to triploidy." J.C.C.P. 56 Suppl. p. 1-16.

_____. 1961. "Sex chromatin in embryonic and fetal tissue." Acta Cytol. 6:61-68.

_____. 1961. "Early reactions of the rodent egg to spermatozoon penetration." J. Exp. Biol. 33:358-365, June.

_____. 1961. "Significance of sperm capacitation." Proc. IV. Cong. Am. Reprod., Hague.

_____. 1961. "The mammalian egg." Blackwell Scientific Pub., Oxford.

_____. 1963. "Fertilization and transport of the ovum." in "Conference on Physiological Mechanisms Concerned with Conception," pp. 285-320. Pergamon Press.

_____. 1965. "Fertilization." 145 pps. Prentice-Hall, Englewood, N.J.

_____. 1965. "Ultrastructural changes in the egg during fertilization and the imitation of cleavage" (G. E. W. Wolstenholme and M. O'Connor, eds.) "Preimplantation Stages of Pregnancy" pps. 3-28. Little, Brown & Co., Boston.

_____, and M. W. H. Bishop. 1957. "Fertilization in mammals." Biol. Rev. 32:296.

_____. 1958. "Capacitation of mammalian spermatozoon." Nature (Lond.). 181:851

_____. 1958. "Some features of the acrosome and perforatorium in mammalian spermatozoon." Proc. Royal Soc. (Lond.). B. 149:241-248.

_____. 1958. "Role of the rodent acrosome and perforatorium in fertilization." Proc. Royal Soc. (Lond.). B. 149:241-248.

Austin, C. R., and A. W. H. Braden. 1953. "An investigation of polyspermy in the rat and rabbit." Austr. J. Biol. Sci. 6:674-692.

_____. 1953. "Polyspermy in mammals." Nature (Lond.). 172:82-3.

_____. 1953. "The distribution of nucleic acid in rat eggs in fertilization and early segmentation." Austr. J. Biol. Sci. 6:324-333.

_____. 1954. "Anomalies in rat, mouse, and rabbit eggs." Austr. J. Biol. Sci. 7: 537-542.

_____. 1955. "Observation on nuclear size and form in living rat and mouse eggs." Exp. Cell. Res. 8:163-172.

Austin, C. R., and A. W. H. Braden. "Early reactions of the rodent egg to spermatozoon penetration." J. Exp. Biol. 33:358-365.

Austin, C. R., and H. M. Bruce. 1956. "Effect of continuous oestrogen administration on oestrous ovulation and fertilization in rat and mice." J. Endocrin. 13:376.

Austin, C. R., and J. Similis. 1948. "Phase-contrast microscopy in the study of fertilization and early development of the rat egg." J. Roy. Microscop. Soc. 68:13-19.

Averill, R. L. W., C. E. Adams, and L. E. Rawson. 1955. "Transfer of mammalian ova between species." Nature (Lond.). 176:167.

Bacisch, P., and G. M. Wyburn. 1945. "Parthenogenesis of atretic ova in the rodent ovary." J. Anat. (Lond.) 79:177-179.

Backman, G. 1939. Das Wachstum der weissen Maus. Lunds Univ. Arsskr. Ard. Z. 35 (12):1-26.

Badtke, G., and K. H. Degenhardt. 1963. "Uber die Enstehung kombinierter mis bildungen der Linse und der Hornkaut in den Augen Neugeborener des embryonalen Linsenblaschens vom Oberflackenektoderm." Klin. Monat. Augenheild. 142:62-89.

Bader, R. S. 1965. "Fluctuating asymmetry in the dentition of the house mouse." Growth. 29:291-300.

Ball, W. D. 1963. "A quantitative assessment of mouse thymus differentiation." Exp. Cell. Res. 31:82-88.

———, and R. Auerbach. 1960. "In vitro formation of lymphocytes from embryonic thymus." Exp. Cell. Res. 20:245.

Ball, Z. B., R. H. Barnes, and Mr. B. Visscher. 1947. "The effects of dietary caloric restriction on maturity and senescence, with particular reference to fertility and longevity. "Am. J. Physiol. 150:511-519.

Balmain, J. H., J. D. Biggers, and P. J. Claringbold. 1956. "Glycogen, wet weight and dry weight changes in the vagina of the mouse." Austr. J. Biol. Sci. 9: 147-158.

Barnett, S. A. 1962. "Total breeding capacity of mice at two temperatures." J. Reprod. Fert. 4:327-335.

———, and E. M. Coleman. 1959. "The effect of low environmental temperature on the reproductive cycle of female mice." J. Endocrin. 19:232-240.

Barnett. S. A., and M. J. Little. 1965. "Maternal performance in mice at -3°C: food consumption and fertility." Proc. Royal Soc. (Lond.) B. 162:492-501.

Barraclough, C. A. 1955. "Influence of age on the response of preweaning female mice to testosterone propionate." Am. J. Anat. 97:493-521.

Barrowman, J., and M. Craig. 1961. "Haemoglobins in foetal C57 B1/6 mice." Nature (Lond.). 189:409-410.

———. 1961. "Haemoglobins of foetal CBA mice." Nature (Lond.). 190:818-819.

Bateman, N. 1954. "The measurement of milk production of mice through preweaning growth of suckling young." Physiol. Zool. 27:163-173.

———. 1954. "Bone growth: a study of the gray lethal and microthalmic mutants in the mouse." J. Anat. 88:212-262.

———. 1960. "Selective fertilization at the locus T of the mouse." Genet. Res. Cambridge. 1:226.

Beatty, R. A. 1951. "Transplantation of mouse eggs." Nature (Lond.). 168:995.

———. 1951. "Heteroploidy in mammals." Animal Breed. (Abstr.) 19:283-292.

———. 1953. "Haploid rodent eggs." Proc. 9th Int. Cong. Genetics Bellogio-Caryologia. 6:784.

———. 1957. "Parthenogenesis and polyploidy in mammalian development." 132 pps. Cambridge U. Press.

Beatty, R. A., and M. Fischberg. 1949. "Spontaneous and induced triploidy in preimplantation mouse eggs." Nature (Lond.). 163:807-808.

———. 1951. "Heteroploidy in mammals. I. Spontaneous heteroploidy in preimplantation mouse eggs." J. Genet. 50:345-359.

Beatty, R. A., and M. Fischberg. 1951. "Cell number in haploid, diploid and polyploid mouse embryos." J. Exp. Biol. 28:541-552.

————. 1952. "Heteroploidy in mammals. III. Induction of tetraploidy in pre-implantation mouse eggs." J. Genet. 50:471-479.

Beatty, R. A., and K. N. Sharma. 1960. "Genetics of gametes. III. Strain differences in spermatozoa from eight inbred strains of mice." Proc. Royal Soc. (Edinb.) B. 68:27-53.

Bennett, D. 1956. "Developmental analysis of a mutation in the pleiotropic effects in the mouse." J. Morphol. 98:199-234.

————. 1964. "Abnormalties associated with a chromosome region in the mouse. II. Embryological effects of lethal alleles in the T-region." Science. 144: 263-267.

————, S. Badenhausen, and L. C. Dunn. 1959. "The embryological effects of four lethal t-alleles in the mouse, which affect the neural tube and skeleton." J. Morphol. 105:105-143.

Bennett, D., and L. C. Dunn. 1958. "Effects on embryonic development of a group of genetically similar lethal alleles derived from different populations of wild house mice." J. Morphol. 103:135-157.

Beaumont, H. M., and A. M. Mandl. 1963. "A quantitative study of primordiol germ cells in the male rat." J. Emb. Exp. Morphol. 11:715-740.

Ben-or, Sarah. 1963. "Morphological and functional development of the ovary of the mouse." J. Emb. Exp. Morphol. 11:1-11.

Berry, R. J. 1960. "Genetical studies on the skeleton of the mouse." Genet. Res. Cambridge.

Biancifiori, C., and F. Caschera. 1963. "The effect of olfactory lobectomy and induced pseudopregnancy on the incidence of methylcholanthrene-induced mammary and ovarian tumors in C3Hb mice. Brit. Jour. Cancer. 17:116-118.

Bierwolf, D. 1958. "Die Embryogenese des Hydrocephalus und der Kleinhirnmiss biklungen bein Dreherstamm der Hausmans." Morphol. Jahrb. 99:542-612.

Biggers, D. 1964. "The biology of cells and tissues in culture." Ed. G. N. Willmer Academic Press.

Biggers, J. D. 1953. "The carbohydrate components of the vagina of the normal and ovariectomized mouse during oestrogenic stimulation." J. Anat. 87327-336.

————, M. R. Ashoub, A. McLaren, and D. Michie. 1958. "The growth and development of mice in three climatic environments. J. Exp. Biol. 35:144-155.

————, and R. L. Brinster. 1964-5. "Biometrical problems in the study of early mammalian embryos in vitro." J. Exp. Zool. 158:39-47.

————, and R. N. Curnow, C. A. Finn, and A. McLaren. 1963. "Regulation of the gestation period in mice." J. Reprod. Fert. 6:125-138.

————, and C. A. Finn, and A. McLaren. 1961. "Long term reproductive performance of female mice." J. Reprod. Fert. 3:303 & 3:313.

————. 1962. "Long term reproductive performance of female mice. II. Variation of litter size with parity." J. Reprod. Fert. 3:315-330.

Biggers, J. D., A. B. L. Gwatkin, and R. L. Brinster. 1962. "Development of mouse embryo in organ cultures of Fallopian tubes on a chemically defined medium." Nature (Lond.). 194:747-749.

————, and A. McLaren. 1958. "Test tube animals - the culture and transfer of early mammalian embryos." Discovery. p. 423.

————, B. Morre, B. Dianne, and D. G. Whittingham. 1965. "Development of mouse embryos in vivo after cultivation from two-cell ova to blastocysts in vitro." Nature (Lond.). 206:734-735.

————, and L. M. Rinaldini. 1957. "The study of growth factors in time culture." Symp. Soc. Exp. Biol. XI. p. 264297.

Bishop, D. H., and J. H. Leathem. 1946. "Response of prepuberal male mice to equine gonadotropin. Anat. Rec. 95:313-319.

————. 1948. "Effect of equine gonadotrophin on prepuberal male mice. Exp. Med. Surg. 6:28-30.

Bishop, D. W., and A. Tyler. 1956. "Fertilizin in mammalian eggs." J. Exp. Zool. 132:575-595.

Bishop, M. W. H., and C. R. Austin. 1957. "Mammalian spermotozoa." Endeavor. 16:137.

_____, and A. Walter. 1960. "Spermatogenesis and the structure of mammalian spermatozoa." Marshall's Physiology of Reproduction 3rd ed. vol. 1 pt. 2 p. 1, A. S. Parkes Ed. Longman Green & Co., Publishers, London.

Bittner, J. J. 1936. "Differences observed in an inbred albino strain of mice following a change in diet. I. Litter size." Jackson Mem. Lab. Nutr. Bull. 1:3-9.

_____. 1936. "Differences observed in an albino strain of mice following a change in diet. II. Mortality." Jackson Mem. Lab. Nutr. Bull. 2:3-11.

Blandau, R. J. 1961. "Biology of eggs and implantation from "Sex and Internal Secretions." Ed. W. C. Young, Pub. Williams & Wilkins, Baltimore.

Blandau, Richard J. 1965. "Observations on the migration of living primordial germ cells in the mouse." Anat. Rec.

Blandau, R., L. Jensen, and R. Rumery. 1958. "Determination of the pH values of the reproductive tract fluids of the rat during heat." Fert. & Steril. 9:207-224.

Blandau, R. J., and D. L. Odor. 1949. "The total number of spermatozoa reaching various segments of the reproductive tract in the female albino rat at intervals after insemination. Anat. Rec. 103:93-109.

_____. 1952. "Observations on sperm penetration into the oöplasm and change in the cytoplasmic components of the fertilizing spermatozoon in rat ova." Fert. & Steril. 3:13-26.

Blandau, R. J., and R. F. Rumery. 1964. "The relationship of swimming movement of epididymal spermatozoa to their fertilizing capacity." Fert. & Steril. 15:571-579.

_____, E. Warrick, and R. E. Rumery. 1965. "*In vitro* cultivation of fetal mouse ovaries." Fert. & Steril. 16:705-715.

_____, B. J. White, and R. E. Rumery. 1962. "*In vitro* observations on the movements of the primordial germ cells of the mouse." Anat. Rec. 142:297.

_____. 1963. "Observations on the movements of the living primordial germ cells in the mouse." Fert. & Steril. 14:482-489.

Bloch, S. 1939. "Contributions to research on the female sex hormones. The implantation of the mouse egg." J. Endocrin. 1:399-408.

_____. 1958. "Beobachtungen uber Falle von fruh zeitiger Trachtigheit der Albino-Maus. Experientia. 14:141-142.

Bluhm, A. 1932. "Uber einen Fall von beeinflussung des geschlechtsverhaltnisses der Albino-Hausmaus durch behandlung des weibchens." Z. ind. Abst. in Vereb. 62:88-89.

Bodermann, E. 1935. "A case of uniovular twins in the mouse." Anat. Rec. 62:291-294.

Boell, E. 1948. "Respiratory metabolism of the mammalian egg." J. Exp. Zool. 109-267.

Bogart, R., R. W. Mason, H. Nicholson, and H. Krueger. 1958. "Genetic aspects of fertility in mice." Int. J. Fert. 3:86-104.

Bomsel-Helmreich, O. 1965. "Heteroploidy and embryonic death" in "Preimplantation Stages in Pregnancy." G. E. W. Wolstenholme & M. O'Connor, eds. p. 246-269, Little, Brown & Co., Boston.

Bonnevie, K. 1950. "New facts on mesoderm formation and proamnion derivatives in the normal mouse embryo." J. Morphol. 86:495-545.

_____, and A. Brodal. 1946. "Hereditary hydrocephalus in the house mouse. IV. The development of the cerebullar anomalies during foetal life with notes on the normal development of the mouse cerebellum." Norske Videns.-Akad. I. Oslo I. Mat. Natur. Klasse no. 4.

Boot, L. M., and O. Muhlboch. 1953. "Transplantation of ova in mice." Acta Physiol. pharm-neerl.

Boot, L. M., and O. Muhlboch. 1957. "The ovarian function in old mice. Acta Physiol. Pharmacol. Neer. 3:463.

Borghese, E., and A. Cassini. 1963. "Cleavage of mouse eggs" in "Cinemicography in Cell Biology." p. 263, Ed. E. Rose, Academic Press, N.Y.

Borum, K. 1961. "Oögenesis in the mouse: a study of the meiotic prophase." Exp. Cell. Res. 24:495-507.

Bosshardt, K. K., W. J. Paul, K. E. O'Doherty, and R. H. Barnes. 1949. "Mouse growth assay procedures for the 'animal protein factor'." J. Nutr. 37:1.

Boving, B. G. 1954. "Blastocyst-uterine relationships." Cold Spring Harbor Symp. Quant. Biol. 19:9.

_____. 1959. "The biology of the trophoblast." Am. N. Y. Acad. Sci. 80:21-43.

_____. 1960. "Implantation." Am. N. Y. Acad. Sci. 75:700-725.

_____. 1960. "Invasion mechanisms in implantation." Anat. Rec. 136:p. 168.

_____. 1963. "Implantation mechanisms" in "Mechanisms concerned with conception." E. G. Hartman. ed. Pergamon Press.

Bowman, J. C., and D. S. Falconer. 1960. "Inbreeding depression and heterosis of litter size in mice." Genet. Res. 1:262-274.

_____, and R. C. Roberts. 1958. "Embryonic mortality in relation to ovulation rate in the house mouse." J. Exp. Biol. 35:138.

Boyd, J. D., and W. J. Hamilton. 1952. "Cleavage, early development and implantation of the egg" pps. 1-126 (A. S. Parkes ed.) "Marshall's physiology of reproduction," 3rd ed. vol. 2. Longman, Green, London.

Boyer, C. C. 1953. "Chronology of development for the golden hamster. J. Morphol. 92:1.

Bracish, P., and G. W. Wyburn. 1945. "Parthenogenesis of atretic ova in the rodent ovary." J. Anat. (Lond.) 79:177.

Bradbury, J. T., and R. G. Bunge. 1958. "Oöcytes in seminiferous tubules." Fert. & Steril. 9:18-25.

Braden, A. W. H. 1953. "The distribution of nucleic acid in rat eggs in fertilization and early segmentation." Austr. J. Biol. Sci. 6:665-673.

_____. 1954. "Reactions of unfertilized mouse eggs to some experimental stimuli." Exp. Cell. Res. 7:277-280.

_____. 1954. "The fertile life of mouse and rat eggs." Science 120:361.

_____. 1957. "The relationship between the diurnal light cycle and the time of ovulation in mice." J. Exp. Biol. 34:177-188.

_____. 1957. "Variation between trains in the incidence of various abnormalties of egg maturation and fertilization in the mouse." J. Genet. 55:476-486.

_____. 1957. "Differences between inbred strains of mice in the morphology of the gametes." Anat. Rec. 127:270-271.

_____. 1958. "Variation between strains of mice in phenomena associated with sperm penetration and fertilization." Genetics. 56:37-47.

_____. 1958. "Influence of time of mating on the segregation rates of alleles of the T locus in the house mouse." Nature (Lond.). 181:786-787.

_____. 1959. "Strain differences in the morphology of the gametes of the mouse." Austr. J. Biol. Sci. 12:65-71.

_____. 1959. "Are nongenetic defects of the gametes important in the etiology of prenatal mortality?" Fert. & Steril. 10:285-298.

_____. 1959. "Spermatozoa penetration and fertilization in the mouse." Proc. Int. Symp. Exp. Biol. Spollanzani, Paira, May.

_____. 1960. "Genetic influences on the morphology and function of the gametes" in "Mammalian genetics and reproduction." J. Cell & Comp. Physiol. 56: (Suppl. 1) 17-29.

_____. 1962. "Spermatozoon penetration and fertilization in the mouse." Symp. Genet. Biol. Ital. 9:1-8.

_____, and C. R. Austin. 1954. "The fertile life of mouse and rat eggs." Science. 120:361.

REFERENCES 371

Braden, A. W. H., and C. R. Austin. 1954. "Fertilization of the mouse eggs and the effect of delayed coitus and of hot-shock treatment." Austr. J. Biol. Sci. 7: 552-565.

———. 1954. "The number of sperms about the eggs in mammals and its significance for normal fertilization." Austr. J. Biol. Sci. 7:543-551.

Braden, A. W. H., C. R. Austin, and H. A. David. 1954. "The reaction of the zona pellucida to sperm penetration." Austr. J. Biol. Sci. 7:391-409.

———, and S. Glueckisohn-Waelsch. 1958. "Further studies of the effects of the T locus in the house mouse on male fertility." J. Exp. Zool. 138:431-452.

Brambell, F. W. R. 1927. "The development and morphology of the gonads of the mouse. I. The morphogenesis of the indifferent gonad and the ovary." Proc. Royal Soc. (Lond.) B. 101:391-409; 102:206-222; 103:258-272.

———. 1928. "The development and morphology of the gonads of the mouse. III. The growth of the follicles." Proc. Royal Soc. (Lond.) B. 103:258-272.

———. 1937. "The influence of lactation on the implantation of the mammalian embryo." Am. J. Obstet. & Gynec. 33:942-953.

———. 1948. "Prenatal mortality in mammals." Biol. Rev. 23:370-407.

———, and A. S. Parkes. 1927. "The normal ovarian cycle in relation to oestrus production." Quart. J. Exp. Physiol. 18:185-198.

Brinster, R. L. 1963. "A method for *in vitro* cultivation of mouse ova from two cell to blastocyst." Exp. Cell. Res. 32:205-208.

———. 1964. "Possible energy sources for the development of the early mouse embryo." Jour. Cell Biol. 23:14A.

———. 1965. "Studies on the development of mouse embryos *in vitro*. I. The effect of osmolarity and hydrogen ion concentration." J. Exp. Zool. 158:49-57.

———. 1965. "Studies on the development of mouse embryos *in vitro*. II. The effect of energy source." J. Exp. Zool. 158:59-68.

———. 1965. "Studies on the development of mouse embryos *in vitro*. III. The effect of fixed nitrogen source." J. Exp. Zool. 158:69-77.

———. 1965. "Lactate dehydrogenase activity in the pre-implanted mouse embryos." Biochem. Biophys. Acta. 110:439-441.

———. 1965. "Studies of the development of mouse embryos *in vitro*: energy metabolism" in "Preimplantation Stages in Pregnancy." G. E. W. Wolstenholme & M. O'Connor, eds. p. 60-81, Little, Brown & Co., Boston.

———. 1966. "Glucose 6-phosphate-dehydrogenase activity in the preimplantation mouse embryo." Biochem. J. 101:161-163.

———, and J. D. Biggers. 1965. "*In vitro* fertilization of mouse ova within the explanted Fallopian tube." J. Reprod. Fert. 10:277-279.

Briody, B. A. 1959. "Response of mice to ectomelia and vaccinia virus." Bact. Rev. 23:61.

Briones, H., and R. A. Beatty. 1954. "Interspecific transfers of rodent eggs." J. Exp. Zool. 125:99.

Bruce, H. M. 1954. "Feeding and breeding of laboratory animals. XIV. Size of breeding group and production of mice." J. Hyg. 52:60-66.

———. 1959. "An exteroceptive block to pregnancy in the mouse." Nature (Lond.). 184:105.

———. 1960. "Further observations on pregnancy block in mice caused by the proximity of strange males." J. Reprod. Fert. 1:311-312.

———. 1960. "A block to pregnancy in the mouse caused by proximity of strange males." J. Reprod. Fert. 1:96-103.

———. 1961. "An olfactory block to pregnancy in mice. Part I. Characteristics of the block." Proc. 4th Congr. Anim. Reprod. (The Hague). 159-162.

———. 1965. "Effect of castration on the reproductive phenomena of male mice." J. Reprod. Fert. 10:141-143.

———, and J. East. 1956. "Number and viability of young from pregnancies concurrent with lactation in the mouse. J. Endocrin. 14:19-27.

Bruce, H. M., and D. M. V. Parrott. 1960. "Role of olfactory sense in pregnancy block by strange males." Science. 131:1526.

Brumby, P. J. 1960. "The influence of maternal environment on growth in mice." Heredity. 14:1-18.

Bryson, D. L. 1964. "Development of mouse eggs in diffusion chambers." Science. 144:1351-1353.

Bryson, V. 1944. "Spermatogenesis and fertility in Mus musculus as affected by factors at the T-locus." J. Morphol. 74:131-187.

––––––. 1945. "Development of the sternum in screw tail mice." Anat. Rec. 91:119-141.

Bullough, W. S. 1942. "Oögenesis and its relation to the oestrous cycle in the adult mouse." J. Endocrin. 3:141-149.

––––––. 1946. "Mitotic activity in the adult female mouse. A study of its relation to the oestrous cycle in normal and abnormal conditions." Philos. Trans. Royal Soc. (Lond.) B. 231:451-576.

––––––, and H. F. Gibbs. 1941. "Oögenesis in adult mice and starlings." Nature (Lond.). 148:439-440.

Bulmer, D., and A. D. Dickson. 1960. "Observations on carbohydrate materials in the rat placenta." J. Anat. 94:46-58.

Burckhart, G. 1901. "Die Implantation des Eies der maus in die Uterusochleimhaut und die Umbildung derselben zur Decidua." Arch. Mikroscop. Anat. 57:528-569.

Burdick, H. O., B. B. Emmerson, and R. Whitney. 1940. "Effects of testosterone proprionate on pregnancy and on passage of ova through the oviducts of mice." Endocrin. 26:1081-1086.

––––––, and G. Pincus. 1935. "The effect of oestrin injections upon the developing ova of mice and rabbits." Am. J. Physiol. 111:201-208.

––––––, and R. Whitney. 1937. "Acceleration of the rate of passage of fertilized ova through the Fallopian tubes of mice by massive injections of an estrogenic substance." Endocrin. 21:637.

––––––, R. Whitney, and B. Emerson. 1942. "Observations on the transport of tubal ova." Endocrin. 31:100-108.

––––––, R. Whitney, and G. Pincus. 1937. "The fate of mouse ova tube-locked by injections of oestrogenic substances." Anat. Rec. 67:513-519.

Burns, E. L., M. Maskop, V. Suntzeff, and L. Loeb. 1936. "On the relation between the incidence of mammary cancer and the nature of the sexual cycle in various strains of mice." Am. J. Cancer. 26:56-68.

Burrows, H. 1935. "Pathological conditions induced by oestrogenic compounds in the coagulating gland and prostate of the mouse." Am. J. Cancer. 23:490-512.

Burstone, M. S. 1950. "The effect of radioactive phosphorus upon the development of the teeth and mandibular joint of the mouse." J. Am. Dental Assn. 41:1-18.

Calarco, Patricia G. 1965. "The histology and fine structure of the murine yolk sac." (Abstr.) Am. Zool. 5:224.

––––––, and Frank H. Moyer. 1966. (Univ. Illinois, Urbana) "Structural changes in the murine yolk sac during gestation: cytochemical and electron microscope observations." J. Morphol. 119:341-356.

Callas, G., and B. E. Walker. 1963. "Palate morphogenesis in mouse embryos after X-irradiation." Anat. Rec. 145:61-68.

Carsner, C. L., and E. G. Rennels. 1960. "Primary site of gene action in anterior pituitary dwarf mice." Science. 131:829.

Carter, T. C. 1954. "The genetics of luxate mice. IV. Embryology." J. Genet. 52:1-35.

––––––. 1959. "Embryology of the Little and Bagg X-rayed mouse stock." J. Genet. 56:401-435.

––––––, M. F. Lyon, and R. Y. S. Phillips. 1955. "Gene-tagged chromosome translocations in eleven stocks of mice." J. Genet. 53:155-166.

Castle, W. E., W. H. Gates, S. C. Reed, and G. D. Snell. 1936. "Identical twins in a mouse cross." Science. 84:581.

Cattanach, B. M., and R. G. Edwards. 1958. "The effects of triethylenemelamine on the fertility of male mice." Proc. Royal Soc. (Edin.). 67:54.

Center, E. M. 1955. "Postaxial polydactyly in the mouse." J. Heredity. 46:144-148.

Cerey, K., J. Elis, and H. Raskova. 1965. "Studies on 6-azuridine and 6-azactidine. VI. Influence of 6-azactidine on prenatal development in mice." Biochem. Pharmacol. 14:1549-1556.

Cerruti, R. A., and W. R. Lyons. 1960. "Mammogenic activities of the mid-gestation mouse placenta." Endocrin. 67:884.

Chai, C. K. "Life span in inbred and hybrid mice." J. Heredity. 50:203-208.

_____. 1966. "Characteristics in inbred mouse populations plateaued by directional selection." Genetics. 54:743-753.

Chandhuri, A. C. 1928. "The effect of the injection of alcohol into the male mouse upon the secondary sex ratio among the offspring." Brit. Jour. Exp. Biol. 5:185-186.

Chang, M. C. 1950. "Cleavage of unfertilized ova in immature ferrets." Anat. Rec. 108:31-44.

_____. 1951. "Fertilizing capacity of spermatozoa deposited into the Fallopian tubes." Nature (Lond.). 168-697.

_____, and D. M. Hunt. 1962. "Morphological changes of sperm head in the oöplasm of mouse, rat, hamster, and rabbit." Anat. Rec. 142:417-426.

Chang, T. K. 1939. "The development of polydactylism in a special strain of Mus musculus." Peking Nat. Hist. Bull. 14:119-132.

_____. 1940. "Cellular inclusions and phagocytosis in normal development of mouse embryos." Peking Nat. Hist. Bull. 14 part 3.

Chapekar, T. N., G. V. Nayak, and K. J. Ranadive. 1966. "Studies on the functional activity of organically cultured mouse ovary." J. Emb. Ex. Morphol. 15:133-141.

Chardard-Ramboult. 1949. "Development de la glande thyroide chez la souris pendant la vie intra-uterine." Compt. Rend. Soc. Biol. 143:40-41.

Charlton, H. H. 1917. "The fate of unfertilized egg in the white mouse." Biol. Bull. 33:321-333.

Chase, E. B. 1941. "Studies on an anophthalmic strain of mice. II. Effect of congenital eyelessness on reproductive phenomena. Anat. Rec. 80:33-36.

Chase, H. B. 1951. "Inheritance of polydactyly in the mouse." Genetics. 36:697-710.

_____. 1954. "Growth of hair." Phys. Rev. 34:113-126.

_____. 1958. "The behaviour of pigment cells and epithelial cells in the hair follicle" in "The biology of hair growth." Chap. I. Academic Press, N.Y.

_____, and E. Chase. 1941. "Studies on an anophthalmic strain of mice. I. Embryology of the eye region." J. Morphol. 68:279-301.

_____, H. Ranch, and V. W. Smith. 1951. "Critical stages of hair development and pigmentation in the mouse." Physiol. Zool. 24:1-8.

Chesley, P. 1935. "Development of the short-tailed mutant in the house mouse." J. Exp. Zool. 70:429-459.

Chiquoine, A. 1954. "The identification, origin, and migration of the primordial germ cells in the mouse embryo." Anat. Rec. 118:135-146.

Chiquoine, A. D. 1958. "The distribution of polysaccharides during gastrulation and embryogenesis in the mouse embryo." Anat. Rec. 129:495-516.

_____. 1959. "Electron microscopic observations on the developmental cytology of the mammalian ovum." Anat. Rec. 133:258.

_____. 1960. "The development of the zona pellucida of the mammalian ovum." Am. J. Anat. 106:149.

Christian, J. J. 1964. "Effect of chronic ACTH treatment on maturation of intact female mice." Endocrin. 74:669-679.

_____, and C. D. Lemunyan. 1958. "Adverse effects of crowding on lactation and reproduction of mice and two generations of their progeny." Endocrin. 63:517-529.

Christy, N. P., M. Dickie, and G. W. Woolley. 1950. "Estrus and mating in gonadec-
    tomized female mice with adrenal cortical abnormalities. Endocrin. 47:
    129-130.

Clark, S. L. 1959. "The ingestion of proteins and colloidal material by columnar ab-
    sorptive cells of the small intestine in suckling rats and mice." J. Biophys.
    Biochem. Cytol. 5:41-50.

Clauberg, C. 1931. "Genitalcyclus and Schwangershoft bei der weissen Maus. Dauer
    de Genital cyclus." Arch. Gynech. 147:549-596.

Clermont, Y., and E. Bustos. 1966. "Identification of five classes of type A sperma-
    togonia in rat seminiferous tubules mounted 'in toto'." Anat. Rec. 154:332.

_____, and C. P. Leblond. 1953. "Renewal of spermatogonia in the rat." Am. J.
    Anat. 93:475-501.

_____, and C. Huckins. 1961. "Microscopic anatomy of the sex cords and seminifer-
    ous tubules in growing and adult albino rats." Am. J. Anat. 108:79-97.

_____, and B. Perey. 1957. "Quantitative study of the cell population of the seminif-
    erous tubule in immature rats." Am. J. Anat. 100:241-267.

Cohen, A. I. 1961. "Electron microscopic observations of the developing mouse eye.
    I. Basement membranes during early development and lens formation." De-
    velop. Biol. 3:297-316.

Cohn, S. A. 1957. "Development of the molar teeth in the albino mouse." Am. J.
    Anat. 101:295-320.

_____. 1966. "Lack of effect of acute neonatal anoxia on development of molar teeth
    of mice." Anat. Rec. 154:333.

Cole, H. A. 1933. "The mammary gland of the mouse during the estrous cycle, preg-
    nancy and lactation." Proc. Royal Soc. (Lond.) B. 114:136-160.

Cole, R. J., and J. Paul. 1965. "Preimplantation stages of pregnancy." Ciba Founda-
    tion Symposium ed. G. E. W. Wolstenholme & M. O'Connor, p. 86 - Churchill,
    Ltd., London.

Coleman, D. L. 1966. "Purification and properties of δ-aminolevulinate dehydratase
    from tissues of two strains of mice." J. Biol. Chem. 241:5511-5517.

Connell, R. S. 1966. "Nucleoside phosphatases of the rat chorioallantoic placenta with
    increasing gestational age: electron microscope study." Anat. Rec. 154:334.

Cook, M. J. 1965. "The anatomy of the laboratory mouse." Academic Press, N.Y.
    143 pps.

Cooper, George W. 1965. "Induction of somite chondrogenesis by cartilage and noto-
    chord: a correlation between inductive activity and specific stages of cytodif-
    ferentiation." Dev. Biol. 12:185-212.

Coppenger, C. J. 1964. "Effects of prenatal chronic gamma irradiation on the pre-
    natal and postnatal development of the albino rat." Thesis, College Station,
    Tex., Agricultural and Mechanical Univ., 125 p.

Cox. D. F., J. E. Legates, and C. C. Cockerham. 1959. "Maternal influence on body
    weight." J. Animal Sci. 18:519-527.

Cox, F. K. 1926. "The chromosomes of the house mouse." J. Morphol. & Physiol.
    43:45-54.

Crabtree, C. 1940. "Sex difference in the structure of Bowman's capsule in the mouse."
    Science. 91:299.

Craig, M. L., and F. S. Russell. 1963. "Electrophoretic patterns of hemoglobin from
    fetal mice of different inbred strains." Science. 142:398-399.

_____. 1964. "A developmental change in hemoglobins correlated with an embryonic
    red cell population in the mouse." Develop. Biol. 10:191-201.

Cranston, E. M. 1945. "The effect of lithospermum ruderale on the estrous cycle of
    mice." J. Pharmacol. Exp. Therap. 83:130-142.

Crelin, E. S., and J. Levin. 1955. "The prepuberal public symphsis and uterus in the
    mouse: their response to estrogen and relaxin." Endocrin. 57:730-747.

Crew, A. E., and P. Ch. Koller. "The sex incidence of chiasma frequency and genetical
    crossing over in the mouse." J. Genet. 26:359-383.

Crew, F. A. E., and L. Mirshaia. 1930. "Mating during pregnancy in the mouse." Nature (Lond.). 125: 564.

———. 1930. "The lactation interval in the mouse." Quart. J. Exp. Physiol. 20:105-110.

———. 1930. "On the effect of removal of the litter upon the reproductive rate of the female mouse." Quart. J. Exp. Physiol. 20:263-266.

Crippa, M. 1964. "The mouse karyotype in somatic cells cultured *in vitro*." Chromosoma. 15:301-311.

Crisp, T. M. 1965. "Observations on the fine structure of lutein cells in mice." Anat. Rec.

Curry, G. A. 1959. "Genetical and developmental studies on droopy-eared mice." J. Emb. Exp. Morphol. 7:39-65.

Cutright, P. R. 1952. "Spermatogenesis in the mouse." J. Morphol. 54:197-220.

Dagg, C. P. 1960. "Sensitive stages for the production of developmental abnormalties in mice with 5-flurourocil." Am. J. Anat. 106:89-96.

———. 1963. "The interaction of environmental stimuli and inherited susceptibility to congenital deformity." Am. Zool. 3:223-233.

Dagg, C. P., *et al.* 1966. "Polygenic control of the teratogenicity of 5-fluorouracil in mice." Genetics. 53:1101-1117.

Dagg, Charles P. 1966. (Jackson Lab., Bar Harbor) "Teratogenesis," pp. 309-328. In E. L. Green (ed.) Biol. Lab. Mouse, 2nd ed. N. Y., McGraw-Hill.

Dalcq, A. M. 1955. "Processes of synthesis during early development of rodent's eggs and embryo." Studies on Fertility. 7:113.

———. 1956. "Effects du reactif de Schiff sur les oeufs en segmentation du rat et de la souris." Exp. Cell. Res. 10:99-119.

Dalcq, A., and J. Pasteels. 1955. "Determination photo metrique de la Teneur relative en DNA des noyaux dans le oeufs en segmentation du rat et de la souris." Exp. Cell. Res. Suppl. 3:72-97.

Danforth, C. H. 1930. "Developmental anomalies in a special strain of mice." Am. J. Anat. 45:275-288.

———, and S. B. de Aberle. 1927. "Distribution of foetuses in the uteri of mice." Anat. Rec. 35:33.

———. 1927. "The functional interrelation of certain genes in the development of the mouse." Genetics. 12:340-347.

———. 1928. "The functional interrelation of the ovaries as indicated by the distribution of foetuses in mouse uteri." Am. J. Anat. 41:65-74.

Daniel, J. F. 1910. "Observations on the period of gestation in white mice." J. Exp. Zool. 9:865-870.

Daoust, R., and Y. Clermont. 1955. "Distribution of nucleic acid in germ cells during the cycle of the seminiferous epithelium in the rat." Am. J. Anat. 96:255-279.

Davenport, C. B. 1925. "Regeneration of ovaries in mice." J. Exp. Zool. 42:1-11.

Dawson, A. B. 1935. "The influence of hereditary dwarfism in the differentiation of the skeleton of the mouse." Anat. Rec. 61:485-493.

DeAberle, S. B. 1927. "A study of the hereditary anemia of mice." Am. J. Anat. 40:219-247.

Deane, H. W., B. L. Rubin, E. C. Driks, B. L. Lobel, and G. Peipsner. 1962. "Trophoblastic giant cells in placentas of rats and mice and their probable role in steroid-hormone production." Endocrin. 70:407-419.

Deanesly, R. 1930. "The corpora lutea of the mouse, with special reference to fat accumulation during the oestrous cycle." Proc. Royal Soc. (Lond.) B. 106:578-595.

———. 1930. "The development and vascularization of the corpus luteum in the mouse and rabbit." Proc. Royal Soc. (Lond.) B. 107:60-76.

———, and A. S. Parkes. 1933. "Size changes in the seminal vesicles of mouse during development and after castration." J. Physiol. 78:442-450.

DeFeo, V. 1965. "Temporal aspect of the distribution of ova, in the rat uterus." Anat. Rec. 151:392.

DeFries, J. C. 1964. "Prenatal maternal stress in mice." J. Heredity. 55:289-295.

———. 1965. "Blocking of pregnancy in mice as a function of stress." Psychol. Rep. 17:96-98.

De Haan, R. L., and H. Ursprung, eds. 1965. "Organogenesis." Holt, Rinehart & Winston, 814 pps.

De Long, B. R., and R. L. Sidman. 1962. "Effects of eye removal at birth on histogenesis of the mouse superior colliculus: an autoradiographic analysis with tritiated thymidine." J. Comp. Neur. 118:205-224.

Deno, R. A. 1937. "Uterine macrophages in the mouse and their relation to involution." Am. J. Anat. 60:433-471.

Deol, M. S. 1962. "Genetical Studies on the skeleton of the mouse. XXVIII. Tailshort." Proc. Royal Soc. (Lond.) B. 155:78-95.

———. 1963. "The development of the inner ear in mice homozygous for shaker-with-syndactylism." J. Emb. Exp. Morphol. 11:493-512.

———. 1964. "The origin of the abnormalities of the inner ear in Dreher mice." J. Emb. Exp. Morphol. 12:727-733.

———. 1964. "The abnormalities of the inner ear in Kreisler mice." J. Emb. Exp. Morphol. 12:475-490.

———. 1966. "Influence of the neural tube on the differentiation of the inner year in the mammalian embryo." Nature (Lond.). 209:219-220.

———, and Margaret C. Green. 1966. "Snell's waltzer, a new mutation affecting behaviour and the inner ear in the mouse." Genet. Res. 8:339-345.

———, H. Grunberg, H. G. Searle, and G. M. Truslove. 1960. "How pure are inbred strains of mice." Genet. Res. 1:50-58.

———, and Priscilla W. Lane. 1966. "A new gene affecting the morphogenesis of the vestibular part of the inner ear in the mouse." J. Emb. Exp. Morphol. 16: 543-558.

Detwiter, S. R. 1932. "Experimental observations upon the developing rat retina." J. Comp. Neur. 55:473-492.

Dewar, A. D. 1957. "Body weight changes in the mouse during the oestrous cycle and pseudopregnancy." J. Endocrin. 15:230-233.

Dickmann, Z. 1965. "Sperm penetration into and through the zona pellucida of the mammalian egg" in "Preimplantation Stages in Pregnancy." G. E. W. Wolstenholme & M. O'Connor, eds. p. 169-182, Little, Brown & Co., Boston.

———, and R. W. Noyes. 1960. "The fate of ova transport into the uterus of the rat." J. Reprod. Fert. 1:197-212.

———. 1961. "The zona pellucida at the time of implantation." Fert. & Steril. 12: 310-318.

Dickson, A. D. 1963. "Trophoblastic giant cell transformation of mouse blastocysts." J. Reprod. Fert. 6:465-466.

———. 1964. "Delay of implantation in superovulated mice subjected to crowded conditions." Nature (Lond.). 201:839-840.

———. 1965. "The structure and size of the mouse blastocyst." Anat. Rec. 151:343.

———. 1966. "The form of the mouse blastocyst." J. Anat. 100:335-348.

———. 1966. "The form of the mouse blastocyst undergoing delay of implantation." Anat. Rec. 154:338.

———. 1966. "Induction of the trophoblastic giant cell transformation after ovariectomy in the mouse." (In press.)

———. 1966. "The size of the inner cell mass in blastocysts recovered from normal and ovariectomized mice." Int. J. Fert. 11:231-234.

———. 1966. "Observations on blastocysts recovered from ovariectomized mice." Int. J. Fert. 11:227-230.

———. 1967. "Variations in development of mouse blastocysts." J. Anat. (Lond.) in press.

Dickson, A. D., and H. B. Aranjo. 1966. "Dissociation of implantation and tropho-blastic giant cell transformation of mouse blastocysts." (In press.)

_____, and D. Bulmer. 1960. "Observations on the placental giant cells of the rat." J. Anat. 94:418-425.

_____. 1961. "Observations on the origin of metrial gland cells in the rat placenta." J. Anat. 95:262-273.

Dietert, S. E. 1966. "Fine structure of the formation and fate of the residual bodies of mouse spermatozoa with evidence for the participation of lysosomes." Anat. Rec. 114:338.

Di Paola, J. A., H. Gatzek, and J. Pichren. 1964. "Malformations induced in the mouse by thalidomide." Anat. Rec. 149:149-155.

Dominic, C. J. 1965. "The origin of the phermones causing pregnancy block in mice." J. Reprod. Fert. 10:469-472.

Doyle, L. L., A. H. Gates, and R. W. Noyes. 1963. "Asynchronous transfer of mouse ova." Fert. & Steril. 11:215-225.

_____, A. H. Gates, and R. W. Noyes. 1963. "Asynchronous transfer of mouse ova." Fert. & Steril. 11:215-225.

Drasher, M. L. 1952. "Morphological and chemical observations on the mouse uterus during the estrous cycle and under hormonal treatment." J. Exp. Zool. 119:333-354.

_____. 1953. "Aging changes in nucleic acid and protein-forming systems of the vir-gin mouse uterus." Proc. Soc. Exp. Biol. & Med. 84:596-601.

_____. 1953. "Uterine and placental nucleic acids and protein during pregnancy in the mouse." J. Exp. Zool. 122:388-408.

_____. 1955. "Strain differences in the response of the mouse uterus to estrogens." J. Heredity. 46:190-192.

Dronkert, A., M. Ota, and A. H. Gates. 1965. "Gonadotropin-inhibiting substances in human urine." Proc. for VI Panamerican Congress of Endocrinology, Mexico City, Excerpta Medica International Congress Series, 99:78.

Dry, F. W. 1926. "The coat of the mouse." J. Genet. 16:287-340.

Duboo, R. J., and R. W. Schaedler. 1960. "The effect of the intestinal flora on the growth rate of mice, and on their susceptibility to experimental infections." J. Exp. Med. 111:407-417.

Dunn, T. B. 1954. "Normal and pathologic anatomy of the reticular tissue in laboratory mice." J. Nat. Cancer Inst. 14,6:1281-1433.

Duplan, J. F., and N. S. Wolf. 1962. "Age-related factors which influence the value of the mouse embryo for post-irradiation restoration of the adult." Int. J. Rad. Biol. 5:597-607.

Dzuick, P. J., and M. N. Runner. 1960. "Recovery of blastocysts and induction of implantation following artificial insemination of immature mice." J. Reprod. Fert. 1:321-333.

Eagle, H. 1956. "The salt requirements of mammalian cells in tissue culture." Arch. Biochem. Biophys. 61:356.

Eaton, G. J., and L. D'Aloisio. 1966. "Effects of progesterone on embryo implantation in the mouse." Abstr. Amer. Zool. 6:316.

Eaton, O. N. 1941. "Crosses between inbred strains of mice." J. Heredity. 32:393-395.

_____, and M. M. Green. 1963. "Giant cell differentiation and lethality of homozygous yellow mouse embryos." Genetics. 34:155-161.

Ebert, J. D. 1965. "Interacting systems in development." Holt, Rinehart & Winston, N.Y. 227 pps.

Eckstein, P. 1959. "Implantation of ova." Mem. Soc. Endocrin. #6 Cambridge U. Press.

Eddo, M. U. 1956. "Immunology and development." Univ. Chicago Press, 59 pps.

Edidin, M. 1964. "Transplantation antigen levels in the early mouse embryo." Trans-plantation. 2:627-637.

Edwards, R. G. 1954. "The experimental induction of pseudogamy in early mouse embryos." Experientia. 10:499.

_____. 1954. "Colchicine-induced heteroploidy in early mouse embryos." Nature (Lond.). 174:276-277.

_____. 1955. "Selective fertilization following the use of sperm mixture in the mouse." Nature (Lond.). 175:215.

_____. 1955. "Colchicine-induced heteroploidy in the mouse. III. The induction of tetraploidy and other types of heteroploidy." J. Exp. Zool. 137:349.

_____. 1957. "The experimental induction of gynogenesis in the mouse." Proc. Royal Soc. (Lond.) Ser. B. 146:469;488;149:117 (1958).

_____, and R. E. Fowler. 1958. "The experimental induction of superfoetation in the mouse." J. Endocrin. 17:223-236.

_____, and A. H. Gates. 1958. "Radioactive tracers and fertilization in mammals." Endeavor. 17:47.

_____. 1959. "Timing of stages of maturation divisions, ovulation, fertilization and the first cleavage of eggs of adult mice treated with gonadotrophins." J. Endocrin. 18:272-304.

_____. 1959. "Embryonic development in superovulated mice not receiving the coital stimulus." Anat. Rec. 135:291.

Edwards, R. G., and J. L. Sirlin. 1956. "Labelled pronuclei in mouse eggs fertilized by labelled sperm." Nature (Lond.). 177:429.

_____. 1957. "Studies in gametogenesis, fertilization and early development in the mouse, using radioactive tracers." Int. J. Fert. 2:185,376.

_____. 1959. "Identification of the C^{14} labelled male chromatin at fertilization in colchicine-treated mouse eggs." J. Exp. Zool. 140:19-27.

Edwards, R. G., E. D. Wilson, and R. E. Fowler. 1963. "Genetic and hormonal influences on ovulation and implantation in adult mice treated with gonadotrophins." J. Endocrin. 26:389-399.

Eguchi, Y., and Y. Hashimoto. 1961. "Histological development of the testis of the mouse during embryonic stages." Bull. Univ. Osoha Prefecture Ser. B. 11:77-83.

Ehling, U. H. 1965. "Dominant skeletal mutations induced by x-irradiation of the mouse spermatogonia." (Abstr.) Genetics. 52:441-442 (see also *ibid* 51:723-732)

Ellinger, F., J. E. Morgan, and F. W. Chambers. 1952. "The use of small laboratory animals in medical radiation biology." NMRI Research Report Proj. NM 006 012.04.43.

Elliott, J. R., and C. W. Turner. "The mammary gland spreading factor in normal pregnant animals." Endocrin. 54:284-289.

Enders, A. C., and S. Schlafke. 1962. "The fine structure of the blastocyst." Anat. Rec. 142:338.

_____. 1965. "The fine structure of the blastocyst: some comparative studies" in "Preimplantation Stages of Pregnancy." G. E. W. Wolstenholme & M. O'Connor, eds., p. 29-59, Little, Brown & Co., Boston.

Engle, E. T. 1927. "A quantitative study of follicular atresia in the mouse." Am. J. Anat. 39:187-203.

_____. 1927. "Polyovular follicles and polynuclear ova in the mouse." Anat. Rec. 35:341-343.

_____. 1931. "Prepubertal growth of the ovarian follicle in the albino mouse." Anat. Rec. 48:341-350.

_____. 1942. "Female mating behavior shown by male mice after treatment with different substances." Endocrin. 30:623.

_____, and J. Rosasco. 1927. "The age of the albino mouse at normal sexual maturity." Anat. Rec. 36:383-388.

Enzmann, E. V. 1933. "Milk-production curve of the albino mouse." Anat. Rec. 56:345-358.

Enzmann, E. V. 1935. "Intrauterine growth of albino mice in normal and in delayed pregnancy." Anat. Rec. 62:31-45.

_____, N. R. Sapkin, and G. Pincus. 1932. "Delayed pregnancy in mice." Anat. Rec. 54:325-342.

Epifanova, O. I. 1963. "A radiographic analysis of the mitotic cycle and kinetics of a cell population in the uterine epithelium of mice." Dokl. Acad. Nauk. USSR 149:424-427.

Espinasse, P. G. 1935. "The oviduccal epithelium of the mouse." J. Anat. 69:363-368.

Evans, E. P., G. Breckon, and C. E. Ford. 1964. "An air-drying method for meiotic preparations from mammalian testes." Cytogenetics. 3:289-294.

Everett, N. B. 1943. "Observational and experimental evidences relating to the origin and differentiation of the definitive germ cells in mice." J. Exp. Zool. 92: 49-91.

Fainstat, T. D. 1951. "Hereditary differences in ability to conceive following coitus in mice." Science. 114:524.

Falconer, D. S. 1955. "Patterns of response in selection experiments with mice." Cold Spring Harbor Symp. Quant. Biol. 20:178-196.

_____. 1960. "The genetics of litter-size in mice." J. Cell. Comp. Physiol. 56:153-168 (Supp. #1).

_____, R. G. Edwards, R. E. Fowler, and R. C. Roberts. 1961. "Analysis of differences in the number of eggs shed by the two ovaries of mice during natural oestrus or after superovulation." J. Rep. Fert. 2:418.

_____, and R. C. Roberts. 1960. "Effect of inbreeding on ovulation rate and foetal mortality." Genet. Res. 1:422-430.

Faris, E. J. 1950. "The care and breeding of laboratory animals." John Wiley and Sons, N.Y., 515 pps.

Fawcett, D. W. 1950. "The development of mouse ova under the capsule of the kidney." Anat. Rec. 108:71-92.

_____. 1965. "The anatomy of the mammalian spermatozoon with particular reference to the guinea pig." Zeit. f. Zellforsch. 67:279-296.

_____, and R. D. Hollenberg. 1963. "Changes in the acrosome of guinea pig spermatozoa during passage through the epididymis." Zeit. f. Zellforsch. 60:276-292.

_____, and S. Ito. 1965. "The fine structures of bat spermatozoa." Am. J. Anat. 116:567-610.

_____, G. B. Wislocki, and C. M. Waldo. 1947. "The development of mouse ova in the anterior chamber of the eye and in the abdominal cavity." Am. J. Anat. 81:413-443.

Fekete, E. 1940. "Observations on three functional tests in a high-tumor and low-tumor strain of mice." J. Cancer. 38:234-238.

_____. 1947. "Differences in the effect of uterine environment upon development in the dba and C57 black strains of mice." Anat. Rec. 98:409-415.

_____. 1950. "Polyovular follicles in the C58 strain of mice." Anat. Rec. 108:699.

_____. 1954. "Gain in weight of pregnant mice in relation to litter size." J. Hered. 45:88-98.

_____, O. Bartholemew, and G. D. Snell. 1940. "A technique for the preparation of sections of early mouse embryos." Anat. Rec. 76:441-447.

_____, and A. B. Griffin. 1954. "Significance of recent developments in nuclear cytology and cytogenetics of the mouse." J. Nat. Cancer Inst. 15:801-808.

_____, and C. C. Little. 1942. "Observations on the mammary tumors incidence in mice born from transferred ova." Cancer Res. 2:525-530.

_____, and L. B. Newman. 1944. "A case of hermaphroditism in the mouse." Yale J. Biol. & Med. 17:395-396.

Fell, H. B., and A. F. Hughes. 1949. "Mitosis in the mouse: a study of living and fixed cells in tissue cultures." Quart. J. Mic. Sci. 90:355-380.

Feller, W. F., and J. Boretos. 1967. "Semiautomatic apparatus for milking mice." Jour. Nat. Cancer Inst. 38:11-17.

Fenner, F. 1949. "Mouse pox (infections ectromelia) in mice. A review." Jour. Immunology. 63:641.

Finkel, M. P. 1947. "The transmission of radio-strontium and plutonium from mother to offspring in laboratory animals." Physiol. Zool. 20:405-421.

_____, and G. M. Hirsch. 1952. "The influence of low, continuous doses of aureoymycin on CF-1 female mice." ANL. 4745:48.

Finn, C. A., and J. R. Hinckliffe. 1964. "Reaction of the mouse uterus during implantation and deciduous formation as demonstrated by changes in the distribution of alkaline phosphatase." J. Reprod. Fert. 8:331-338.

Firlit, C. F., and J. R. Davis. 1965. "Morphogenesis of the residual body of the mouse testis." Quart. J. Mic. Sci. 106:93-98.

Fischberg, M., and R. A. Beatty. 1950. "Aufange einer genetischen Analyse der spontanen Heteroploidie bei Mausen." Arch. Klaus. Stift. Vereb. Forsch. 25:22-27.

_____. 1950. "Experimentelle Herstellung von polyploiden Mausblastulae." Arch. Klaus. Stift. Vereb. Forsch. 25:54-55.

_____. 1951. "Spontaneous heteroploidy in mouse embryos up to mid-term." J. Exp. Zool. 118:321-335.

_____. 1952. "Heteroploidy in mammals. II. Induction of triploidy in pre-implantation mouse eggs." J. Genet. 50:455-470.

_____. 1952. "Heteroploidy in the mouse embryo due to crossing of inbred strains." Evolution. 6:316-324.

Fitch, M. 1957. "A mutation in mice producing dwarfism, brachycephaly, cleft palate, and micromelia." J. Morphol. 109:141-149.

_____. 1957. "An embryological analysis of two mutants in the house mouse, both producing cleft palate." J. Exp. Zool. 136:329-361.

Fitzgerald, M. J. T. 1966. "Perinatal changes in epidermal innervation in rat and mouse." J. Comp. Neurol. 126:37-41.

Flax, M. H. 1953. "Ribose nucleic acid and protein during oögenesis and early embryonic development." Ph. D. Thesis, Columbia U. 44 pps.

Flexner, B., and A. Gellhorn. 1900. "The transfer of water and sodium to the amniotic fluid of the guinea pig." Am. J. Physiol. 136:757-761.

Flynn, R. J. 1955. "Ectoparasites of mice." Proc. Animal Care Panel. 6:75-91.

_____, L. Greco, and P. B. Jinkins. 1963. "A disease of mice characterized by lung congestion." Lab. An. Care. 13:499-501 (1962 UAC-6762).

Fogg, L. C., and R. F. Cowing. 1952. "Cytologic changes in the spermatagonial nuclei correlated with increased radio-resistance." Exp. Cell. Res. 4:107-115.

Forbes, T. R. 1957. "Progestin in mouse embryos placenta and amniotic fluid." Endocrin. 61:593-594.

_____, and C. W. Hooker. 1957. "Plasma levels of progestin during pregnancy in the mouse." Endocrin. 61:281-286.

Ford, C. E., and G. L. Woollam. 1963. "A colchicine, hypotonic citrate, air drying sequence for fetal mammalian chromosomes." Stain Techn. 38:271.

_____. 1964. "Selection pressure in mammalian cell populations." Symp. of the Int. Soc. for Cell Biology. Vol. 3. Academic Press, Inc., New York City.

Forsberg, J. G. 1965. "Origin of vaginal epithelium." Obstet. & Gynec. 25:787-791.

_____. 1965. "Mitotic rate and auto-radiographic studies in the derivation and differentiation of the epithelium in the mouse vaginal anlagé." Acta Anat. (Basel). 62:266-282.

_____, and H. Olivicrona. 1965. "Further studies on the differentiation of the epithelium in the mouse vaginal anlagé." Z. Zellforsch. 66:867-877.

Forsthoefel, P. F. 1958. "The embryological development of the skeletal effects of the luxoid gene in the mouse, including its interactions with the luxate gene." J. Morphol. 104:81-142.

_____. 1963. "The embryological development of the effects of Strongs Luxoid gene in the mouse." J. Morphol. 113:427-452.

_____. 1963. "Observations on the sequence of blastemal condensations in the limbs of the mouse embryo." Anat. Rec. 147:129-138.

Fowler, R. E., and R. G. Edwards. 1957. "Induction of superovulation and pregnancy in mature mice by gonado-trophins." J. Endocrin. 15:374-384.

Fowler, R. E., and R. G. Edwards. 1960. "The fertility of mice selected for large or small body size." Genet. Res. 1:393-407.

Francke, C. 1948. "Some observations on the morphological structure of explanted immature ovaries and on 'ovariohypophyses'." Acta Neer. Morphol. Pathol. 6:129-140.

Frankenberger, Z. 1926. "Sur la morphologie et le développement des voies biliaires chez le genre Mus." Arch. Anat. Histol. et Embryol. 6:201-216.

Fraser, F. C., T. D. Fainstat, and H. Kalter. 1953. "The experimental production of congenital defects with particular reference to cleft palate" in "Neo-natal Studies." Int. Children's Center. 2:43-58. McGill University.

Freye, H. 1954. "Anatomische und entwicklungs geschicht-liche Untersudningen und oligs dactyler Mause." Z. Martin-Luther-Univ. Math. Naturw. Reike 3 (1+4) 801-824.

Fridhandler, L., S. E. Hafez, and G. Pincus. 1956. "Respiratory metabolism of mammalian eggs." Proc. Soc. Exp. Biol. & Med. 92:127-129.

Friedrich, F. 1964. "The development of the so-called yolk sac diverticulum in the placenta of white mice." Z. Znat. Entwickl. 124:153-170.

Fritze, C. 1956. "Statistical contributions to the study of human fertility." Fert. & Steril. 87:88-95.

Friz, M., and R. May. 1959. "Early embryonal death before implantation." Int. J. Fert. 4:306.

Frommer, J. 1964. "Prenatal development of the mandibular joint in mice." Anat. Rec. 150:449-461.

Froud, M. D. 1959. "Studies on the arterial system of three inbred strains of mice." J. Morphol. 104:441-478.

Fruhman, G. J. 1966. "Shunting of erythropoiesis in mice." Anat. Rec. 154:346.

Fujita, S. 1963. "Matrix cell and cytogenesis of the developing central nervous system." J. Comp. Neur. 120:37-40.

_____, 1964. "Analysis of neuron differentiation in the central nervous system by tritiated thymidine autoradiography." J. Comp. Neur. 122:311-327.

Fuller, John L. 1967. "Effect of drinking schedule upon alcohol preference in mice." Quart. J. Studies Alcohol. 28:22-26.

Fulton, J. D., A. C. Arnold, and R. B. Mitchell. "The antibody response of animals exposed to X-radiation. III. The protective effect of chemotherapeutic agents on a specific immune status of X-radiated mice." USAF School of Aviation Medicine, Project No. 21-47-002, Report No. 3.

Fuxe, K., and O. Nilsson. 1963. "The mouse uterine surface epithelium during the estrous cycle." Anat. Rec. 145:541-548.

Galton, M., and S. F. Holt. 1965. "Asynchronous replication of the mouse sex chromosomes." Exp. Cell. Res. 37:111-116.

Gardner, P. J. 1966. "Fine structure of the seminiferous tubule of the Swiss mouse. The spermatid." Anat. Rec. 155:235-250.

_____, and E. A. Holyoke. 1963. "Observations on the fine structure of the seminiferous tubules of the Swiss mouse." Anat. Rec. 145:320.

_____. 1964. "Fine structure of the seminiferous tubule of the Swiss mouse. I. The limiting membrane, Sertoli cell, spermatagonia, and spermatocytes." Anat. Rec. 150:391-404.

Gates, Allen. 1956. "Viability and developmental capacity of eggs from immature mice treated with gonadotrophins." Nature (Lond.). 177:754-5.

Gates, A. H. 1959. "Early embryology of the mouse as studied by transplantation of ova." Ph. D. Thesis, University of Edinburgh.

_____. 1963. "Postnatal growth following asynchronous development of the egg and endometrium." Proc. XVI. Int. Cong. Zool. 2:94.

_____. 1965. "Rate of ovular development as a factor in embryonic survival." Ciba Found. Symp. on "Preimplantation Stages of Pregnancy." p. 270-297. Ed. G. E. W. Wolstenholme & M. O'Connor. Churchill Ltd., London.

Gates, A. H., and R. A. Beatty. 1954. "Independence of delayed fertilization and spontaneous triploidy in mouse embryos." Nature (Lond.). 174:356-357.

————, L. L. Doyle, and R. W. Noyes. 1961. "A physiological basis for heterosis in hybrid mouse fetuses." (Abstr.). Am. Zool. 1:449.

————, and A. Dronkert. 1965. "Gonadotropin preparations and supervulation in the mouse." Am. Zool. 5:(Abstr.). #148.

————, and M. Karasek. 1965. "Hereditary absence of sebaceous glands in the mouse." Science. 148:1471-1473.

————, and M. N. Runner. "Factors affecting survival of transplanted ova of the mouse." Anat. Rec. 113:555.

————. 1957. "Influence of prepuberal age on number of ova that can be superovulated in the mouse." Anat. Rec. 128:554.

Gates, R. R. 1953. "Polyploidy and sex chromosomes." Acta Biochem. (Leiden). 11: 27-44.

Gates, W. H. 1925. "Litter size, birth weight, and early growth rate of mice." Anat. Rec. 29:183-193.

————. 1930. "The effect of polygamy on the sex ratio in mice." J. Exp. Biol. 7:235-240.

Gaunt, W. A. 1963. "An analysis of the growth of the cheek teeth of the mouse." Acta Anat. 54:220-259.

————. 1964. "Changes in the form of the jaws of the albino mouse during ontogeny." Acta Anat. (Basel). 58:37-61.

————. 1965. "Growth of the facial region of the albino mouse as revealed by the mesh diagram." Acta Anat. (Basel). 61:574-588.

Gaunt, W. S. 1956. "The development of enamel and dentin on the molars of the mouse, with an account of the enamel-free areas." Acta Anat. 28:111-134.

Geyer-Duszynska, I. 1964. "Cytological investigations on the T-locus in Mus musculus L." Chromosoma. 15:478-502.

Gihys, R. 1959. "Comparative study of the lethal effects of cobalt-60 gamma rays and 200 Kv X-rays on C57 mice." Onocologia. 12:279-94 (French).

Glass, L. 1963. "Transfer of native and foreign serum antigens to oviduccal mouse eggs." Am. Zool. 3:135-156.

Glass, L. E., and T. P. Lin. 1961. "Development of x-irradiated and non-irradiated mouse oöcytes transplanted to x-irradiated and non-irradiated recipient females." J. Cell. Comp. Physiol. 61:53-60.

Gluecksohn-Waelsch, S. 1953. "Lethal factors in development." (y, Sd, T, ki, Fu). Quart. Rev. Biol. 28:115-135.

————. 1954. "Some genetic aspects of development." Cold Spring Harbor Symp. 19: 41-49.

————. 1954. "Genetic control of embryonic growth and differentiation." J. Nat. Cancer Inst. 15:629-634.

————. 1965. "Genetic control of mammalian differentiation." Proc. XI. Int. Cong. Genet. 2:209-219.

Goodman, J. W., and L. H. Smith. 1961. "Erythrocyte life span in normal mice and in radiation bone marrow chimeras." Am. J. Physiol. 200:764-770.

Goss, C. M. 1940. "First contractions of the heart without cytological differentiation." Anat. Rec. 76:19-27.

Graff, Ralph J., Willys K. Silvers, Rupert E. Billingham, W. H. Hildemann, and George D. Snell. 1966. "The cumulative effect of histocompatibility antigens." Transplantation. 4:605-617.

Gray, A. P. 1954. "Mammalian hybrids: A check list with bibliography." Tech. Com. #11 Common. Bur. An. Breed, Gen. Edinburgh.

Green, E. L. ed. 1966. "Biology of the laboratory mouse." McGraw-Hill, N.Y.

————, and M. C. Green. 1946. "The effect of the uterine environment on the skeleton of the mouse." J. Morphol. 78:105.

Green, E. L., and M. C. Green. 1953. "Modification of difference in skeletal types tween reciprocal hybrids by transplantation of ova in mice." Genetics. 38: 666.

———. 1959. "Transplantation of ova in mice." J. Heredity. 50:109-114.

Green, J. A. "Ovarian weight and responsiveness to gonadotrophins throughout the estrous cycle of mice." Proc. Soc. Exp. Biol. & Med. 95:504-506.

Green, M. C. 1952. "A rapid method for clearing and straining specimens for the demonstrations of bone." Ohio J. Science. 52:31-33.

———. 1955. "Luxoid-a new hereditary leg and foot abnormality in the house mouse." J. Heredity. 46:90-99.

Green, Margaret C. 1966. (Jackson Lab., Bar Harbor). "Genes and development," p. 329-336. In E. L. Green (ed.) "Biol. Lab. Mouse," 2nd ed. N.Y., Mc-Graw-Hill.

———. 1967. "A defect of the splanchnic mesoderm caused by the mutant gene dominant hemimelia in the mouse." Develop. Biol. 15:62-89.

Green, M. C., and R. L. Sidman. 1962. "Tottering, a new neurological mutant in the mouse." J. Heredity. 53:233-237.

Greenwald, G. S. 1956. "The reproductive cycle of the field mouse." J. Mammology. 37:213.

———, and N. B. Everett. 1959. "The incorporation of S^{35} methionine by the uterus and ova of the mouse." Anat. Rec. 134:171-184.

Gresson, R. A. R. 1933. "A study of the cytoplasmic inclusions and nuclear phenomena during oögenesis of the mouse." Quart. J. Mic. Sci. 75:697-721.

———. 1940. "Presence of the sperm middle-piece in the fertilized egg of the mouse." Nature (Lond.). 145:425.

———. 1940. "A cytological study of the centrifuged oöcyte of the mouse." Quart. J. Mic. Sci. 81:569-583.

———. 1941. "A study of the cytoplasmic inclusions during the maturation, fertilization and first cleavage divisions of the egg of the mouse." Quart. J. Mic. Sci. 83:35-59.

———. 1948. "Fertilization, parthenogenesis, and the origin of the primitive germ cell of some animals" in "Essentials of General Cytology." p. 64-75. Edinburgh U. Press.

Griffen, A. B., and M. C. Bunker. 1964. "Three cases of trisomy in the mouse." Proc. Nat. Acad. Sci. 52(5):1194-1198.

Grillo, T. A. T. 1964. "The occurrence of insulin in the pancreas of foetuses of some rodents." J. Endocrin. 31:67-73.

Grobstein, C. 1949. "Behavior of components of the early embryo of the mouse in culture and in the anterior chamber of the eye." Anat. Rec. 105:490.

———. 1953. "Morphogenetic interaction between embryonic mouse tissues separated by a membrane filter." Nature (Lond.). 172:869-871.

———. 1955. "Inductive interaction in the development of the mouse metanephros." J. Exp. Zool. 130:319-339.

———. 1956. "Trans-filter induction of tubules in mouse metanephrogenic mesenchyme." Exp. Cell. Res. 10:424-440.

Grüneberg, H. 1937. "The relations of endogenous and exogenous factors in bone and tooth development. The teeth of the gray-lethal mouse." J. Anat. (Lond.). 71:246-244.

———. 1941. "The growth of the blood of the suckling mouse." J. Path. Bact. 52: 323-329.

———. 1942. "Inherited macrocytic anaemias of the house mouse. II. Dominance relationships." J. Genet. 43:285-293.

———. 1943. "The development of some external features in mouse embryos." J. Heredity. 34:88-92.

———. 1950. "Embryology of the mammalian genes." Rev. suisse Zool. 57. Suppl. 1:729-739.

Gruneberg, H. 1952. "The genetics of the mouse." 2nd ed. Martinus Nizhoff The Hague (Bibliogrophia Genet. Vol. 15 complete)

_____. 1954. "Genetical studies on the skeleton of the mouse." XII. The development of undulated." J. Genet. 52:441-455.

_____. 1956. "Hereditary lesions of the labyrinth in the mouse." Brit. Med. Bull. 12:153-157.

_____. 1956. "A ventral ectodermal ridge of the tail in mouse embryos." Nature (Lond.). 177:787-788.

_____. 1960. "Developmental genetics in the mouse, 1960." J. Cell. Comp. Physiol. Suppl. Vol. 56:49-60.

_____. 1963. "The pathology of development." John Wiley & Sons, N.Y. 301 pps.

Guttenberg, I. 1961. "Plasma levels of 'Free' progestin during the estrous cycle in the mouse." Endocrin. 68:1006-1009.

Gwatkin, R. B. A. 1964. "Effect of enzyme and acidity on the zona pellucida of the mouse egg before and after fertilization." J. Reprod. Fert. 7:99-105.

Halberg, F., and M. B. Visscher. 1952. "A difference between the effects of dietary calorie restriction on the estrous cycle and on the 24-hour adrenal cortical cycle in rodents." Endocrin. 51:329-335.

Hall, B. U. 1935. "The reactions of rat and mouse eggs to hydrogen ions." Proc. Soc. Exp. Biol. & Med. 32:747-748.

_____. 1936. "Variations in acidity and oxidation reduction potentials of rodent uterine fluids." Physiol. Zool. 9:471.

Hamburgh, M., L. Nebel, and G. Greenhouse. 1966. "Penetration and uptake of trypan blue in the yolk sac placenta of the mouse." Am. Zool. 6:

Hammond, J. 1949. "Survival of mouse ova *in vitro* and induced multiple pregnancies in cattle." Proc. 1st. Nat. Egg Transfer Breed Conf. Texas. p. 22.

_____. 1949. "Recovery and culture of tubal mouse ova." Nature (Lond.). 163:28-29.

Hampton, J. C. 1966. "The effects of ionizing radiation on absorptive and secretory cells in the intestinal epithelium of the mouse." Anat. Rec. 154:353.

Hancock, R. L. 1966. "S-adenosylmethionine-synthesizing activity of normal and neoplastic mouse tissues." Cancer Res. 26:2425-2430.

Hanna, C. 1965. "Changes in DNA, RNA and protein synthesis in the developing lens." Invest. Ophthal. 4:480-495.

Hargitt, C. T. 1926. "The formation of the sex glands and germ cells of mammals. II. The history of the male germ cells in the albino rat." J. Morphol. 42: 253-305.

_____. 1930. "The formation of the sex glands and germ cells of mammals. III. The history of the female germ cells in the albino rat to the time of sexual maturity." J. Morphol. 49:277-331.

Haring, O. M. 1965. "Effects of prenatal hypoxia on the cardiovascular system in the rat." A. M. A. Arch. Path. 80:351-356.

_____, and F. J. Lewis. 1961. "The etiology of congenital developmental anomalies." Surgery Gynec. & Obstet. 113:1-18.

Harkness, R. A., A. McLaren, and E. J. Roy. 1964. "Oestrogens in mouse placenta." J. Rep. Fert. 8:411-413.

Harper, M. J. K. 1965. "Transport of eggs in cumulus through the ampulla of the rabbit oviduct in relation to day of pseudopregnancy." Endocrin. 77:114-123.

Harrington, F. E. 1965. "Transportation of ova and zygotes through the genital tract of immature mice treated with gonadotrophins." Endocrin. 77:635-640.

Harris, R. G. 1927. "Effect of bilateral ovariectomy upon the duration of pregnancy in mice." Anat. Rec. 37:83.

Harvey, E. B., R. Yanagimachi, and M. C. Chang. 1961. "Onset of estrus and ovulation in the golden hamster." J. Exp. Zool. 146:231-236.

Harvey, S. C., and Y. Clermont. 1962. "The duration of the cycle of the seminiferous epithelium of normal, hypophysectomized and hypophysectomized hormone-treated albino rats." Anat. Rec. 142:

Hashima, H. 1956. "Studies on the prenatal growth of the mouse with special references to the site of implantation of the embryo." Tohoku J. Agric. Res. 6:307.

Haushka, T. S. 1959. "The chromosomes in ontogeny and oncology." Cancer Res. 21:957-974.

Healy, M., A. McLaren, and D. Michie. 1960. "Foetal growth in the mouse." Proc. Roy. Soc. (Lond.) B. 153:367.

Heinecke, H., and H. Grimm. 1958. "Untersuchungen zur Offnungszeit der Vaginalmembran bei verschiedenen Mausestammen." Endokrinologie. 35:205-213.

Hemmingsen, A. M., and N. B. Krareys. 1937. "Rhythmic diurnal variations in the oestrous phenomena of the rat and their susceptibility to light and dark." Levin and Munksgaard, Kobenhaven.

Henderson, N. D. 1964. "Behavioural effects of manipulation during different stages in the development of mice." J. Comp. & Physiol. Psych. 57:284-289.

Henin, A. 1941. "Etude des modification de l'oviducte au cours du cycle oestral (souris)." Arch. Biol. 52:97-115.

Henricson, B., and A. Nilsson. 1964. "Chromosome investigations on the embryo progeny of male mice treated with ^{90}Sr." Royal Veterinary Coll., Stockholm & Res. Inst. Nat'l Defence, Sunndbyberg, Sweden. Acta Radiol., Therapy, Phys. Biol. (N.S.) 2:315-320.

Herrmann-Erlee, M. P. 1964. "Quantitative histochemistry of the embryonic mouse radius: influence of parathyroid extract on the activity of lactic dehydrogenase." J. Histochem. Cytochem. 12:481-482.

Hinricksen, K. 1959. "Morphologische Untersuchungen zum Topagenese der Mandibularen Nogezakneder Maus." Anat. Anz. 107:59-74.

Hitzerman, Sister J. W. 1962. "Development of enzyme activity in the Leydig cells of the mouse testis." Anat. Rec. 143:351-361.

Hoag, W. G. 1961. "Oxyuriosis in laboratory mouse colonies." Am. J. Vet. Res. 22:150-153.

_____. 1964. "Animal health control for inbred mouse colonies of the Jackson Laboratory." Laboratory & Animal Care. 14:253-259.

_____, and M. M. Dickie. 1962. "Studies of the effect of various dietary protein fat levels on inbred laboratory mice." Proc. Anim. Care Panel. 12:7-10.

_____, and J. Rogers. 1961. "Techniques for the isolation of Salmonella tryphimcuricum from laboratory mice." Jour. Bact. 82:153-154.

Hollander, W. F. 1959. "The problem of superfetation in the mouse." Heredity. 50:71-73.

_____. 1959. "Sperm abnormality of a mutant type involving the "p" locus in the mouse." Proc. X Int. Cong. Gen. Vol. 2, Toronto. p. 123.

_____. 1960. "Genetics in relation to reproductive physiology in mammals." J. Cell. Comp. Phys. Vol. 56, Suppl. p. 61-72.

_____, J. W. Gowen, and J. Stadler. 1956. "A study of 25 gynandromorphic mice of the Bagg albino mice." Anat. Rec. 124:223-243.

_____, and J. W. Gowen. 1959. "A single-gene antagonism between mother and fetus in the mouse." Proc. Soc. Exp. Biol. & Med. 101:425-428.

_____, and L. C. Strong. 1950. "Intra-uterine mortality and placental fusions in the mouse." J. Exp. Zool. 115:131-150.

Holmes, R. L. 1953. "Nuclear studies on the thalamus of the mouse." J. Comp. Neural. 99:377-414.

Hooker, C. H., and W. L. Williams. 1940. "Retardation of mammary involution in the mouse by irritation of the nipples." Yale J. Biol. Med. 12:559-564.

Hoshino, K. 1962. "Influences of estrogen upon pregnancy in mice." Anat. Rec. 142:p. 242. Gardner through Nat'l. Cancer Inst. U.S.P.H.S. C 343-C14 and Jane Coffin Childs Memorial Fund.

_____. 1964. "Regeneration and growth of quantitatively transplanted mammary glands of normal female mice." Anat. Rec. 150:221-235.

Hoshino, K. 1965. "Development and function of mammary glands of mice prenatally exposed to testosterone proprionate." Endocrin. 76:789-794.

_____. 1966. "Development and growth of mammary glands of CBA mice prenatally exposed to progesterone." Anat. Rec. 154:360.

_____. 1967. "Transplantability of mammary gland in brown fat pads of mice." Nature (Lond.). 213:194-195.

_____, and W. U. Gardner. 1967. "Transplantability and life span of mammary gland during serial transplantation in mice." Nature (Lond.). 213:193-194.

Howard, A., and S. R. Pelc. 1950. "p^{32} autoradiographs of mouse testis. Preliminary observations of the timing of sperm at ogenic stages." Brit. Jour. Radiol. 33:634-641.

Hummel, Katharine P., Margaret M. Dickie, and Douglas L. Coleman. 1966. "Diabetes, a new mutation in the mouse." Science. 163:1127-1128.

Hungerford, D. A. 1955. "Chromosome number of ten day fetal mouse cells." J. Morphol. 97:497-509.

Hunter, R. L. 1951. "Distribution of esterase in the mouse embryo." Proc. Exp. Biol. & Med. 78:56-57 (A).

Hupp, E. W., H. B. Pace, E. Furchtgott, and R. L. Murphree. 1960. "Effect of fetal irradiation on mating activity in male rats." Psychological Reports. 7:289-294.

Hussey, K. L. 1957. "Syphacia muris vs. Sobvelata in laboratory rats and mice." J. Paristoloty. 43:555-559.

Ibery, P. L. T. 1958. "Evidence for a direct mechanism in leukaemogenesis." Australian Atomic Energy Symposium. 681-4.

Ingalls, Th., E. F. Ingenito, and F. J. Curley. 1964. "Acquired chromosomal anomalies induced in mice by known teratogens." J.A.M.A. 187:836-838.

_____, G. Keleman, and F. J. Curley. 1957. "Development of the inner ear after maternal hypoxia." A.M.A. Arch. Otolaryngology. 65:558-566.

Jackson, R. C. 1964. "Genotype and sex drive in intact and in castrated male mice." Science. 145:514-515.

Jacobs, R. M. 1964. "S^{35} - liquid scintillation count analysis of morphogenesis and teratogenesis of the palate in mouse embryos." Anat. Rec. 150:271-277.

_____. 1964. "Histochemical study of morphogenesis and teratogenesis of the palate in mouse embryos." Anat. Rec. 149:691-697.

Jagrillo, G. M. 1965. "A method for meiotic preparations of mammalian ova." Cytogenetics. 4:245-250.

Johnson, M. L. 1933. "The time and order of appearance of ossification centers in the albino mouse." Am. J. Anat. 52:241-271.

Jollie, W. P. 1961. "The incidence of experimentally produced abdominal implantations in the rat." Anat. Rec. 141:159.

Jolly, J., and M. Ferester-Tadie. 1935. "1. Sur la disposition de l'embryon dans l'oeuf et ses rapports avec les membranes ovulaires chez le rat et souris. 2. La formation du mesoderme dans l'oeuf de la souris." Comp. Rend. Soc. Biol. 119:1055-1058.

_____. 1936. "Recherches sur l'oeuf du rat et de la souris." Arch. Anat. Micr. 32:323-390.

Jones, E. C., and P. L. Krohn. 1961. "The relationships between age, numbers of oöcytes and fertility in virgin and multiparous mice." J. Endocrin. 21:469-496.

Jones, G. E. S., and E. B. Astwood. 1942. "The physiological significance of the estrogen: progesterone ratio on vaginal cornification in the rat." Endocrin. 30:295-300.

Jones, N., and G. A. Harrison. 1958. "Genetically determined obesity and sterility in the mouse." Proc. Soc. Study Fert. 9:51-64.

Jones-Seaton, A. 1950. "A study of cytoplasmic basophily in the egg of the rat and some other mammals." Ann. Soc. Roy. Zool. (Belgium). 80:76-86.

Jones-Seaton, A. 1950. "Etude de l'organization cytoplasmique de l'oeuf des rongeurs principalment quant a la basophilie ribonucleique." Arch. Biol. 61:291-444.

Kallen, B. 1953. "Notes on the development of the neural crest in the head of *Mus musculus*." J. Embryol. Exp. Morphol. 1:393-398.

_____. 1953. "Formation and disappearance of neuromery in *Mus musculua*." Acta Anat. 18:273-282.

Kalter, H. 1954. "The inheritance of susceptibility to teratogenic action of cortisone in mice." Genetics. 39:185-196.

Kaneko, K. 1940. "Ueber die Entwicklung der Thalamuskerne der Maus." Fol. Anat. Japan. 19:557-596.

Keeley, K. 1962. "Prenatal influence in behaviour of offspring of crowded mice." Science. 135:44-45.

Keighley, G. H., P. Lowy, Elizabeth S. Russell, and Margaret W. Thompson. 1966. "Analysis of erythroid homeostatic mechanisms in normal and genetically anaemic mice." Brit. Jour. Haematol. 12:461-477.

Kelemen, G. 1947. "The junction of the nasal cavity and the pharyngeal tube in the rat." Arch. Otolaryngol. 45:159-168.

_____. 1955. "Experimental defects in the ear and the upper airways induced by radiation." Arch. Otolaryngol. 61:405-418.

_____. 1955. "Aural changes in the embryo of a diabetic mother." A.M.A. Arch. Otolaryngol. 62:357-363.

_____. 1963. "Hemorrhage: a specific poison to tissue of the ampullar cupulae." A.M.A. Arch. Otolaryngol. 4:365-379.

_____. 1963. "Radiation and ear." Acta-Oto-Laryngologia Suppl. 184:5-48.

Kent, H. A. 1960. "Polyovular follicles and multinucleate ova in the ovaries of young mice." Anat. Rec. 137:521.

Kerr, T. 1946. "The development of the pituitary of the laboratory mouse." Quart. J. Micr. Sci. 87:3-29.

Kile, J. C. 1950. "An improved method for the artificial insemination of mice." O.R.N.L.-808., Anat. Rec. 109:109-117, 1951.

King, J. W. B. 1950. "Pigmy, a dwarfing gene in the house mouse." J. Heredity. 41:249-252.

_____. 1955. "Observations of the mutant 'pigmy' in the house mouse." J. Genet. 53:487-497.

Kingery, H. M. 1914. "So-called parthenogenesis in the white house." Biol. Bull. 27:240.

_____. 1917. "Oögenesis in the white mouse." J. Morphol. 30:261-315.

Kirby, D. R. S. 1960. "Development of mouse eggs beneath the kidney capsule." Nature (Lond.). 187:707-708.

_____. 1962. "Reciprocal transplantation of blastocysts between rats and mice." Nature (Lond.). 194:785.

_____. 1962. "The influence of the uterine environment on the development of the mouse egg." J. Emb. Exp. Morphol. 10:496-506.

_____. 1962. "The development of mouse blastocysts transplanted to the scrotal and cryptorchid testis." J. Anat. (Lond.). 97:119-130.

_____. 1963. "Development of the mouse blastocyst transplanted to the spleen." J. Reprod. Fert. 5:1.

_____. 1965. "The role of the uterus in the early stages of mouse development" in "Preimplantation Stages in Pregnancy." pps. 325-344, Little, Brown & Co., Boston.

_____, W. D. Billington, S. Bradbury, and D. J. Goldstein. 1964. "Antigen barrier of the mouse placenta." Nature (Lond.). 204:548-549.

_____, and S. K. Malhotra. 1964. "Cellular nature of the invasive mouse trophoblast." Nature (Lond.). 201:520.

Kirkham, W. B. 1906. "The maturation of the mouse egg." Biol. Bull. 12:259-265.

_____. 1910. "Ovulation in mammals with special reference to the mouse and rat." Biol. Bull. 18:245-251.

———. 1916. "The germ cell cycle in the mouse." Anat. Rec. 10:217-219.

———. 1916. "The prolonged gestation period in suckling mice." Anat. Rec. 11:31-40.

Kliman, B., and H. A. Salhanick. 1952. "Relaxation of public symphysis of the mouse during the estrous cycle and pseudopregnancy." Proc. Soc. Exp. Biol. Med. 81:201-202.

Knowlton, N. P., Jr., and W. R. Widner. 1950. "The use of x-rays to determine the mitotic and intermitotic time of various mouse tissues." Cancer Res. 10:59-63.

Knudsen, P. A. 1964. "Mode of growth of the choroid plexus in mouse embryos." Acta Anat. 57:172-182.

———. 1964. "The surface area of choroid plexus in normal mouse embryos." Acta Anat. 58:355-367.

———. 1965. "Congenital malformations of upper incisors in exencephalic mouse embryos, induced by hypervitaminosis A. Types and frequency." Acta Odont. (Scand.) 23:71-89.

———. 1965. "Congenital malformations of upper incisors in exencephalic mouse embryos, induced by hypervitaminosis A. II. Morphology of fused upper incisors." Acta Odont. (Scand.) 23:391-408.

———. 1965. "Fusion of upper incisors at bud or cap stage in mouse embryos with exencephaly induced by hypervitaminosis A." Acta Odont. (Scand.) 23:549-565.

Koch, W. E. 1965. "In vitro development of tooth rudiments in embryonic mice." Anat. Rec. 152:513-524.

———. 1965. "The interaction of embryonic tooth tissues growing in vitro." Anat. Rec.

———. 1966. "In vitro studies of developmental associations in the embryonic mouse incisor." Anat. Rec. 154:370.

Kochi, T. 1936. "Ueber die Entwicklungstudien bei der Hypophysis cerebri von Mus musculus (in Japanese)." Akayama-Igakkai-Zasshi. 48:2213-2239.

Koller, P. C., and C. A. Auerback. 1941. "Chromosome breakage and sterility in the mouse." Nature (Lond.). 148, p. 501.

Konigsmark, B. W., and R. N. Sidman. 1963. "Origin of brain macrophages in the mouse." J. Neuropath. Exp. Neurol. 22:643-676.

Krabbe, K. H. 1944. "Studies on the morphogenesis of the brain of Rodentia, Prosimiae, and Edentates." Munksgaard, (Copenhagen).

Krebhiel, R. H., and J. C. Plagge. 1962. "Distribution of ova in the rat uterus." Anat. Rec. 143:239-241.

Kremer, J. 1924. "Das Verhalten der Vorkerne in befruckteten Eider Ratte und der Mause mit besonderer Beruchsichtgung ihren Nucleolen." Z. Mikr. Anat. Forsch. 1:353.

Kroc, R. L., B. G. Steinetz, and V. L. Beach. 1959. "The effect of estrogens progestrogens, and relaxin in pregnant and non-pregnant laboratory rodents." Am. N.Y. Acad. Sci. 75:942.

Krzanowska, H. 1960. "Studies on heterosis. II. Fertilization rate in inbred lines of mice and their crosses." Folia Biol. 8:269-279.

———. 1960. "Early embryonal mortality in inbred lines of mice and their crosses." Bull. Soc. Roy. Belge de Gynec. et Obstetr. 30:719-728.

———. 1962. "Sperm quantity and quality in inbred lines of mice and their crosses." Acta Biologica (Cracoviensia). 5:279-291.

———. 1964. "Studies on Heterosis. III. The course of the sexual cycle and the establishment of pregnancy in mice, as affected by the type of mating." Folia Biologica. 12:415-426.

Kuhl, W., and H. Friedrich-Freksa. 1936. "Richtungs korperbildung und Furchung des Eies sourie das verholten des das Trophoblasten der weiben maus." Zool. Anz. 9:187 (Suppl. 2).

Ladman, A. J., and M. N. Runner. 1951. "Comparison of sensitivities of the immature and pregnant mouse for estimation of gonadotropin." Endocrin. 48:358-364.

_____. 1959. "Correlation of the maternal pituitary weight with the number of uterine implantation sites in pregnant mice." Endocrin. 65:580-585.

_____. 1960. "Induction of ovulation in normal and hypophysectomized immature mice with purified pituitary extracts of F.S.H. and LH." 1st Int. Cong. Endocrin. Abst. #100.

Laguchev, S. S. 1959. "Comparison of the estrous cycles in mice of high and low cancer lines (English trans.)." Bull. Exp. Biol. Med. 48:1149-1152.

Laird, A. K. 1966. "Dynamics of embryonic growth." Growth. 30:263-275.

Lamond, D. R. 1958. "Infertility associated with extirpation of the olfactory bulbs in female albino mice." Austral. J. Exp. Biol. Med. Sci. 36:103-108.

_____. 1959. "Effect of stimulation derived from other animals of the same species on oestrous cycles in mice." J. Endocrin. 18:343-349.

_____, and C. W. Emmens. 1959. "The effect of hypophysectomy on the mouse uterine response to gonadotrophins." J. Endocrin. 18:251-261.

Lams, H., and J. Doorme. 1907. "Nouvelles recherches sur la maturation et la fecondation de l'oeuf des mammiferes." Arch. Biol. 23:259-363.

Lane, P. W. 1959. "The pituitary - gonad response of genetically obese mice in parabiosis with thin and obese siblings." Endocrin. 65:863-868.

Larrson, K. S. 1960. "Studies on the closure of the secondary palate. II. Occurrence of sulpho-mucopolysaccharides in the palatine processes of the normal mouse embryo." Exp. Cell. Res. 21:498-503.

_____, H. Bostrom, and S. Carlsoo. 1959. "Studies on the closure of the secondary palate. I. Autoradiographic study in the normal mouse embryo." Exp. Cell. Res. 16:379-383.

_____, et al. 1966. (Dept. Ped. Pathol., Univ. Uppsala) "A microradiographic study of salicylate-induced skeletal anomalies in mouse embryos." Acta Pathol. Microbiol. Scand. 66:560.

Lash, J. W. 1966. "Chemical embryogenesis of skeletal tissues." Birth Defects, Nat'l. Foundation. 11:56-57.

Lataste, F. 1887. "Recherches de Zooethique sur les Mammiferes de l'orde des Rongeuro." Acta Soc. Linneus Bordeau. 40:202.

Leblond, C. P., and Y. Clermont. 1952. "Spermiogenesis of rat, mouse, hamster and guinea-pig as revealed by the periodic acid-fuchsin sulfurous acid technique." Am. J. Anat. 90:167-216.

_____. 1952. "Definition of the stages of the cycle of the seminiferous epithelium in the rat." Ann. N.Y. Acad. Sci. 55:548-573.

Leduc, E. H., J. W. Wilson, and D. H. Winston. 1949. "The production of biotin deficiency in the mouse." J. Nutrition. 38:73-86.

Lee, S. van der., and L. M. Boot. 1955. "Spontaneous pseudopregnancy in mice." Acta Physiol. Pharmac. Neerland. 4:442-444.

Lenz, W. 1962. "How can the physician lessen a hazard to offspring?" (Universitat, Hamburg) Med. Welt. 48:2554-8 (in German).

Leuchtenberger, C. 1960. "The relation of the deoxyribose-nucleic acid (DNA) of sperm cells to fertility." J. Dairy Sci. Suppl. 43:31-53.

_____, and F. Schrader. 1950. "The chemical nature of the acrosome in the male germ cells." Proc. Nation. Acad. Sci. Washington. 36:677-683.

Lewis, W. H., and E. S. Wright. 1935. "On the early development of the mouse egg." Contrib. to Embryol. pub. #429, no. 148, Carnegie Inst. pps. 113-143.

Leziak, K. 1958. "Studies on retarded pregnancy in mice from inbred matings (sibmatings) I. Effect of inbreeding on the course of the sexual cycle in mice." Folia Biol. 6:63-70.

_____. 1959. "Studies on retarded pregnancy in mice from inbred matings (sibmatings) II. Resorption of foetuses." Folia Biol. 7:267-275.

Lin, T. P. 1956. "Dl-methionine (S^{-35}) for labelling unfertilized mouse eggs in transplantation." Nature (Lond.) 178:1175-1176.

Lin, T. P. 1966. "Microinjection of mouse eggs." Science. 151:333-337.
_____, and D. W. Bailey. 1965. "Difference between two inbred strains of mice in ovulatory response to repeated administration of gonadotrophins." J. Reprod. Fert. 10:253-253-259.
_____, and L. A. Glass. 1962. "Cause of pre-implantation death of mouse oöcytes x-irradiated *in vitro*." Anat. Rec. 142:253.
_____, J. K. Sherman, and E. L. Willett. 1957. "Survival of unfertilized mouse eggs in media containing glycerol and glycine." J. Exp. Zool. 134:275.
Lindop, P. J. 1961. "Growth rate, life span, and causes of death in SAS/4 mice." Gerontologia. 5:193-208.
Lipkow, J. 1959. "Die Bedentung des Vaginal-propfes bei der weissen Maus." Naturwiss. 49:63.
Lloyd, C. W. (ed.). 1959. "Recent progress in the endocrinology of reproduction." Academic Press, N.Y.
Locke, M. 1964. "The Role of chromosomes in development." 290 pps. Academic Press, N.Y.
Loevy, H. 1962. "Developmental changes in the palate of normal and cortisone-treated strong A mice." Anat. Rec. 142:375-589.
Loewenstein, J. E., and A. I. Cohen. 1964. "Dry mass, lipid content and protein content of the intact and zona-free mouse ovum." J. Emb. Exp. Morphol. 12:113-121.
Long, J. A. 1912. "The living eggs of rats and mice with a description of apparatus for obtaining and observing them." Minn. Col. Publ. Zool. 9:105-136.
_____. 1912. "Studies on early stages of development in rats and mice." Univ. Calif. Pub. Zool. 9:105.
_____, and E. L. Mark. 1911. "The maturation of the egg of the mouse." Carnegie Inst. Wash. #142:1-72.
Lugo, F. P. 1959. "The effect of a diet supplemented with hens' eggs upon the estrus cycle of mice." Anat. Rec. 133:408. (Abstr.)
Lyon, M. F. 1966. "Lack of evidence that inactivation of the mouse X-chromosome is incomplete." Genetic Res., Camb. 8:197-203.
_____, and R. Meredith. 1966. "Autosomal translocations causing male sterility and viable aneuploidy in the mouse." Cytogenetics. 5:335-354.
Macaluso, M. C. 1965. "The fine structure of the corpus luteum of the Swiss mouse after parturition." Anat. Rec. (Abstr.).
MacDowell, E. C. 1924. "A method of determining the prenatal mortality in a given pregnancy of a mouse without affecting its subsequent reproduction." Anat. Rec. 27:329-336.
_____. 1928. "Alcohol and sex ratios in mice." Am. Nat. 62:48-54.
_____, and E. Allen. 1927. "Weight of mouse embryos 10-18 days after conception, a logarithmic function of embryo age." Proc. Soc. Exp. Biol. & Med. 24:672-674.
_____, E. Allen, and C. G. MacDowell. 1927. "The prenatal growth of the mouse." J. Gen. Physiol. 11:57-70.
_____. 1929. "The relation of parity, age, and body weight to the number of corpora lutea in mice." Anat. Rec. 41:267-272.
MacDowell, E. C., and E. M. Lord. 1925. "Data on the primary sex ratio in the mouse." Anat. Rec. 31:143-148.
_____. 1926. "The relative viability of male and female mouse embryos." Am. J. Anat. 37:127-140.
_____. 1927. "Reproduction in alcohol mice." Arch. Entwickl. Org. 109:549-583. (Quoted by Strong and Fuller, 1958).
Makino, S. 1941. "Some attempts to induce chromosome doubling in germ cells of mice." Bot. & Zool., Tokyo. 9:424-426.
_____. 1941. "Studies on the murine chromosomes. I. Cytological investigations of mice, included in the genus Mus." J. Fac. Sci. Hokkaido Univ. 7:305.

Makino, S. 1951. "An atlas of chromosome numbers in animals." Iowa State College Press.

Mandl, A. 1963. "Pre-ovulatory changes in the oöcyte of the adult rat." Proc. Roy. Soc. B. 158:105-118.

Mann, S. J. 1912. "Prenatal formation of hair follicle types." Anat. Rec. 144:135-141.

Mark, F. L., and J. A. Long. 1912. "Studies in early stages of development in rats and mice." Univ. Calif. Pub. Zool. 9:105.

Markert, C. L. 1959. "Biochemical embryology and genetics." N.C.I. Monog. #2 Symp. on "Normal and Abnormal Differentiation and Development."

_____, and W. K. Silvers. 1956. "The effects of geno-type and cell environment on melanoblast differentiation in the house mouse." Genetics. 41:429-450.

Marsden, H. M., and F. H. Bronson. 1964. "Estrous synchrony in mice" in "Alteration by exposure to male urine." Science. 144:1469.

_____. 1965. "The synchrony of oestrus in mice: relative roles of the male and female environments." J. Endocrin. 32:313-319.

_____. 1965. "Strange male block to pregnancy: its absence in inbred mouse strains." Nature (Lond.). 207:878.

Marston, J. H., and M. C. Chang. 1964. "The fertilizable life of ova and their morphology following delayed insemination in mature and immature mice." J. Exp. Zool. 155:237-252.

_____. 1964. "Action of intra-uterine foreign bodies in the rat and rabbit." Excerpta Med. Int. Congress Ser. #86.

Martin, L., and P. J. Claringbold. 1960. "Sensitive oestrogen assay in mice." J. Endocrin. 20:173.

Martinoosteh, P. W. 1939. "The effect of subnormal temperature on the differentiation and survival of cultivated *in vitro* embryonic and infantile rat and mouse ovaries." Proc. Royal Soc. (Lond.) B. 128:138-143.

Matthey, R. 1951. "Chromosomes in muridae." Experientia. 7:340-341.

_____. 1952. "Chromosomes de Muridae." Chromosoma. 5:113-118.

_____. 1953. "Les chromosomes des muridae." Rev. Suisse Zool. 60:225-283.

_____. 1954. "Nouvelles recherches sur les chromosomes des muridae." Caryologia. 6:1-44.

_____. 1954. "Un cas nouveau de chromosomes sexuels multiples dans le genre Gerbillus." Experientia. 10:464-465.

Mayer, T. C. 1965. "The development of piebold spotting in mice." Develop. Biol. 11:319-334.

_____, and E. L. Maltby. 1964. "An experimental investigation of pattern development in lethal spotting and belted mouse embryos." Develop. Biol. 9:269-286.

McCafferty, R. E., and H. P. Mack. 1964. "Tissue porphyrin in pregnant and non-pregnant mice injected with hematoporphyrin." Quart. J. Exp. Physiol. 49:394-407.

_____, and M. L. Wood. 1963. "Intra-amniotic pressure of the mouse." Anat. Rec. 145:285-289. (Abstr.).

_____, M. L. Wood, and W. H. Knisley. 1964. "Uterine contractions and intraamniotic pressures of gravid mice." Am. J. Obstet. & Gynec. 90:120-127.

_____. 1965. "Morphological and physiological effects of thalidomide and trypan blue on uteri and concepti of gravid mice." Am. J. Obstet. & Gynec. 91:260-269.

McCarthy, J. C. 1965. "Effects of concurrent lactation on litter size and prenatal mortality in an inbred strain of mice." J. Reprod. Fert. 9:29-39.

_____. 1965. "Genetic and environmental control of foetal and placental growth in the mouse." Animal Produc. 7:347-361.

McLaren, A. 1963. "The distribution of eggs and embryos between sides in the mouse." J. Endocrin. 27:157-181.

_____. 1965. "Genetics and environmental effects on foetal and placental growth in mice." J. Reprod. Fert. 9:79-98.

McLaren, A. 1965. "Maternal factors in nidation" in "Symposium on the Early Conceptus, Normal and Abnormal," ed. W. W. Park, Dundee.

———. 1965. "Placental weight loss in late pregnancy." J. Reprod. Fert. 9:343-346.

———, and J. D. Biggers. 1958. "Successful development and birth of mice cultivated *in vitro* as early embryos." Nature (Lond.). 182:877-878.

———, and D. Michie. 1954. "Transmigration of unborn mice." Nature (Lond.). 174: 844.

———. 1956. "The spacing of implantations in the mouse uterus" in "Implantation of Ova." P. Echstein, ed. p. 65-75. Cambridge Univ. Press.

———. 1956. "Studies on the transfer of fertilized mouse eggs to uterine foster-mothers. I. Factors affecting the implantation and survival of native and transferred eggs." J. Exp. Biol. 33:394-416.

———. 1958. "An effect of the uterine environment upon skeletal morphology in the mouse." Nature (Lond.). 181:1147.

———. 1958. "Factors affecting vertebral variation in mice. IV. Experimental proof of the uterine basis of a maternal effect." J. Emb. Exp. Morphol. 6:645-659.

———. 1959. "The spacing of implantations in the mouse uterus." Endocrin. 6:65.

———. 1959. "Experimental studies on placental fusion in mice." J. Exp. Zool. 141: 47-73.

———. 1959. "Superpregnancy in the mouse. I. Implantation and the foetal mortality after induced superovulation in females of various ages." J. Exp. Biol. 36: 281.

———. 1959. "Superpregnancy in the mouse. II. Weight and gain during pregnancy." J. Exp. Biol. 36:301.

———. 1959. "Studies on the transfer of fertilized mouse eggs to uterine foster-mothers. II. The effect of transferring large numbers of eggs." J. Exp. Biol. 36:40.

———. 1960. "Control of pre-natal growth in mammals." Nature (Lond.). 187:363.

———. 1960. "Congenital runts." Cibia Edition Symp. Congen. Malform. p. 178-194.

———. 1963. "Nature of the systemic effect of litter size on gestation period in mice." J. Reprod. Fert. 6:139-141.

McLaren, A., and A. K. Tarkowski. 1963. "Implantation of mouse eggs in the peritoneal cavity." J. Reprod. Fert. 6:385-392.

McChere, T. J. 1962. "Infertility in female rodents caused by temporary inanition at or about the time of implantation." J. Reprod. Fert. 4:241.

McPhail, M. K., and H. G. Read. 1942. "The mouse adrenal. I. Development, degeneration and regeneration of the x-zone." Anat. Rec. 84:51-73.

Meier, Hans. 1967. "The neuropathy of teetering, a neurological mutation in the mouse." Arch. Neurol. 16:59-66.

Melissinos, K. 1907. "Die Entwicklung des Eies der mause von den ersten Furchungsphenomenen bis zur Festoetzung der Allantois an der Ectoplacentorplatte." Arch. Miki. Anat. 70:577-628.

Meller, K., W. Breipohl, and P. Glees. 1966. "Early cytological differentiation in the cerebral hemispheres of mice." Zeits. Zellforsch. 72:525-553.

Menefee, M. S. 1955. "The differentiation of keratin-containing cells in the epidermis of embryo mice." Anat. Rec. 122:181-191.

Menzies, J. I. 1957. "Gene-controlled sterility in the African mouse (Mastomys)." Nature (Lond.). 179:1142.

Merritt, G. C. 1959. "The histochemical demonstration of nucleic acid with pyronin-methyl green." Med. Techn. 1:36-41.

Merklin, R. J. 1957. "Pregnancy in mice immediately after parturition." Anat. Rec. 127:333.

Merton, H. 1938. "Studies on reproduction in the albino mouse. I. The period of gestation and the time of parturition." Proc. Roy. Soc. (Edinburgh). 58:80-96.

Merton, H. 1939. "Reproduction in the albino mouse. III. Duration of life of sperm in the female reproduction tract." Proc. Roy. Soc. (Edinburgh). 59:207.

Miale, I. L., and R. L. Sidman. 1961. "An autoradiographic analysis of histogenesis in the mouse cerebellum." Exptl. Neurol. 4:277-296.

Michie, D., and A. McLaren. 1955. "The importance of being cross-bred." New Biol. 19:48.

Michie, D., and M. E. Wallace. 1953. "Affinity: a new genetic phenomenon in the house mouse." Nature (Lond.). 171:26.

Midgley, A. R., and G. B. Pierce. 1963. "Immunohistochemical analysis of basement membranes of the mouse." Am. J. Anat. 63:929-944.

Milaire, J. 1963. "Etude morphologique et cytochimique du development des membres chez la souris et chez la taupe." Arch. Biol. 74:129-317.

Miller, J. F., and J. S. Davies 1964. "Embryological development of the immune mechanism." Am. Rev. Med. 15:23-36.

Mintz, B. 1957. "Embryological development of primordial germ cells in the mouse. Influence of a new mutation." J. Emb. Exp. Morphol. 5:396-406.

———. 1957. "Interaction between two allelic series modifying primordial germ cell development in the mouse embryo." Anat. Rec. 128:591.

Mintz, B. 1957. "Germ cell origin and history in the mouse: genetic and histochemical evidence." Anat. Rec. 127:335-6.

———. 1958. "Environmental influences on prenatal development." 87 pps. ed. B. Mintz, N. A. S. - N. R. C. Univ. Chicago Press.

———. 1959. "Continuity of the female germ cell line from embryo to adult." Arch. Anat. Mic. Morphol. Exptl. 48, Suppl. 155-172.

———. 1960. "Embryological phases of mammalian gametogenesis." J. Cell. Comp. Phys. 56, Suppl. 1:31-48.

———. 1962. "Experimental study of the developing mammalian egg. Removal of the zona pellucida." Science. 138:594-595.

———. 1962. "Formation of genotypically mosaic mouse embryos." Am. Zool. 2:432.

———. 1962. "Incorporation of nucleic acid and protein precursors by developing mouse eggs." Am. Zool. 2:432.

———. 1962. "Experimental recombination of cells in the developing mouse egg normal and lethal mutant genotypes." Am. Zool. 2:541-542.

———. 1963. "Growth *in vivo* of t^{12}/t^{12} lethal mutant mouse eggs." Am. Zool. 3:550-551.

———. 1964. "Synthetic processes and early development in the mammalian egg." J. Exp. Zool. 157:85-100.

———. 1964. "Gene expression in the morula stage of mouse embryos, as observed during development of T^{12}/T^{12} lethal mutants *in vitro*." J. Exp. Zool. 157:267-272.

———. 1964. "Formation of genetically mosaic mouse embryos, and early development of lethal (T^{12}/T^{12}) normal mosaics." J. Exp. Zool. 157:273-292.

———. 1965. "Genetic mosaicism in adult mice of quadri-parental lineage." Science. 148:1232-3.

———. 1965. "Nucleic acid and protein synthesis in the developing mouse embryo" in "Preimplantation Stages in Pregnancy." G. E. W. Wolstenholme & O'Connor, eds. p. 145-168, Little, Brown & Co., Boston.

———, and E. S. Russell. 1955. "Developmental modifications of primordial germ cells, induced by the W-series genes in the mouse embryo." Anat. Rec. 122:443.

———. 1957. "Gene-induced embryological modifications of primordial germ cells in the mouse." J. Exp. Zool. 134:207-230.

Mirskaia, L., and F. A. E. Crew. 1930. "On the genetic nature of the time of attainment of puberty in the female mouse." Quart. J. Exp. Physiol. 20:299-304.

———. 1930. "Maturity in the female mouse." Proc. Roy. Soc. (Edinburgh). 50:179-186.

Mirskaia, L., and F. A. E. Crew. 1931. "On the pregnancy rate in the lactating mouse and the effect of suckling on the duration of pregnancy." Proc. Roy. Soc. (Edinburgh). 51:1-7.

Mitchison, N. A. 1952. "The effect on the offspring of maternal immunization in mice." J. Genet. 51:406-420.

Miyazaki, T. 1940. "Uber die Entwicklung der Lymphknoten bei Maus." Soc. Path. Jap. Trans. 30:29-34.

Monesi, V. 1962. "Autoradiographic study of DNA synthesis and the cell cycle in spermatogonia and spermatocytes of mouse testis using tritiated thymidine." J. Cell. Biol. 14:1-18.

Moog, F. 1951. "The functional differentiation of the small intestine. II. The differentiation of alkaline phosphomonesterase in the duodenum of the mouse." J. Exp. Zool. 118:187-207.

Morgan, W. 1964. "Bipaternity in mice." Proc. S. D. Acad. Sci. 43:81-84.

Mori, A. 1961. "The difference in sperm morphology in different strains of mice." Tohoku J. Agric. Res. 12:107-118.

Mowry, R. W., and R. C. Millican. 1952. "A histochemical study of the distribution and fate of dextran in tissues of the mouse." Am. J. Path. 28:522.

Muhlbock, O. 1947. "On the susceptibility of different strains of mice for oestrone." Acta Brev. Neer. 15:18-20.

Mukeyee, H., J. S. Ram, and G. B. Pierce. 1965. "Basement membranes. V. Chemical composition of neoplastic basement membrane muco-protein." Am. J. Path. 46:49-57.

Mulnard, J. 1955. "Contributions a la connaissance des enzymes dans l'ontogenese, les phosphomonesterases acide et alcoline dans le developpement du Rat et de la Souris." Arch. d. Biol. 66:525-688.

Mulnard, J. G. 1965. "Studies of regulation of mouse ova in vitro" in "Preimplantation Stages in Pregnancy." G. E. W. Wolstenholme and M. O'Connor, eds. p. 123-144, Little, Brown & Co., Boston.

Munford, R. E. 1963. "Changes in the mammary glands of rat and mice during pregnancy, lactation and biochemical changes." J. Endocrin. 28:35-44.

Murakami, U. 1963. "Developmental disturbances of the central nervous system-mechanisms of their formation in experimental teratology." Animal Rep. of The Research Institute of Environmental Medicine. 11:63-75. (Japan).

———. 1963. "Studies on mechanisms manifesting congenital anomalies." (Japan). J. Human Genetics. 8:202-226.

———, Y. Kameyama, and T. Kato. 1956. "A pathologic process in the initial phase of maldevelopments of the central nervous system." Am. Rep. Res. Inst. Environ. Med. Nagoya Univ. (Japan). 67-80.

Muthukkauppen, V. 1965. "Inductive tissue interaction in the development of the mouse lens in vitro." J. Exp. Zool. 159:269-287.

Nadamitsu, S. 1961. "Studies in vitro ovulation in vertebrates. IV. In vitro ovulation in the mouse." J. Sc. Hiroshima U. Ser. B. Div. 1. Vol. 20:27-32.

Naeye, R. L. 1966. "Organ and cellular development in mice growing at simulated high altitude." Lab. Invest. 15:700-706.

Nakum, L. H. 1964. "Science tools. VI. The lethal gene." Conn. Med. 28:163-164.

Nakamura, H. 1926. "Etudes experimentale sur la duree de gestation de la souris." Ann. Inst. Pasteur. 40:303-308.

Nakamura, T. 1957. "Cytological studies on abnormal ova in mature ovaries of mice observed at different phases of oestrous cycle." J. Fac. Fish. Anim. Husbandry, Hiroshima Univ. 1:343-362.

Nandi, S. 1959. "Hormonal control of mammogenesis and lactogenesis in the C_3H/HeCrgl mouse." Univ. Calif. Pub. Zool. 65:1-128.

Nebel, B. R., A. P. Amarose, and E. M. Hackett. 1961. "Calender of gametogenic development in the prepuberal mouse." Science. 134:832-833.

Nebel, B. R., and E. Hackett. 1961. "Synaptinemal complexes in primary sperma-
    tocytes of the mouse. The effect of elevated temperatures and some ob-
    vations on the structure of these complexes in control material." Z.
    Zullforsch. 55:556-565.

Nebel, L., and M. Hamburgh. 1966. "Observations on the penetration and uptake of
    trypan blue in embryonic membranes of the mouse." Z. Zullforsch. 75:129-
    137.

Nelson, A., S. Ullberg, H. Kristoffersson, and C. Ronnback. 1962. "Distribution of
    radioruthenium in mice." Acta Radiologica. 58:353-360.

Nelson, J. B., and G. R. Collins. 1961. "The establishment and maintenance of a spe-
    cific pathogen-free colony of Swiss mice. Proc. Animal Care Panel. 11:65-
    72.

New, D. A. 1963. "Effects of excess vitamin A on cultures of skin and buccal epithelium
    of embryonic rat and mouse." Brit. Jour. Derm. 75:320-325.

_____, and K. F. Stein. 1963. "Cultivation of mouse embryos *in vitro*." Nature
    (Lond.). 199:279-299.

_____. 1964. "Cultivation of post-implantation mouse and rat embryos on plasma
    dots." J. Emb. Exp. Morphol. 12:101-111.

Newton, W. H. 1935. "Pseudo-parturition in the mouse and the relation of the placenta
    to post partem oestrus." J. Physiol. 84:196.

_____, and N. Beck. 1939. "Placental activity in the mouse in the absence of the pi-
    tuitary gland." J. Endocrin. 1:65.

Nicholas, J. S. 1934. "Experiments on developing rats. I. Limits of foetal regenera-
    tion; behaviour of embryonic material in abnormal environments." Anat.
    Red. 58:387.

_____. 1934. "Mechanisms affecting embryonic growth." Cold Spring Harbor Symp.
    Quart. Biol. 19:36-40.

_____. 1949. "The problems of organization" in "The Chemistry and Physiology of
    Growth." A. K. Parpart, ed. Princeton U. Press.

Niimi, K., and I. Harada. 1957. "On the ontogenetic development of the massa inter-
    media of the mouse (in Japanese)." Arb. II. Abt. Anat. Inst. Tokushima
    2:97-128.

_____, I. Harada, Y. Kusaka, and S. Kishi. 1961. "The ontogenetic development of
    the diencephalon of the mouse." Tokushima J. Exp. Med. 8:203-238.

Nogami, H. 1964. "Digital malformations in the mouse foetus caused by x-radiation
    during pregnancy." J. Emb. Exp. Morphol. 12:637-650.

Noyes, R. W. 1953. "The fertilizing capacity of spermatozoa." West. J. Surg. 61:342.

_____, and Z. Dickmann. 1960. "Relationship of ovular age to endometrial develop-
    ment." J. Reprod. Fert. 1:186-196.

_____, L. L. Doyle, A. H. Gates, and D. L. Bentley. 1961. "Ovular maturation and
    fetal development." Fert. & Steril. 12:405-416.

_____, Z. Dickmann, L. L. Doyle, and A. H. Gates. 1963. "Ovum transfers, syn-
    chronous and asynchronous, in the study of implantation." pps. 197-211,
    A. C. Enders, (Ed.) "Delayed Implantation." Univ. Chicago Press, Chicago.

Oakberg, E. F. 1956. "A description of spermiogenesis in the mouse and its use in
    analysis of the cycle of the seminiferous epithelium and germ cell renewal."
    Am. J. Anat. 99:391-413.

_____. 1956. "Duration of spermatogenesis in the mouse and timing of stages of the
    cycle of the seminiferous epithelium." Am. J. Anat. 99:507-516.

_____. 1957. "Duration of spermatogenesis in the mouse." Nature (Lond.). 180:1137-
    1139, 1497.

_____. 1965. "The mammalian oöcyte." W. H. O. Conf. Chemistry & Physiology of
    the Gametes, Geneva, Switzerland.

Odor, D. L. 1955. "The temporal relationship of the first maturation division of rat
    ova to the onset of heat." Am. J. Anat. 97:461-492.

Ohno, S., L. C. Christian, and C. Stenius. 1963. "Significance in mammalian oogenesis
    of non-homologous association of bivalents." Exp. Cell. Res. 32:590-592.

Okuda, H., B. Haga, T. Kawachi, S. Fujii, and Y. Yamamura. 1960. "Studies on liver catalase, beta-glucuromidose and plasma iron during development of mice." Gann. 51:231-234.

Orsini, M. W. 1962. "Study of ova-implantation in the hamster, rat, mouse, guinea-pig, and rabbit in cleared uterine tracts." J. Reprod. Fert. 3:288-293.

———. 1963. "Morphological evidence on the intrauterine career of the ovum" (ed. Enders) in "Delayed Implantation." Univ. Chicago Press.

Otis, E. M. 1949. "Intra-uterine death-time in semisterile mice." Anat. Rec. 105:533.

———, and R. Brent. 1952. "Equivalent ages in mouse and human embryos." U.R.-194 and Anat. Rec. 120:33-64. (1954)

Overton, J. 1965. "Fine structure of the free cell surface in developing mouse intestinal mucosa." J. Exp. Zool. 159:195-202.

Padykula, H. A., J. J. Deren, and T. H. Wilson. 1966. "Development of structure and function in the mammalian yolk sac. I. Developmental morphology and vitamin B_{12} uptake of the rat yolk sac." Dev. Biol. 13:311-348.

———, and D. Richardson. 1963. "A correlated histochemical and biochemical study of glycogen storage in the rat placenta." Am. J. Anat. 112:215-242.

Pai, A. C. 1965. "Developmental genetics of a lethal mutation, muscular dysgenesis, in the mouse. II. Developmental analysis." Dev. Biol. 11:93-109.

Parkes, A. S. 1924. "Studies on the sex ratio and relayed phenomena. I. Foetal retrogression in mice." Proc. Royal Soc. (Lond.) 95:551-8.

———. 1925. "The age of attainment of sexual maturity of the albino mouse." J. Roy. Microscop. Soc. 315-319. (Quoted by Engle and Rosasco, 1927).

———. 1926. "Studies on the sex-ratio and related phenomena: observations on fertility and sex ratio in mice." Brit. Jour. Exp. Biol. 4:93-104.

———. 1926. "Observations on the oestrous cycle of the albino mouse." Proc. Royal Soc. (Lond.) B. 100:151-170.

———. 1928. "The length of the oestrous cycle in the unmated normal mouse, records of 1,000 cycles." Brit. Jour. Exp. Biol. 5:371-377.

———. 1929. "The internal secretions of the ovary." Longmans Green & Co., London.

———. 1929. "The functions of the corpus luteum. II. The experimental production of placentomata in the mouse." Proc. Royal Soc. (Lond.) B. 104:183-188.

———. 1953. "Prevention of fertilization by hyaluronidase inhibitor." Lancet. 265:1285-1287.

———. 1960. "The biology of spermatozoa and artificial insemination" in "Marshall's physiology of reproduction." pps. 161-263. A. S. Parkes (ed.). 3rd ed. Vol. 1, Part 2, Longmans Green, London.

———, and C. W. Bellerby. 1926. "The mammalian sex-ratio." Biol. Rev. 2:1-51.

———, and H. M. Bruce. 1961. "Olfactory stimuli in mammalian reproduction." Science. 134:1049-1054.

———. 1962. "Pregnancy block in female mice placed in boxes soiled by males." J. Reprod. Fertil. 4:303-308.

Parkes, A. S., W. Fielding, and F. W. R. Brambell. 1927. "Ovarian regeneration in the mouse after complete double ovariotomy." Proc. Royal Soc. (Lond.) B. 101:328-354.

Patten, B. M. 1957. "Varying developmental mechanisms in teratology." Pediatrics. 19:734-748.

———. 1964. "Foundations of Embryology." McGraw-Hill Book Co. 622 pps.

Penchez, R. I. 1929. "Experiments concerning ovarian regeneration in the white rat and white mouse." J. Exp. Zool. 54:319-339.

Pennycink, R. R. 1965. "The effects of acute exposure to high temperatures on prenatal development in the mouse with particular reference to secondary vibrissae." Austr. J. Biol. Sci. 18:97-113.

Perrotta, C. A. 1962. "Initiation of cell proliferation in the vaginal and uterine epithelia of the mouse." Am. J. Anat. 111:195-204.

Perrotta, C. A. 1966. "Effect of x-irradiation on DNA synthesis in the uterine epithelium." Rad. Res. 28:232-242.

Pesonen, S. 1946. "Abortive egg cells in the mouse." Hereditas. 32:93-96.

_____. 1949. "On abortive eggs. III. On the cytology of fertilized ova in the mouse." Ann. Clin. Gynec. Fenn. Suppl. 3 Vol. 38:337-352.

Peters, H., and K. Borum. 1961. "The development of mouse ovaries after low-dose irradiation at birth." Int. J. Rad. Biol. 3:1-16.

_____, E. Levy, and M. Crane. 1962. "Deoxy-ribonucleic acid synthesis in oöcytes of mouse embryos." Nature (Lond.). 195:915-916.

Pfeiffer, C. A., and C. W. Hooker. 1942. "Early and late effects of daily treatment with pregnant mare serum upon the ovary of mice of the A strain." Anat. Rec. 84:311-330.

Pierce, E. T. 1965. "An autoradiographic study of mouse brain stem histogenesis." Anat. Rec. (Abstr.) #151, p. 400.

_____. 1966. "Histogenesis of the nuclei griseum pontis, corporis pontobulbaris and reticularis tegmenti pointis (Bechterew) in the mouse." J. Comp. Neurol. 126:219-240.

Pierce, G. B., and T. F. Beals. 1964. "The ultrastructure of primordial germinal cells of the fetal testes and of embryonal carcinoma cells of mice." Cancer Res. 24:1553-1567.

_____, A. R. Midgley, J. Sri Ram, and J. D. Feldman. 1962. "Parietal yolk sac carcinoma: due to the histogenesis of Reicherts' membrane of the mouse embryo." Am. J. Path. 41:549-566.

Pierce, L. J. 1965. "Hereditary eye defects in the mouse." (Abstr.). Genet. 52:467.

Pincus, G. 1936. "The eggs of mammals." The Macmillan Co., N.Y.

_____. 1951. "Fertilization in mammals." Scientific American. 184:44-47.

Pinsky, L., and A. M. DiGeorge. 1965. "Cleft palate in the mouse. A teratogenic index of glucocorticoid potency." Science. 147:402-403.

Porter, David G. 1966. "Observations on the yolk sac and Reichert's membrane of ectopic mouse embryos." Anat. Rec. 154:847-860.

Potter, R. G. 1958. "Artificial insemination by donors." Fert. & Steril. 9:37-53.

Price, D. 1963. "Comparative aspects of development and structure in the prostate." Nat. Cancer Inst. Monogr. 12:1-27.

Purdy, D. M., and H. Hillemann. 1950. "Prenatal growth in the golden hamster." Anat. Rec. 106:591-598.

Purshottam, N., M. M. Mason, and G. Pincus. 1961. "Induced ovulation in the mouse and the measurement of its inhibition." Fert. & Steril. 12:346-352.

_____, and G. Pincus. 1961. "*In vitro* cultivation of mammalian eggs." Anat. Rec. 140:51.

Ravn, E. 1894. "Ueber die Arteria omphalomesenterica der Ratten und Mause." Anat. Anz. 9:420-424.

_____. 1894. "Zur Entwicklung des Nobelstranges der weissen Maus." Arch. Anat. u. Physiol. Anat. Abt. 293-312.

_____. 1895. "Ueber das Proamnion, besonders bei der Maus." Arch. Anat. u. Physiol. Anat. Abt. 189-224.

Rawles, M. E. 1947. "Origin of pigment cells from the neural crest in the mouse embryo." Physiol. Zool. 20:248-266.

Raymond, J. 1960. "Controles hormonal de la glande sous-maxillaire de la souris." Bull. Biol. 94:399-523.

Raynaud, A. 1941. "Reaction du sinus urogenital des embryons de souris aux hormones genitales injectees a la mere en gestation." Compt. Rend. Acad. Sci. 213:187-189.

_____. 1942. "Developpement des glandes annexes du tractus genital de la souris." Compt. Rend. Soc. Biol. 136:292-294.

_____. 1948. "Retards de developpement, observes chez des foetus de souris, au voisenage du terme de la gestation." Notes et Rev. Arch. Zool. Exp. et Gen. 85:83-99.

Raynaud, A. 1950. "Recherches experimentales sur le developpement de l'appareil endocrines et le fonctionnement des glandes endocrines des foetus de souris et de mulot." Arch. d'Anat. Mic. et de Morphol. Exp. 39:518-576.

_____. 1957. "Sur le developpement et la differenciation sexuelle de l'appareil guber- naculaire du foetus de souris." C. R. Acad. Sci. 245:2101-2123.

Reading, A. J. 1966. "Effects of parity and litter size on the birth weight of inbred mice." J. Mammology. 47:111-114.

Reamer, G. R. 1963. "The quantity and distribution of nucleic acids in the early cleavage stages of the mouse embryo." Ph. D. Thesis, Brown University. 108 pps.

Reed, S. C. 1933. "An embryological study of harelip in mice." Anat. Rec. 56:101- 110.

_____. 1938. "Uniovular twins in mice." Science. 88:13.

Reinius, S. 1965. "Morphology of the mouse embryo, from the time of implantation to mesoderm formation." Zeits. Zellforsch. 68:711-723.

_____. 1967. "Ultrastructure of epithelium in mouse oviduct during egg transport." (In press, Excerpta Medica).

_____. 1967. "Ultrastructure of blastocyst attachment in the mouse." Z. Zellforsch. 77:257-266.

Ressler, R. H. "Parental handling in two strains of mice reared by foster parents." Science. 137:129-130.

Reynaud, J., and A. Reynaud. 1947. "Observations sur la structure du tractus genital des souris femelles castrees a la naissance." Ann. Endocrin. 8:81-86.

Richardson, F. 1953. "The mammary gland development in normal and castrate male mice at nine weeks of age." Anat. Rec. 117:449-465.

Rietsckel, P. E. 1929. "Zur Morphologie der Genital-ausfukrungs gange in Individual cyclus der weissen Maus." Z. Wissensch. Zool. 135:428-494.

Ring, J. R. 1944. "The estrogen-progesterone induction of sexual receptivity in the spayed female mouse." Endocrin. 34:269-275.

Ritter, W., and K. H. Degenhardt. 1963. "Clefts of the lips, jaws and palate induced in mice by means of x-rays." Int. Dental J. 13:489-494.

Roberts, R. C. 1960. "The effect of litter size of crossing lines of mice inbred without selection." Genetic Res. 1:239-252.

Robertson, G. G. 1940. "Ovarian transplantations in the house mouse." Proc. Soc. Exp. Biol. & Med. 44:302-304.

_____. 1942. "An analysis of the development of homozygous yellow mouse embryos." J. Exp. Zool. 89:197-231.

Robson, J. M. 1934. "Uterine reactivity and activity in the mouse at various stages of the sex cycle." J. Physiol. 82:105.

_____. 1935. "The effect of oestrin on the uterine reactivity and its relation to exper- imental abortion and parturition. J. Physiol. 84:21 and 86:171.

Roosen-Runge, E. C. 1962. "The process of spermatogenesis in mammals." Biol. Rev. 37:343-377.

Ross, L. 1965. "In vivo palatine shelf movement in mice after maternal treatment with several teratogens." Anat. Rec. (Abstr.).

Rothschild, L. 1954. "Polyspermy." Quart. Rev. Biol. 29:332.

_____. 1965. "Fertilization." Methuen, London.

Rowe, W. P. 1961. "The epidemiology of mouse polyoma virus infection." Bact. Rev. 25:18.

Rudali, G., and J. Reverdy. 1959. "Action of very weak doses (5r) of x-rays given at birth on the leukemogenesis of AKR mice." Compt. Rend. Soc. Biol. 248: 1248-9.

Rudeberg, S. I. 1964. "Topographic distribution of non-specific esterase during cere- bellar development in mouse." Z. Anat. Entwicklungs gesch. 124:226-233.

Rudkin, G. T., and A. A. Griech. 1962. "On the persistance of oöcyte nuclei from fetus to maturity in the laboratory mouse." J. Cell. Biol. 12:169-176.

Rugh, R. 1964. "The mouse: a discoid placentate," p. 236-303, in R. Rugh, "Verte-brate embryology; the dynamics of development." Harcourt, Brace & World, N.Y.

Runner, M. N. 1947. "Attempts at *in vitro* semination of the mouse eggs." Anat. Rec. 99:564.

_____. 1947. "Development of mouse eggs in the anterior chamber of the eye." Anat. Rec. 98:1-17.

_____. 1949. "Limitation of litter size in the mouse following transfer of ova from artificially induced ovulations." Anat. Rec. 103:585.

_____. 1951. "Differentiation of intrinsic and maternal factors governing intrauterine survival of mammalian young." J. Exp. Zool. 116:1-20.

_____. 1954. "Ovulation in the prepuberal mouse: a delicate bioassay." Anat. Rec. 128:514.

_____. 1954. "Inheritance of susceptibility to congenital deformity-embryonic insta-bility." J. Nat. Cancer Inst. 15:637-649.

_____. 1959. "Embryocidal effect of handling pregnant mice and its prevention with progesterone." Anat. Rec. 133:330-331. (Abstr.).

_____, and A. J. Ladman. 1950. "The time of ovulation and its diurnal regulation in the post-parturitional mouse." Anat. Rec. 108:343-361.

_____, and A. H. Gates. 1954. "Conception in prepuberal mice following artificially induced ovulation and mating." Nature (Lond.). 174:222-223.

_____. 1954. "Sterile, obese mothers." J. Heredity. 45:51.

Runner, M. N., and J. Palm. 1952. "Length of life of the unfertilized ovum of the mouse." Anat. Rec. 112:383.

_____. 1953. "Transplantation and survival of unfertilized ova of the mouse in rela-tion to post-ovulatory age." J. Exp. Zool. 124:303-316.

Russell, E. S., and E. Fekete. 1958. "Analysis of W-series pleiotropism in the mouse: Effect of W W substitution on definitive germ cells and on ovarian tumorigen-esis." J. Nat. Cancer Inst. 21:365-381.

_____, and F. A. Lawson. 1959. "Selection and inbreeding for longevity of a lethal type." J. Heredity. 50:19-25.

_____, and E. C. McFarland. 1965. "Erythrocyte populations in fetal mice with and without two hereditary anaemias." Fed. Proc. 24:240 (Abstr.).

Russell, L. B., S. K. Badgett, and C. L. Saylors. 1959. "Comparison of the effect of acute continuous, and fractionated irradiation during embryonic develop-ment." from "Immediate and Low Level Effects of Ionizing Radiations." Int. J. Rad. Biol. Suppl. 343-359.

_____, and W. L. Russell. 1954. "An analysis of the changing radiation response of the developing mouse embryos." J. Cell. Comp. Physiol. 43:104-145.

Russell, Liane B., and Florence N. Woodiel. 1966. "A spontaneous mouse chimera formed from separate fertilization of two meiotic products of oögenesis." Cytogenetics. 5:106-119.

Russell, W. L. 1963. "The effect of radiation dose rate and fractionation on mutation in mice." from Sobels "Repair from Genetic Radiation." Pergamon Press.

Rutter, W. J., N. K. Wessells, and C. Grobstein. 1964. "Control of specific synthesis in the developing pancreas." Nat. Cancer Inst. Monogr. 13:51-65.

Sanel, F. T., and W. M. Copenhaver. 1965. "Histogenesis of mouse thymus studied with the light and electron microscope." Anat. Rec. 151:410.

Sapsford, C. S. 1957. "The development of the Sertoli cell." J. Endocrin. Proc. 15:lv, lvi.

_____. 1962. "Changes in the cells of the sex cords and seminiferous tubules during the development of the testis of the rat and mouse." Austral. J. Zool. 10: 178-192.

Sato, K. 1936. "Uber die Entwicklungs geschichte des Maus eies. I. Die intratubase Entwicklung derselben." Okayama-Igakkai-Zosski. 48:423-441.

_____. "Uber die Entwicklungs gehichte des Maus eies. II. Die intrauterine Ent-wicklung derselben, besonders der Emostetekungs-mechanismus des Am-mions." Okayama-Igakkai-Zosaki. 48:792-832.

Sawada, T. 1957. "An electron microscope study of spermatid differentiation in the mouse." Okayamas Fol. Anat. Japan. 30:73-80.

Schlafke, S., and A. C. Enders. 1963. "Observations on the fine structure of the rat blastocyst." J. Anat. 97:353-360.

Schlager, Gunther. 1966. "Systolic blood pressure in eight inbred strains of mice." Nature (Lond.). 212:519-520.

Schlesinger, M. 1964. "Serologic studies of embryonic and trophoblastic tissues of the mouse." J. Immun. 93:253-263.

———. 1965. "Immune lysis of thymus and spleen cells of embryonic and neonatal mice." J. Immun. 94:358-364.

Searle, A. G. 1954. "The influence of maternal age on development of the skeleton of the mouse." Ann. N.Y. Acad. Sci. 57:558-563.

———. 1964. "Effects of low-level irradiation on fitness in an inbred mouse strain." Genetics. 50:1159-1178.

Selye, H., and T. McKeown. 1934. "The effect of mechanical stimulation of the nipple on the ovary and the sexual cycle." Surg. Gynec. & Obstet. 59:886-890.

———. 1934. "Production of pseudo-pregnancy by mechanical stimulation of the nipples." Proc. Soc. Exp. Biol. & Med. 31:683-687.

Sharma, K. N. 1960. "Genetics of gametes. IV. The phenotype of mouse spermatozoa in four inbred strains and their F_1 crosses." Proc. Royal Soc. (Edinburgh) B. 68:54.

Shelesnyak, M. C., and A. M. Davies. 1953. "Relative ineffectiveness of electrical stimulation of the cervix for inducing pseudopregnancies in the mouse." Endocrin. 52:362-363.

Sherman, J. K., and T. P. Lin. 1958. "Survival of unfertilized mouse eggs during freezing and thawing." Proc. Soc. Exp. Biol. & Med. 98:902-905.

———. 1959. "Temperature shock and cold-storage of unfertilized mouse eggs." Fert. & Steril. 10:384.

Shintani, Y. K. 1959. "The nuclei of the prectectal region of the mouse brain." J. Comp. Neur. 113:43-60.

Shoji, R., and E. Ohzu. 1965. "Breeding experimental of white rats and mice. XI. Notes on implantation rate, prenatal mortality and spontaneous abnormality in eight strains of inbred mice." (Jap., Eng. summ.) Zool. Mag. (Dobutsugaku Sasshi). 74:115-118.

Sidman, R. L. 1961. "Histogenesis of mouse retina studies with thymidine-H^3" in "The Structure of the Eye." pps. 487-506, ed. G. Smelser, Academic Press, N.Y.

———. 1963. "Organ culture analysis of inherited retinal degeneration in rodents." J. Nat. Cancer Inst. 11:227-246.

———, and J. B. Angevine, Jr. 1962. "Audoradiographic analysis of time of origin of nuclear versus cortical components of mouse telencephalon." (Abstr.). Anat. Rec. 142:326.

———, S. H. Appel., and J. F. Fuller. 1965. "Neurological mutants of the mouse." Science. 150:513-516.

———, M. M. Dickie, and S. H. Appel. 1964. "Quaking and Jumpy, mutant mice with deficient myclination in the central nervous system." Science. 144:309-311.

———, and M. C. Green. 1965. "Retinal degeneration in the mouse. Location of the rd locus in linkage group XXII." J. Heredity. 56:23-29.

———, P. Lane, and M. Dickie. 1962. "Stagger, a new mutation in the mouse affecting the cerebellum." Science. 137:610-612.

———, I. L. Miale, and N. Feder. 1959. "Cell proliferation migration in the primitive ependymal zone. An autoradiographic study of histogenesis in the nervous system." Exp. Neur. 1:322-333.

———, P. A. Mottla, and N. Feder. 1961. "Improved polyester wax embedding for histology." Stain Techn. 36:279-284.

———, and G. B. Wislocki. 1954. "Histochemical observations on rods and cones in retinas of vertebrate." J. Histochem. & Cytochem. 2:413-433.

Silagi, S. 1963. "Some aspects of the relationship of RNA metabolism to development in normal and mutant mouse embryos cultivated *in vitro.*" Exp. Cell. Res. 32:149-152.

Silvers, W. K. 1958. "Origin and identity of clear cells found in hair bulbs of albino mice." Anat. Rec. 130:135-144.

Simkins, C. S. 1923. "Origin and migration of the so-called primordial germ cells in the mouse and rat." Acta Zool. 4:241-278.

Simmons, R. L., and P. S. Russell. 1962. "The antigenicity of mouse trophoblast." Ann. N.Y. Acad. Sci. 99:717-732.

_____, and J. Weintraub. 1965. "Transplantation experiments on placental ageing." Nature (Lond.). 208:82-83.

Simonds, J. P. 1925. "The blood of normal mice." Anat. Rec. 30:99-101.

Sirlin, J. L., and R. G. Edwards. 1959. "Timing of DNA synthesis in ovarian oöcyte nuclei and pronuclei of the mouse." Exp. Cell. Res. 18:190-198.

Sisken, B. F., and S. Gluechoohn-Waelsch. 1959. "A developmental study of the mutation 'phocomelia,' in the mouse." J. Exp. Zool. 142:623-642.

Skalko, R. G. 1963. "Deoxyribonucleic acid synthesis in the irradiated rat embryo." Thesis, Univ. of Florida. 89 pps.

Skoje, R., Ohzu, E. 1965. "Breeding experiments of white rats and mice. XI. Notes on implantation rate, prenatal mortality and spontaneous abnormality in eight strains of inbred mice." Zool. Mag. (Dobutsugaki-Zosski). 74:115-118.

Slizynski, B. M. 1949. "A preliminary pachytene chromosome map of the house mouse." J. Genetics. 49:242-244.

Smith, C. 1965. "Studies on the thymus of the mammal. XIV. Histology and histochemistry of embryonic and early postnatal thymuses of C57BL/6 and AKR strain mice." Am. J. Anat. 116:611-629.

_____, and M. J. Waldron. 1956. "Glycogen in the thymus of the fetal mouse." Anat. Rec. (Abstr.). 108:113.

Smith, L. J. 1956. "A morphological and histochemical investigation of a pre-implantation lethal (T^{12}) in the house mouse." J. Exp. Zool. 132:51-84.

_____. 1964. "The effects of transection and extirpation on axis formation and elongation in the young mouse embryo." J. Emb. Exp. Morphol. 12:787-803.

_____. 1966. "The changing pattern of basophilia in the mouse uterus from mating through implantation." Am. J. Anat. 119:1-14.

_____. 1966. "Metrial gland and other glycogen containing cells in the mouse uterus following mating and through implantation of the embryo." J. Anat. 119:15-23.

Smith, P. E., and E. T. Engle. 1927. "Experimental evidence regarding the role of the anterior pituitary in the development and regulation of the genital system." Am. J. Anat. 40:159-217.

_____. 1927. "Induction of precocious sexual maturity in the mouse by daily homeo and heterotransplants." Proc. Soc. Exp. Biol. Med. 24:561-562.

Smithberg, M. 1953. "The effect of different proteolytic enzymes on the zona pellucida of mouse ova." Anat. Rec. 117:554.

_____, and M. N. Runner. 1956. "The induction and maintenance of pregnancy in prepuberal mice." J. Exp. Zool. 133:441-457.

_____. 1957. "Pregnancy induced in genetically sterile mice." J. Heredity. 48:97-100.

_____. 1960. "Retention of blastocysts in non-pregestational uteri of mice." J. Exp. Zool. 143:21-30.

Smitten, N. A. 1963. "Cytological analysis of catechalamine synthesis in the ontogenesis of vertebrates and problems of melanogenesis." Gen. Comp. Endocrin. 3:362-377.

Snell, G. D. 1944. "Antigenic differences between the sperm of different inbred strains of mice." Science. 100:272-273.

_____, ed. 1941. "Biology of the Laboratory Mouse." Blakiston, Philadelphis. 497 pps.

Snell, G. D., E. Fekete, K. P. Hummel, and L. W. Law. 1940. "The relation of mating, ovulation, and estrous smear in the house mouse to time of day." Anat.
Rec. 76:39-54.

_____, K. P. Hummel, and W. H. Abelmann. 1944. "A technique for the artificial
insemination of mice." Anat. Rec. 87:473 & Anat. Rec. 90:243-253.

Sobel, E. H., M. Hamburgh, and R. Koblin. 1960. "Development of the fetal thyroid
gland." First Intern. Congr. Endocrinology, Copenhagen, Session XIIIc,
No. 608 - Thyroid Exper.

Sobotta, J. 1895. "Die Befruchtung and Furchung des Eies der Maus." Arch. Mikr.
Anat. 45:15-93.

_____. 1903. "Die Entwichlung des Eies du Maus vom Schlusse der Furchungsperiode
bis zum Auftreten der Amniosfolten." Arch. Mikr. Anat. 61:274-330.

_____. 1911. "Die Entwicklung des Eies du Maus vom Ersten Auftreten des Meso-
derms an bis zur Ausbildung den Embryonanlage und dem Auftretern der
Allantois." Arch. Mikr. Anat. 78:271-352.

Solomon, J. B. 1964. "Deoxyribonuclease II in the developing mouse embryo." Nature
(Lond.). 201:618-619.

Soper, E. H. 1963. "Ovarian and uterine responses to gonadotrophin in immature mice
as related to age." Anat. Rec. 145:352-353 (Abstr.).

Soriano, L. 1965. "Differenciation des epitheliums du tube digestif *in vitro*." Jour.
Emb. Exp. Morphol. 14:119-128.

Southard, J. L. 1965. "Artificial insemination of dystrophic mice with mixture of
spermatazoa." Nature (Lond.). 208:1126-1127.

Staats, J. 1964. "Standardized nomenclature for inbred strains of mice, third listing."
Cancer Research. 24:147-168.

Stafford, E. S. 1930. "The origin of the blood of the placental sign." Anat. Rec. 47:
43-57.

Stanley, N. F., D. C. Dorman, and J. Ponsford. 1953. "Studies on the pathogenesis of
a hitherto underscribed virus (hepato-encephalo myelitis) producing unusual
symptoms in suckling mice." Austral. J. Exp. & Med. Sci. 31:147-160.

Staugaard, Burton Christian. 1965. "Enzyme activities in the developing kidney of the
normal mouse and of Sd heterozygotes and homozygotes." Abstr. Ph. D.
Thesis, Univ. Connecticut, 1964.

Steinetz, B. G., V. L. Bead, and R. L. Kroc. 1957. "The influence of progesterone,
relaxin and estrogen on some structure and functional changes in the pre-
parturient mouse." Endocrin. 61:271.

Stevens, J. C., and J. A. Mackensen. 1958. "The inheritance and expression of a mutation in the mouse affecting blood formation, the axial skeleton, and body
size." J. Heredity. 49:153-160.

Stevens, L. C. 1964. "Experimental production of testicular teratomas in mice."
Proc. Nat. Acad. Sci. 52:645-661.

Stevens, W. L. 1937. "Significance of grouping and a test for uniovular twins in mice."
Ann. Eugenics. 8:57-73.

Stoner, R. D., and W. M. Hale. 1953. "A method for eradication of the mite 'Myocop-
tes musculinus' from laboratory mice." J. Econ. Entomal. 46:692.

Stotsenburg, J. M. 1915. "The growth of the fetus of the albino rat from the thirteenth
to the twenty-second day of gestation." Anat. Rec. 9:667-682.

Stowell, R. E. 1941. "A case of probable superfetation in the mouse." Anat. Rec. 81:
215-220.

Strong, L. C. 1942. "The origin of some inbred mice." Cancer Res. 2 No. 8:531.

_____. 1961. "The Springville mouse, further observations on a new 'luxoid' mouse."
J. Heredity. 52:122-124.

_____, and C. A. Fuller. 1958. "Maternal age at time of first litters in mice." J.
Gerontol. 13:236-240.

Sugahara, T. 1964. "Genetic effects of chronic irradiation given to mice through three
successive generations." Genetics. 50:1143-1158.

Sugiyama, M. 1959. "A comparative cytoarchitectural study on the diencephalon of the mouse (in Japanese)." Arb. II. Abt. Anat. Inst. Tokushima. 7:49-104.

Sugiyama, T. 1961. "Morphological studies on the placenta of mice of various ages and strains. I. Variations in fetal and placental weight at term" Acta Med. Med. Univ. Kioto. 37:139.

Suntzeff, V., E. L. Burns, M. Moskop, and L. Loeb. 1938. "On the proliferative changes taking place in the epithelium of vagina and cervix of mice with advancing age and under the influence of experimentally administered estrogenic hormones." Amer. J. Cancer. 32:256-289.

Swyer, G. I. M. 1947. "The release of hyaluronidase from spermatozoa." Biochem. J. 41:413-417.

Szollosi, D. 1965. "Development of "yolky substance" in some rodent eggs." Anat. Rec. (Abstr.). 151:424.

Taber, E. 1963. "Histogenesis of brain stem neurons studied autoradiographically with thymidine H^3 in the mouse." Anat. Rec. 145:291 and Anat. Rec. 148:344.

Takasugi, N., and H. A. Bern. 1962. "Crystals and concretions in the vaginae of persistent-estrous mice." Proc. Soc. Exp. Biol. Med. 109:662-624.

Talbert, G. B., and P. L. Krohn. 1965. "Effect of maternal age on the viability of ova and on the ability of the uterus to support pregnancy." Anat. Rec. (Abstr.).

Tarkowski, A. K. 1955. "Experiments in the development of isolated blastomeres of mouse eggs." Nature (Lond.). 184:1286-1287.

———. 1959. "Experimental studies on regulation in the development of isolated blastomeres of mouse eggs." Acta Theriologica. 3:191.

———. 1959. "Experiments on the transplantation of ova in mice." Acta Theriologica. 2:251.

———. 1960. "The effects of transplantation on the early development of rat eggs." Symp. on "Germ Cells and Development." Inst. Int. d. Emb. & Fondozioni, A., Baselli.

———. 1961. "Mouse chimeras developed from fused eggs." Nature (Lond.). 190:857-860.

———. 1962. "Interspecific transfers of eggs between rat and mouse." J. Emb. Exp. Morphol. 10:476-495.

———. 1963. "Studies on mouse chimeras developed from eggs fused in vitro." Nat. Cancer Inst. Monog. 11:51-67.

———. 1964. "True hermaphroditism in chimaeric mice." J. Emb. Exp. Morphol. 12:735-757.

———. 1964. "Patterns of pigmentation in experimentally produced mouse chimaerae." J. Emb. Exp. Morphol. 12:575-585.

———. 1965. "Embryonic and postnatal development of mouse chimeras" in "Preimplantation Stages in Pregnancy." G. E. W. Wolstenholme & M. O'Connor, eds. pps. 183-193, Little, Brown & Co., Boston.

Taylor, R. B. 1965. "Pluripotential stem cells in mouse embryo liver." Brit. J. Exp. Pathol. 46:376-383.

Tedeschi, C. G., and T. H. Ingalls. 1956. "Vascular anomalies of mouse fetuses exposed to anoxia during pregnancy." Am. J. Obstet. & Gynec. 71:16-28.

Tennant, Judith R., and George D. Snell. 1966. "Some experimental evidence for the influence of genetic factors on viral leukemogenesis." Nat. Cancer Inst. Monogr. 22:61-72.

Theiler, K. 1954. "Die entetehung von spaltwirbeln bei Danforth's short-tail maus." Acta Anatomica. 21:259-283.

Theiler, K. 1957. "Über die differenzierung der rumpfmyotome beim menschen und die herkunft der bauchwandmuskeln." Acta Anat. 30:842-864.

———. 1958. "Zelluntergang in den histersten Rumpfsomiten bei der Maus." Zeits. Anat. Extwickl. 120:274-278.

———. 1959. "Anatomy and development of the 'Truncate' (Boneless) mutation in the mouse." Am. J. Anat. 104:319-343.

Theiler, K., and Salome Gluecksohn-Waelsch. 1956. "The morphological effects and the development of the fused mutation in the mouse." Anat. Rec. 125:83-104.

————, and L. C. Stevens. 1960. "The development of rib fusions, a mutation in the house mouse." Am. J. Anat. 106:171183.

Thiery, M. 1960. "Les variations de la teneur en acid desoxyribonucleique (DNA) des noyaux de l'epithelium vaginal de la souris au cours du cycle oestral." Arch. Biol. 71:389-406.

Thomas, L. J. 1926. "Ossification centers in the petrosal bone of the mouse." Anat. Rec. 33:59-68.

Thomson, J. L., and J. D. Biggers. 1966. "Effect of inhibitors of protein synthesis on the development of preimplantation mouse embryos." Exp. Cell. Res. 41:411-427.

Thompson, W. R., and S. Olian. 1961. "Some effects on offspring behavior of maternal adrenalin injection during pregnancy in three inbred mouse strains." Psychol. Rep. 8:87-90.

Thung, P. J., L. M. Boot, and O. Muhlbock. 1956. "Senile changes in the oestrous and in ovarian structure in some inbred strains of mice." Acta Endocrin. 23:8-32.

Tijo, J. H., and A. Levan. 1954. "Some experiences with acetic orcein in animal chromosomes." An. Estoc. Exp. Anla. Dei. 3:225-228.

Togari, C. 1927. "On the ovulation of the mouse." Nagoya J. Med. (Japan). 2:17-50.

Torrey, T. W. 1945. "The development of the urino-genital system of the albino rat. II. The gonads." Am. J. Anat. 76:375-400.

Trasler, D. G. 1960. "Influence of uterine site on occurrence of spontaneous cleft palate in mice." Science. 152:420-421.

————. 1965. "Aspiring-induced cleft lip and other malformations in mice." Lancet. 606-607.

Traurig, H. H., and C. F. Morgan. 1964. "Autoradiographic studies of the epithelium of mammary gland as influenced by ovarian hormones." Proc. Soc. Exp. Biol. & Med. 115:1076-1080.

Truslove, G. M. 1956. "The anatomy and development of the fidget mouse." J. Genet. 54:64-86.

Turner, C. D., and H. Asakawa. 1964. "Experimental reversal of germ cells in ovaries of fetal mice." Science. 143:1344-1345.

Turner, C. W., and E. T. Gomez. 1932. "The normal development of the mammary gland of the male and female albino mouse." Univ. Miss. Res. Bull. 182.

Tutikawa, Kiyosi and Akira Akahori. 1965. "Strain difference of susceptibility to the teratogenic action of ethylurethane in mice." Abstr. Ann. Rep. Nat. Inst. Genet. Japan (1964).

Tyan, M. L. 1964. "Thymus: role in maturation of fetal lymphoid precursors." Science. 145:934-935.

Tyan, M. L., and L. J. Cole. 1962. "Development of transplantation isoantigens in the mouse embryo plus trophoblast." Transplantation Bull. 30:526-529.

Tyler, A. 1961. "The fertilization process." Sterility - Vol. pp. 25-26.

Umansky, R. 1966. "The effect of cell population density on the developmental fate of reaggregating mouse limb bud mesenchyme." Dev. Biol. 13:31-56.

Uzman, L. 1960. "The histogenesis of the mouse cerebellum as studied by its tritiated thymidine uptake." J. Comp. Neur. 114:137-159.

Valverde, F., and R. L. Sidman. 1965. "Successful Golgi impregnations in brains of mutant mice with deficient myelination." Anat. Rec. 151:479-480.

Van der Lee, S., and L. M. Boot. 1955. "Spontaneous pseudopregnancy in mice." Acta Physiol. Pharmacol. Neer. 4:442-444.

Van der Stricht, O. 1901. "L'atresie ovulaire et l'atresie folliculaire du follicle de De Graaf dans l'ovaire de chauve-souris." Verh. Anat. Ges., Jena 15:108-121.

————. 1902. "Le spermatozoi de dans l'oeuf de chauve-souris." Verh. Anat. Ges. 16:163.

Van Ebbenhorst Tengbergen, W. J. P. R. 1955. "The morphology of the mouse anterior pituitary during the oestrous cycle." Acta Endocrin. 18:213-218.

Velardo, J. T. 1958. "The endocrinology of reproduction." Oxford Univ. Press. p. 340.

Venable, J. H. 1945. "Pre-implantation stages in the golden hamster." Anat. Rec. 94:105-120.

_____. 1946. "Volume changes in the early development of the golden hamster." Anat. Rec. 94:129-138.

Veneroni, G., and A. Bianchi. 1957. "Correcting the genetically determined sterility of W W male mice." J. Emb. Exp. Morphol. 5:422-427.

Vivien, J. H. 1950. "Epoque de la differenciation hypophysaire chez la souris albinos." Compt. Rend. Soc. Biol. 144:284-287.

Von Heyningen, H. E. 1961. "The initiation of thyroid function in the mouse." Endocrin. 69:720-727.

Wada, T. 1923. "Anatomical and physiological studies on the growth of the inner ear of the albino rat." Wistar Mem. 10, Philadelphis.

Waddington, C. H. 1956. "Principles of embryology." Macmillan Co. 510 pps.

Walker, B. E. 1954. "Genetics - embryological studies on normal and cleft palates in mice." Ph.D. Thesis, McGill Univ., Montreal.

_____. 1958. "Polyploidy and differentiation in the transitional epithelium of mouse urinary bladder." Chromosoma. 9:105-118.

_____. 1960. "A special component of embryonic mesenchyme." Anat. Rec. 136:298.

_____. 1961. "The association of muco-polysaccharides with morphogenesis of the palate and other structures in mouse embryos." J. Emb. Exp. Morphol. 9:22-31.

_____, and B. Crain. 1959. "The lethal effect of cortisone on mouse mbryos with spontaneous cleft lip-cleft palate." Texas Rep. Biol. Med. 17:637-644.

_____. 1960. "Effects of hypervitaminosis A on palate development in two strains of mice." Am. J. Anat. 107:49-58.

_____. 1961. "Abnormal palate morphogenesis in mouse embryos induced by riboflavin deficiency." Proc. Soc. Exp. Biol. & Med. 107:404-406.

Walker, B. E., and F. C. Fraser. 1956. "Closure of the secondary palate in three strains of mice." J. Emb. Exp. Morphol. 4:176-189.

_____. 1957. "The embryology of cortisone induced cleft palate." J. Emb. Exp. Morphol. 5:201-209.

Ward, M. C. 1946. "A study of the estrous cycle and the breeding of the golden hamster." Anat. Rec. 94:139-162.

_____. 1948. "The early development and implantation of the golden hamster and the associated endometrial changes." Am. J. Anat. 82:231-276.

Waring, H. 1935. "The development of the adrenal gland of the mouse." Quart. J. Mier. Sci. 78:329-366.

Washburn, W. W. 1951. "A study of the modification in rat eggs observed *in vitro* following tubal retention." Arch. Biol. 62:439.

Wegelius, O. 1959. "The dwarf mouse - an animal with secondary myxedema." Proc. Sn. Exp. Biol. & Med. 101:225-227.

Weir, J. A. 1958. "Sex ratio related to sperm source in mice." J. Heredity. 49:223-227.

Weir, M. W., and J. C. De Fries. 1963. "Blocking of pregnancy in mice as a function of stress." Psychol. Rep. 13:365-366.

_____. 1964. "Prenatal maternal influence on behaviour in mice: Evidence of a genetic basis." J. Cell. Comp. Phys. 58:412-417.

Weitlauf, H. M., and G. C. Greenwold. 1965. "Comparisons of S^{35} methionine incorporation by the blastocysts of normal and delayed implanting mice." J. Rep. Fert. 10:203-208.

Wessells, N. K., and K. D. Roessner. 1965. "Nonproliferation in dermal condensations of mouse vibrissae and pelage hairs." Develop. Biol. 12:419-433.

West, W. T., H. Meier, and W. G. Hoag. 1966. "Hereditary mouse muscular dystrophy with particular emphasis on pathogenesis and attempts at therapy." Ann. N.Y. Acad. Sci. 138:4-13.

Whitten, W. K. 1956. "Modification of the oestrous cycle of the mouse by external stimuli associated with the male." J. Endocrin. 13:399-404.

_____. 1956. "The effect of removal of the olfactory bulbs on the gonads of mice." J. Endocrin. 14:160-163.

_____. 1956. "Culture of tubal mouse ova." Nature (Lond.). 177:96. Nature (Lond.). 179:1081-1082, 1957.

_____. 1957. "The effect of progesterone on the development of mouse eggs *in vitro*." J. Endocrin. 16:80-85.

_____. 1957. "Effect of exteroceptive factors on the oestrous cycle of mice." Nature (Lond.). 180:1436.

_____. 1958. "Endocrine studies on delayed implantation in lactating mice: Role of the pituitary in implantation." J. Endocrin. 16:435-440.

_____. 1959. "Occurrence of anoestrus in mice caged in groups." J. Endocrin. 18: 102-107.

_____, and C. P. Dagg. "Influence of spermatozoa on the cleavage rate of mouse eggs." J. Exp. Zool. 148:173-183.

Wilson, E. D., and M. X. Zarrow. 1958. "Induction of superovulation with HCG in immature mice primed with PMS." Proc. Am. Soc. Zool. #19.

Wilson, I. B. 1960. "Implantation of tissues in the uteri of pseudopregnant mice." Nature (Lond.). 185:553-554.

_____. 1963. "New factor associated with the implantation of the mouse egg." J. Reprod. Fert. 5:281-282.

_____. 1963. "A tumour tissue analogue of the implanting mouse embryo." Proc. Zool. Soc. (Lond.). 141:137-151.

Wilson, J. W., C. S. Gwat, and E. H. Leduc. 1963. "Histogenesis of the liver." Ann. N.Y. Acad. Sci. 111:8-24.

_____, and E. H. Leduc. 1948. "The occurrence and formation of binucleate and multinucleate cells and polyploid nuclei in the mouse liver." Am. J. Anat. 82:3, 353-392.

_____. 1950. "Abnormal mitosis in mouse liver." Am. J. Anat. 86:51-74.

_____, and H. Winston. 1949. "The production of biotin deficiency in the mouse." J. Nutrition, v. 38: no. 1.

Wimsatt, W. A., and C. M. Waldo. 1945. "The normal occurrence of a peritoneal opening in the bursa ovarii of the mouse." Anat. Rec. 93:47-57.

Winick, M., and R. G. Greenberg. 1965. "Appearance and localization of a nerve growth-promoting protein during development." Pediatrics. 35:221-228.

Wirsen, C. 1964. "Histochemical heterogeneity of muscle spindle fibers." J. Histochem. Cytochem. 12:308-309.

_____, and K. S. Larsson. 1964. "Histochemical differentiation of skeletal muscle in foetal and newborn mice." J. Emb. Exp. Morphol. 12:759-767.

Wirtschafter, Z. T. 1960. "Genesis of the mouse skeleton." C. C. Thomas, Springfield.

Wischnitzer, S. 1966. "The maturation of the ovum and growth of the follicle in the mouse ovary; a phase contrast microscope study." Growth. 30:239-255.

Wislocki, G. B., H. W. Deane, and E. W. Dempsey. 1964. "The histochemistry of the rodent's placenta." Am. J. Anat. 78:281-345.

_____, and R. L. Sidman. 1954. "Chemical morphology of the retina." J. Comp. Neur. 101:53-100.

Witschi, E. 1956. "Development of Vertebrates." W. B. Saunders, Phila. 588 pps.

Wolf, M. K. 1964. "Differentiation of neuronal types and synopses in myelinating cultures of mouse cerebellum." J. Cell. Biol. 22:259-279.

Wolfe, H. Glenn. 1966. "Plasma proteins in fetal mice." Abstr. Genetics. 54:369.

Wolstenholme, G. E. W., and M. O'Connor. 1965. "Preimplanation stages of pregnancy." (Ciba Foundation Symposium) Little, Brown & Co., Boston, 430 pps.

Wood, M. L. 1967. "*In-vivo* observations on normal and experimentally induced variations in the electrical activity of the uterus of the mouse." (in press)

_____, and R. E. McCafferty. 1964. "Relationship of intra-amniotic pressure changes to electrical activity of surrounding uterine muscle." Anat. Rec. 148:351-352. (Abstr.).

_____. 1966. "*In-vivo* observations of spontaneous electrical activity in uteri of normally cycling, pregnant and ovariectomized mixe." Anat. Rec. 154:443 (Abstr.).

Woollam, D. H. M. 1964. "The effect of environmental factors on the foetus." Jour. Col. Gen. Pract. Suppl. #2, 8:35-46.

Yamada, E. 1955. "The fine structure of the gall bladder epithelium of the mouse." J. Biophys. Biochem. Cytol. 1:445-458.

_____. 1955. "The fine structure of the renal glomerulus of the mouse." J. Biophys. & Biochem. Cytol. 1:551.

_____, T. Muta, and A. Matamura. 1957. "The fine structure of the oöcyte in the mouse ovary studied with electron microscope." Kurume Med. J. 4:148-171.

Yanagimachi, R., and M. C. Chang. 1961. "Fertilizable life of golden hamster ova and their morphological changes at the time of losing fertilizability." J. Exp. Zool. 148:185-197.

_____. 1962. "Fertilizable life of the golden hamster ova and the cortical change at the time of losing fertilizability." Anat. Rec. 142:334.

Yoon, C. H. 1955. "Homeostasis associated with heterozygosity in the genetics of time of vaginal opening in the house mouse." Genetics. 40:297-309.

Young, H. B. 1917. "Some phases of spermatogenesis in the mouse." Univ. Calif. Pub. Zool. 16:371-380.

Young, W. C. 1933. "Die Resoystron in den Ductuli efferentes der Maus und ihre Bedentung fur das Problem der Unterbindung un Hoden-Nebenhodensystem." Z. Zellforsch. U. Miki. Anat. 17:729-759.

Zarrow, M. X., and E. D. Wilson. 1961. "The influence of age on superovulation in superovulation in the immature rat and mouse." Endocrin. 69:851-855.

Zeman, W., and J. R. McInnes. 1963. "Craigies Neuroanatomy of the rat." Academic Press, N.Y., 230 pps.

Zimmerman, L. E., and A. B. Eastham. 1959. "Acid mucopolysaccharide in the retinal pigment in the primitive apendymal zone: An autoradiographic study of histogenesis in the nervous system." Exp. Neurol. 1:322-333.

ADDITIONAL REFERENCES

Albert S., et al. 1966. Observations on the origin of lymphocyte-like cells in mouse bone marrow. Nature. 212:1577-1579.

Alescio, Tommaso, and Emilia Colombo Piperno. 1967. A quantitative assessment of mesenchymal contribution to epithelial growth rate in mouse embryonic lung developing *in vitro*. J. Embryol. Exp. Morphol. 17:213-227.

Aldred, John Phillip. 1963. Climatic environmental influences on growth and fertility in the mouse. Abstr. Diss. Abstr. 23:2962-2963.

Baccarini, Iracema M., et al. 1967. An autoradiographic study of vaginal epithelium of mice by serial biopsy. Abstr. Anat. Rec. 157:207-208.

Blackburn, Will R., and J. F. A. P. Miller. 1967. Electron microscopic studies of thymus graft regeneration and rejection I. Syngeneic grafts. Lab. Invest. 16:66-83.

Borum, Kirstine. 1967. Oögenesis in the mouse; a study of the origin of the mature ova. Exp. Cell Res. 45:39-47.

Browning, Henry C., and M. Perley. 1966. Effect of single prenatal dosage of androgen on female mice. Abstr. Amer. Zool. 6:569.

Browning, Henry C., and Wilma D. White. 1967. Relation between the size of ovarian isografts and abnormal reproductive cycles in the mouse. Anat. Rec. 157: 155-162.

Burnet, F. M. 1966. Mast cells in the mouse thymus, p. 335-340. In G. E. W. Wolstenholme & R. Porter (ed.) The thymus. Boston; Little & Brown.

Caruso, R., and N. L. Petrakis. 1966. Studies of the coagulation and prothrombin time in the mouse embryo. Thromb. Diath. Haemorrh. 16:732-737.

Chaudhry, Anand P., and Susan Siar. 1957. *In vitro* study of fusion of palatal shelves in A/Jax mouse embryos. J. Dent. Res. 46:257-260.

Clark, Sam L., Jr. 1966. Cytological evidences of secretion in the thymus, p. 3-30. In G. E. W. Wolstenholme & R. Porter (ed.). The thymus. Boston; Little & Brown.

Daems, W. Th., and E. Wisse. 1966. Shape and attachment of the cristae mitochondriales in mouse hepatic cell mitochondria. J. Ultrastruct. Res. 16:123-140.

Disher, Lenore. 1967. Histogenesis of mouse submandibular salivary gland. Abstr. Anat. Rec. 157:235.

Flower, Michael, and Clifford Grobstein. 1967. Interconvertibility of induced morphogenetic responses of mouse embryonic somites to notochord and ventral spinal cord. Develop. Biol. 15:193-205.

Godowicz, Barbara, and Halina Krzanowska. 1966. DNA content of mouse spermatozoa from inbred strain KE of low male fertility. Folia Biol. (Warsaw) 14:235-242.

Heath, Trevor, and Steven L. Wissig. 1966. Fine structure of the surface of mouse hepatic cells. Amer. J. Anat. 119:97-128.

Hoshino, K. 1967. Comparative study on the skeletal development in the fetus of rat and mouse. Congen. Anomalies (Japan). 7:32-38.

Izard, Jacques. 1967. A class of dense reticular cells with long processes in the mouse thymus. Abstr. Anat. Rec. 157:264.

Jacobs, Richard M. 1966. Effects of cortisone acetate upon hydration of embryonic palate in two inbred strains of mice. Anat. Rec. 156:1-4.

Knudsen, P. A. 1966. Congenital malformations of lower incisors and molars in exencephalic mouse embryos, induced by hypervitaminosis A. Acta Odontol. Scand. 24:55-90.

Krzanowska, Halina. 1966. Fertilization rate in mice after artificial insemination with epididymal or "capacitated" sperm from inbred and crossbred males. Folia Biol. (Warsaw) 14:171-175.

———. 1966. Inheritance of reduced male fertility, connected with abnormal spermatozoa, in mice. Acta Biol. Cracoviensia (Zool.) 9:61-70.

Laird, Anna Kane, and Alma Howard. 1967. Growth curves in inbred mice. Nature. 213:786-787.

Ogawa, T. 1967. Comparative study on development in the stage of organogenesis in the mouse and rat. Congen. Anomalies (Japan). 7:27-31.

Reading, Anthony J. 1966. Influence of room temperature on the growth of house mice. J. Mannal. 47:694-697.

Richardson, Flavia L. 1967. The acinar pattern in the mammary glands of virgin mice at different ages. J. Nat. Cancer Inst. 38:305-315.

Schlesinger, Michael, and Zeev Koren. 1967. Mouse trophoblastic cells in tissue culture. Fertil. Steril. 18:95-101.

Shelton, Emma. 1966. Differentiation of mouse thymus cultured in diffusion chambers. Amer. J. Anat. 119:341-358.

Simnett, J. D., and A. G. Heppleston. 1966. Cell renewal in the mouse lung; the influence of sex, strain, and age. Lab. Invest. 15:1793-1801.

———. 1956. Factors controlling organ growth; a comparison of mitotic activity of newborn and adult mouse lung in organ culture. Exp. Cell Res. 45:96-105.

Smiley, G. R. 1967. A profile cephalometric appraisal of normal growth parameters in embryonic mice. <u>Abstr</u>. Anat. Rec. 157:323.

Smiley, Gary R., and Andrew D. Dixon. 1966. Fine structure of midline epithelium in the developing palate. <u>Abstr</u>. J. Cell Biol. 31:162A.

Stevens, Leroy C. 1967. The development of teratomas from intratesticular grafts of 2-cell mouse eggs. <u>Abstr</u>. Anat. Rec. 157:328.

Stoeckel, M. E., and A. Porte. 1966. Observations ultrastructurales sur la parathyroide de souris. I. Etude chez la souris normale. Zeits. Zellforsch. 73:488-502.

Tarkowski, A. K. 1966. An air-drying method for chromosome preparations from mouse eggs. Cytogenetics. 5:394-400.

Wakasugi, Noboru, et al. 1967. Differences of fertility in reciprocal crosses between inbred strains of mice: DDK, KK and NC. J. Reprod. Fertil. 13:41-50.

Wessells, Norman K., and Julia H. Cohen. 1967. Early pancreas organogenesis: morphogenesis, tissue interactions, and mass effects. Develop. Biol. 15:237-270.

Index*